STATISTICS
FOR BIOLOGISTS

STATISTICS FOR BIOLOGISTS

THIRD EDITION

R. C. CAMPBELL

Fellow of Churchill College, Cambridge

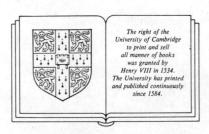

The right of the
University of Cambridge
to print and sell
all manner of books
was granted by
Henry VIII in 1534.
The University has printed
and published continuously
since 1584.

CAMBRIDGE UNIVERSITY PRESS

Cambridge
New York   New Rochelle
Melbourne Sydney

Published by the Press Syndicate of the University of Cambridge
The Pitt Building, Trumpington Street, Cambridge CB2 1RP
32 East 57th Street, New York, NY 10022, USA
10 Stamford Road, Oakleigh, Melbourne 3166, Australia

First published 1967
Reprinted 1972
Second edition 1974
Reprinted 1975, 1979, 1981, 1986, 1987
Third edition 1989

Printed in Great Britain at the University Press, Cambridge

British Library cataloguing in publication data

Campbell, R. C. (Richard Colin)
Statistics for biologists – 3rd ed.
1. Statistical mathematics – For biology
I. Title
519.5′024574

Library of Congress cataloguing in publication data

Campbell, R. C. (Richard Colin)
Statistics for biologists / R. C. Campbell.—3rd ed.
    p.   cm.
Bibliography: p.
Includes index.
ISBN 0 521 36095 1. ISBN 0 521 36932 0 (pbk.)
1. Biometry.   I. Title.
QH323.5.C35 1989
519.5′02474—dc19   88-7297 CIP

ISBN 0 521 36095 1 hard covers
ISBN 0 521 36932 0 paperback

MCP

CONTENTS

CONTENTS

CONTENTS

These tables have mostly been adapted from those already available in the statistical literature, and I am most grateful to the authors and publishers of the following works for their permission to use their material.

Siegel (1956) Tables 3, 4, 6, 7, 8, 9 and 17; *Lindley and Scott* (1984) 1, 5, 11, 13, 15 and 16; *Fisher and Yates* (1974) 12, 14, 18, and 21; *Kendall* (1970) 10; *Nair* (1940) 2; *Pearson and Hartley* (1958) 19, 20 and Fig. A1.

I have given no mathematical definitions nor any guidance as to methods of interpolation; these matters are important, but are out of place in an elementary book. They are well discussed in the larger sets of tables referred to above.

PREFACE

This book sets out to give an account of the principles and more elementary techniques of statistical reasoning, particularly as they are relevant to the biologist. I have tried to keep the use of symbolism and mathematical jargon to a minimum and, except in a very few sections, there is nothing assumed beyond that normally contained in a school elementary mathematics course. I have attempted rather to concentrate on the exposition of ideas, because I believe that when a scientist first realizes his need for statistical tools it is their meaning rather than their procedure which causes him difficulty. Once the logic is clear it is relatively easy to follow the form of the calculations.

The chapters are arranged for consecutive reading; a few sections of a slightly mathematical nature, which can be omitted at a first reading, are marked with an asterisk after the section number. A decimal system is used for this numbering.

The first chapter aims to show the need for statistical procedures in experiments and surveys, and does not contain any techniques. Chapters 2, 3 and 4 deal with the simpler non-parametric methods, with particular emphasis on the logic of statistical reasoning; I believe that this is easier to follow in non-parametric procedures than in normal distribution methods. Chapter 5 is concerned with decision theory ideas, which seem certain to have an increasing effect on statistical practice. Chapters 6, 7 and 8 contain the application to the normal distribution of the ideas developed earlier; these applications, particularly in the analysis of variance, are so powerful that their discussion necessarily occupies a good half of the book.

Many teachers and students have contributed to this exposition, and I am more than grateful for their patience. I would like to acknowledge particularly the help I have had in the preparation of the manuscript from Mrs B. G. Baker, Mrs D. I. Hughes and Mrs V. Partridge.

1967   R. C. CAMPBELL

PREFACE TO THE SECOND EDITION

In this edition I have added a chapter dealing with some elementary aspects of non-normal distributions, and have included a section of exercises at the end of each chapter. For the most part these exercises are direct applications of methods described in the text, but in a few places I have included a little additional discussion. The contexts of the exercises are intended to be realistic, but the 'observations' have been synthesized by computer; by this method an author has more control over the distributions he presents. As in the text, one or two slightly more theoretical exercises are marked with an asterisk.

I have not tried to make the exercises artificially easy; if experience in analysis is the key to statistical understanding, it is important that this experience should not lead to a false impression of what the discipline requires. There is therefore quite a lot of calculation needed in some exercises, and the reader will be wise to obtain the use of a good desk calculator. I believe that the day is still a little way off when the biologist learns his statistics with the help of one of the large computers, but he should certainly have the use of a modern desk machine.

I am very grateful for the many kind and helpful comments the first edition received and I have done my best to act on them. I would also like especially to thank Robert Marrs for the great help he has given me in constructing the exercises.

R. C. CAMPBELL

1973

PREFACE TO THE THIRD EDITION

The main object of the new material in this edition is to illustrate the potential usefulness of computers to the biologist when he or she needs to undertake statistical analyses; to this end I have given computer analyses of a selection of the examples in the text and exercises, using several different statistical languages. I have not attempted to illustrate all the languages available nor to introduce all the facilities of the chosen languages which could be applied even to the limited range of examples in this book. To achieve this would demand a very large book on those topics alone, a book moreover which would be out of date before it was published.

The biologist who wishes to use a statistical language on a computer should be prepared to seek local advice about the facilities available and to go on seeking help as difficulties are encountered in applying the chosen language: but I hope that the examples in this book will show something of what can reasonably be looked for and encourage the reader to make the effort that will be necessary to gain some facility in the use of the very powerful tools that these languages offer.

In incorporating the new material I have tried to disrupt the existing text as little as possible. In particular, I have been careful not to reduce the—sometimes lengthy—existing explanations of how calculations are built up, because I believe that understanding these processes is an important part of comprehending the statistical methods: this comprehension is no less necessary just because we can use a computer to undertake the labour of the calculations. Sections dealing with computing are labelled with (c) after the section number; thus sections 2.1.1 and 2.1.1(c) are different—and appear in that order in the text. New tables and figures are labelled as CTables and CFigures and are numbered independently of those not concerned with computer procedures. I hope that this arrangement will make this edition easy to use both for those biologists who have access to computer facilities and for those who do not.

I have also made some small changes to the text of the second edition, removing a few sections and making alterations in others; one that should be mentioned is a redefinition of *regret* in sections 5.3 *et seq*.

R. C. CAMPBELL

1987

1 WHAT IS STATISTICS ABOUT?

1.1 Random variation

The simplest biological observations are qualitative, but when we try to refine them we usually find it necessary to quantify them in some way. 'Grass is green' sounds a simple and reasonable observation but if we explore its implications we come very soon to further questions, such as 'Is it always green? Is it as green in Cambridge, in the drier east of England, as in Hereford, in the wetter west of the country? Are all varieties equally green?' These queries leave us needing to measure greenness; when we answer them we shall have quantitative observations.

The progression from qualitative to quantitative is found in all the sciences, and is not by itself enough to explain why statistical methods are of special use to the biologist. The extra reason is the variability of the biologists' material. If we take observations to try to answer the first question in the preceding paragraph, we find that even on a single day different samples of grass exhibit different degrees of greenness. If we measure greenness on five samples today and on five more in a week's time, the ten measurements may well all be different and the range of values covered by the first five may overlap that of the second five. This is a simple example of biological variation; the purpose of statistical methods is to reach worthwhile quantitative conclusions in its presence.

Such variation is more than just experimental error. This is the variation we should find if we made repeated measurements on a single sample of grass; it is important but is usually smaller than the differences we find between samples. Experimental error is a part, but no more than a part, of biological variation. This distinction helps us to understand why statistical methods are of more importance in elementary biology than in elementary work in the physical sciences. Experimental error occurs in both but the biologist has, from his first measurements, to cope with important sources of variation that are quite distinct from it—and are usually larger.

The statistical approach is not concerned to any great extent with the causes of variation. In comparing our ten samples of grass, we may be able to decide whether there are important differences in greenness between the two groups of five without knowing why the samples differ within the groups. We consider how much the samples differ without exploring the reasons for the differences. This limitation of approach reminds us that statistical methods are for the biologist a tool, but nothing more. It is indeed useful to be able to compare the greenness of the two groups without knowing why this character varies, but our investigation would be limited if we stopped at this point. We ought to remember that, although the variation can be regarded as random whilst we compare the groups, there must be causes which would explain it if we knew them. If, however, we look for these causes we shall continue to encounter random variation, and so will again need to use statistical tools; very few studies are so simple as to avoid this need.

1.1.1 Surveys and experiments

The investigations in which statistical methods are a useful tool are of two main kinds. We may take nature as we find her, or we may impose changes and observe their consequences; in both circumstances we find that random variation impedes us in drawing quantitative conclusions.

When we made the ten measurements discussed in 1.1, we were careful not to do anything that would alter the greenness of the samples; we wanted to observe the state of nature, not to change it. This condition defines the type of investigation known in statistical terminology as a *survey*.

When we set out to discover why greenness varied we would, at some stage, arrive at a list of possible causes—soil water content, hours of sunshine, and so on—and would then wish to measure and compare their effects. To do this we should have to change the state of nature and observe what effect the changes had on greenness; in other words we would carry out an *experiment*.

1.2 Sample and population

Suppose that in an attempt to clarify the simple statement 'grass is green', we have defined our terms more precisely. Let us now say that on the 1 May 1965, in the City of Cambridge, the grass is green—and suppose that we have defined how we will measure greenness. We will try to discover the meaning of the refined statement.

We agree at once that there is a lot of grass in Cambridge. Have we examined all of it? Most probably not! Does this matter? Is our statement invalidated because we have examined only some of the grass? Or did we perhaps intend our statement to apply only to the grass we had examined? The question of validity may be difficult, but there is no doubt that we meant our statement to apply more widely than to the few specimens we measured on one day; unless it had wider application, it could be of very little interest. The distinction implicit in this argument is fundamental to all scientific—and therefore statistical—reasoning; we take a few observations, but we want the conclusions that we draw to have much wider application. In statistical terminology, we observe a *sample* but we want to apply the conclusions to a *population*. The whole purpose of statistical analysis is to extract from the sample worthwhile information about the population.

The problem of validity, mentioned above, can be rephrased by asking whether our sample is one on which we can reasonably base conclusions that apply to the population in which we are interested. In the example quoted, the population—the grass in the City of Cambridge on 1 May 1965—was adequately defined and we need only consider whether the sample was satisfactory. This is the sequence that we should always seek to adopt; first we specify the population that interests us, then we consider how to take a good sample from it. In all too many cases, however, this sequence is not observed; we take some observations, our

3

sample, and frequently do not even consider whether they form a representative sample of any population, let alone deciding what that population is.

We now consider, slightly more formally, how to ensure that the sample gives useful information about the population. The case we have quoted is a survey type investigation, because we are trying to observe the state of nature rather than to change it, and we will for the moment limit the discussion to that situation.

1.2.1 Subdividing the population

The first precaution that an investigator suggests when considering how to make a sample representative is usually that we must take account of any known subdivisions in the population. Thus in investigating Cambridge grass he may mention the well kept College lawns and contrast these with the rough grazing of the commons or the cultivated leys of the few farms within the city boundary. This is a good starting point; if we have any prior knowledge we should use it in the design of our investigation. We shall see later that it is equally important to use it in the statistical analysis.

In the present case, the simplest way to use this prior knowledge is to take samples from each type of grass; the jargon is that we divide the population into *strata*, and by taking observations from each stratum we construct a *stratified* sample. The idea implicit in this procedure is that we believe that, whilst the separate areas of grass in a particular subgroup may be pretty much alike, the subgroups will differ rather markedly; the separate College lawns may be very similar, but they will differ substantially from the various commons. The prerequisite for stratified sampling is that the variation within the strata is less than that within the population as a whole.

This is however not the only kind of subdivision that can occur. If we consider for the moment only the stratum of College lawns, the further division into separate lawns may not imply any reduction in variation; the parts of a single lawn may differ as much as the parts of several distinct lawns. In this circumstance there is no need to sample every lawn; we can obtain satisfactory information by choosing a few lawns and studying these. This is called a *two-stage* sample; at the first stage we choose certain lawns for study, and at the second stage we sample within these chosen *first-stage units* (lawns). Clearly the procedure can sometimes be extended to more than two stages; we have then a *multi-stage* sample, with the characteristic feature that at each stage we sample only within the units chosen at the previous stage. This type of sample is sometimes of considerable practical convenience; it may save us a lot of trouble to work within a few first-stage units, rather than having to cover all the subgroups, as is necessary in a stratified sample.

In the grass problem we would probably decide to stratify on cultivation types (lawns, commons and leys) and so would sample each of these: we would then take a two-stage sample within each stratum, studying some College lawns, some of the commons and some of the leys. This mixed type of sampling is often convenient.

1.2.2 Random selection

The procedures discussed in the last section are aimed at making the sample more fully representative of the population, but they are not by themselves enough to ensure that it is satisfactory. There still remains the question of how to choose the samples (*individuals*) that we take in a given stratum or first-stage unit. The only fully satisfactory way of doing this is to select them *at random*. This means that at every stage of the sampling process all individuals have the same chance of being chosen and that, if we are selecting a total of n individuals, all possible sets of n are equally likely to be selected. Only by these conditions can we ensure that the information we obtain from the sample is relevant to the population and is not simply a property of the method of selection.

Truly random selection is by no means trivially easy to achieve; 'choosing at random', in the sense of taking those individuals that you think you will, is entirely unsatisfactory and should never be considered. Various quasi-random devices—tossing pennies, shuffling cards, and so on—are sometimes advocated but even these are very liable to introduce non-random elements into the selection process. Any reliable method uses a table of *random numbers*; this is a table in which any of the possible entries is equally likely to occur in any given position. A small table is given for illustrative purposes (Table A1); any serious user should have available a larger table such as that by Fisher and Yates (1974). We will give a simple example of the use of the table by considering how to select at random 5 rats from a group of 20. We first identify the rats in some convenient way and number them from 1 to 20; it does not matter how we arrange them for numbering. We then choose a starting point in the table of random numbers; a convenient rule for this is to begin where we ended our last use of the table. The simplest, but not the most efficient, method of selecting five numbers is to reject all numbers greater than 20 (00 counts as greater than 20) and to take the first five remaining, rejecting any repetitions. Supposing that we begin at the top of the first column, this procedure gives 20, 16, 4, 3, 1; the rats with these numbers are then our random sample. More elaborate and efficient methods of using random number tables are described in the notes to Table A1.

1.3 Experimental design

We consider how far the ideas introduced in the discussion of surveys—sample, population, representativeness, random selection—require to be modified or extended when we apply them to experiments in the presence of random variation. Starting with an example, suppose we wish to compare the effect of three different levels of soil moisture on the greenness of grass.

The most elementary procedure would be to take an area of grass, divide it into three parts and maintain one level of soil moisture in each; after a suitable interval we would measure the greenness of each part. This measuring process would amount to carrying out a survey on each of the three parts; we would expect each part to be fairly small so that no question of any subdivision into strata or first-stage units would arise, but we would certainly require random selection in choosing the sample or samples of grass on which greenness was measured. However, even if the sampling were satisfactory, the *design* of the experiment suffers several sampling defects. First, the method of allocating the *treatments* (moisture levels) to the *plots* (parts of the experimental area) has not been specified. This allocation must be at random so as to ensure as far as possible that the differences observed are due to the treatments and not to the plots themselves; of course, in any single experiment random allocation will not guarantee this, but used regularly it will ensure that it is true 'on average', and this is all that we can hope for.

Secondly, we have no means of knowing whether the differences we observe are due to the treatments or have occurred by chance in our variable material. This uncertainty can be resolved only if we have a measure of the random variation, and this in turn can be obtained only if we have several plots receiving each treatment instead of one as in the proposed design. Thus an experiment in which the test area was divided into twelve plots and these were then allocated at random, four to each of the three treatments, would enable us to compare (by suitable calculations) the observed treatment differences with the observed random variation; we could then decide whether the treatment differences were large enough to be important. This design is known as a *completely randomized* layout; it might result in an arrangement of plots such as that shown in Fig. 1a.

In interpreting the results of this simple experiment we need to remember that our observations are a sample, and that conclusions based upon them apply only to the population from which the sample was taken. Thus if the moisture-level experiment had been carried out, in the Cambridge context of 1.2, on one of the commons, we could reasonably extend our conclusions only to other commons, perhaps only to other Cambridge commons. If moreover we had selected a heavily shaded experimental area, we could not be sure that the results were of relevance to non-shaded areas. One experiment is just a single sample from the population of possible experiments and can no more lead to reliable general conclusions than can one sample from any population.

1.3.1 The idea of blocking

One weakness of the completely randomized design is that it does not enable us to take account of any subdivision that we know of in the experimental material. Thus if the experiment on moisture levels were carried out on an area that was partly shaded and partly exposed to the sun, we might find that the randomization gave a layout like that in Fig. 1b. This would clearly be undesirable because most of the plots receiving treatment A were also shaded so that we could not separate the effects of shade and A; we would say these effects were *confounded*.

A simple way of lessening the effect of shading would be to divide the experimental area into sub-areas (*blocks*) before allocating the treatments, choosing the blocks in such a way that

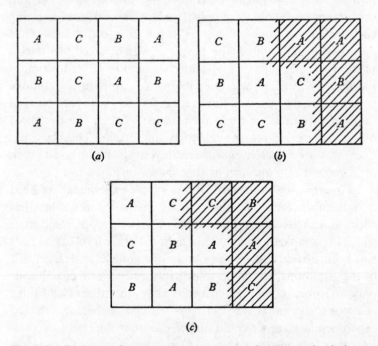

Fig. 1. Possible layouts for an experiment to compare 3 treatments in 12 plots; see 1.3, 1.3.1. Figs 1a and 1b are completely randomized layouts: in 1b part of the experimental area is shaded and part in sun and the randomization has made it difficult to distinguish between the effects of shade and of treatment A. This difficulty is reduced by the randomized block layout in Fig. 1c.

the plots of a block were as nearly alike in their conditions as possible, and to allocate treatments to plots so that each treatment was used in each block. The allocations would, as before, be at random, but with a separate *randomization* for each block; this would give a layout such as that in Fig. 1*c*. The design would be described as a *randomized block*. It is a type which is very widely used; it is the first we consider when we know something about subdivisions in our experimental material, and its importance is a further example of the principle mentioned in 1.2.1, that any prior knowledge must always be used when planning an investigation. The terms block and plot have become an accepted part of statistical jargon and are used even when the experiment does not involve an area of ground. Blocks are subdivisions of the experimental material chosen, like strata, to be more uniform than the material as a whole: plots are the units to which the treatments are applied.

1.4 Exercises

1 A batch of 20 male rats is to be divided at random into 4 groups of 5; the rats have been labelled with numbers 1 to 20. Prepare a suitable allocation.

Let the groups be numbered 1 to 4; corresponding to the rat numbers 1 to 20 we require a set of 20 random digits from 1 to 4, each digit occurring 5 times. Enter the table of random digits (Table A1) at the point where our last use of the table ended; this avoids the bias introduced by using one part of the table time after time. Use the first entry, in the way described in the note to the table, to obtain a random number from 1 to 4: allocate rat 1 to the group with this number. Use the next entry from the table to allocate rat 2 and continue in this way until all the rats are allocated, disregarding any allocation which would give more than 5 rats in a group.

2 Four bacterial cultures are to be applied to each of 10 culture dishes, which have been marked with equally spaced positions numbered 1 to 4. Prepare a suitable allocation.

3 Twenty-five of a general practitioner's patients have agreed to take part in a comparison of 4 sleeping drugs, *A, B, C, D*, which the doctor is carrying out for a pharmaceutical firm. Each patient will use each drug in turn, with a different random order for every patient. Prepare an allocation sheet for the doctor to use in distributing the drugs.

The patients will not be told the name or the code letter of each drug as it is given to them; this is necessary to prevent any 'comparing notes' between patients. The doctor will not know the labelling code of the drugs, so that his questions cannot inadvertently influence the patients' answers. These precautions constitute a double blind *trial.*

4 Describe one situation known to you in which a completely randomized experiment would be appropriate; prepare a suitable random allocation.

5 Fifty individually potted seedlings are to be used in a greenhouse experiment to compare growth under 10 treatments involving gibberellic acid and related substances. The seedlings are to be arranged in 5 rows of 10 pots on the staging in the greenhouse: it is known from past experience that the rows differ somewhat in average growth. Prepare a suitable allocation.

The knowledge that rows differ indicates that rows should form blocks in a randomized block allocation. If there are any differences in apparent vigour between the seedlings these differences should also be removed, as far as possible, as part of the block effect, by allocating the 10 most vigorous plants to one block, the next 10 to another block, and so on. The treatments should be allocated randomly to the pots in a block, with a separate randomization for each block.

6 An experimental piggery is arranged with two rows of pens: the rows are separated by a gangway and each row contains 16 individual pens. An experiment is to be carried out to compare the effects of 4 diets on weight gain and food conversion efficiency. The 32 pigs for the experiment are to be selected from the 5 litters which are available; the sexes and weights (kg) of the pigs in these litters are as follows:

Litter	A		B		C		D		E	
Sex	M	F	M	F	M	F	M	F	M	F
	12.0	15.7	15.1	13.4	17.4	18.4	13.5	13.2	14.4	14.1
	13.0	14.0	9.3	13.2	16.6	14.5	16.3	14.6	11.0	12.2
	15.1	16.6	16.3	13.7	17.4	15.6	15.3	14.5		11.6
	17.7		14.3	12.2		15.1	15.2			13.8
	16.1		14.7	17.1		16.8				14.1
	14.3					16.4				11.4

Prepare a suitable allocation.

First decide how to choose groups of 4 pens that will be treated as blocks. Then choose groups of 4 pigs, as alike as possible in litter, sex and weight, to be allocated to the blocks; it will not be possible to get completely uniform groups and you must use your judgement and knowledge of pigs to do as well as you can. Then allocate treatments to pens at random within blocks.

7 The owner of a boarding kennel wishes to compare the efficiency of 3 proprietary insecticides in keeping dogs free from fleas. Discuss the design of a suitable experiment, specifying the allocation of treatments to individual dogs.

8 A histologist wishes to compare 4 differential stains used in determining the proportion of 'dead' spermatozoa in a sample of bull semen. He can obtain 20 semen samples. Discuss the design of a suitable experiment, specifying the allocation of stains.

9 Describe one situation known to you in which a randomized block experiment would be appropriate; prepare a suitable random allocation.

10 A parasitologist working in a hospital in a large city wishes to study the distribution of intestinal parasites in the population served by the hospital. Discuss the design of a suitable survey, considering particularly the place of random selection in the design.

Consider such questions as the following:
(a) *should he use hospital inpatients only, or include outpatients, or attempt to select from the population as a whole;*
(b) *what groups in the population are relevant;*
(c) *how does he select individuals for study.*

11 A veterinarian is investigating the incidence in cattle of a certain type of tumour, which can be detected by an immunological test. It is believed that the incidence differs between breeds of cattle and between parts of the country. Discuss the design of a suitable survey.

12 A consumer organization wishes to investigate the retail sale of frozen vegetables; they are concerned with such factors as length of time in freeze, care in maintenance of correct temperature, loss of flavour, colour and vitamin content, brand differences and so on. Discuss the design of a suitable survey.

13 An ecologist wishes to investigate the concentration of mercury in the flesh of river fish in Great Britain. Discuss the design of a suitable survey.

14 Describe one situation known to you in which a survey would be a suitable means of obtaining required information. Discuss the design of an appropriate survey.

2 PRESENTING THE INFORMATION CONTAINED IN ONE SAMPLE

2.1 Pictorial presentation

The procedures discussed in chapter 1 had the purpose of making our sample of observations representative of the population we wished to investigate. We need now to consider what we can learn about the population from our sample. We require first to condense the information contained in the sample; even twenty observations are more than we can usefully comprehend without some form of summary.

The first and simplest form of summary should be a suitable pictorial presentation. There are many of these available, but we will consider only the more important types. The choice of the best type is partly a matter of objective judgement of the possible methods, but also depends on what features of the sample we consider important to our argument. This comment applies in varying degree to all statistical investigations and is on the one hand a challenge to our skill and understanding, but is also the origin of many misinterpretations—or worse! As soon as we summarize a sample, as every analysis does in one way or another, we decide what to emphasize and at the same time what to ignore. In presenting our own analyses and understanding other people's, we need always to be just as clear about what aspects of the sample have been passed over as we are about the meaning of the features emphasized. A most enjoyable account of some of the distortions which statistical methods can produce is given by Huff (1986).

2.1.1 Histogram

The commonest form of pictorial presentation is the *histogram*, which shows how often the various values or groups of values of our measurement occur. The methods of its extraction and presentation are illustrated in Table 1a and Fig. 2a; the table gives the weight of 35 rats. The first step in summarizing is to find the *range*—the difference between the largest and smallest values—and to decide how this should be divided for clear presentation. The range in Table 1 is from 369 g to 437 g, a range of 68 g; a convenient division into intervals of 10 g, defined so that they do not overlap, is shown in the table. The end intervals are left open because it is undesirable to add several intervals to cover one or two outlying values: the presentation of these open-ended intervals requires some care and is discussed below. The choice of subdivisions for the range is guided by the considerations that if the intervals are too short they will each contain so few observations that the histogram will appear very irregular, whilst if they are too long only a very coarse presentation of the sample will be achieved. We obtain the number of observations in each interval by going through the table in a systematic manner (each row in turn or each column in turn) and recording each value by a stroke against its interval; it is convenient to group the strokes in fives as shown in Table 1. These strokes are counted to obtain the *frequencies*, and these are totalled to check that no observations have been left out.†

The histogram is now constructed from the frequencies. The horizontal axis is marked off according to the division of the range, and the vertical axis shows the frequency; a block is drawn over each interval, with height given by the frequency (Fig. 2a). It is helpful to observe that frequency is proportional to area as well as to height, because all the intervals have the same width; this comment is taken up later. We notice that blocks are not drawn for the open-ended intervals, but the numbers of observations in them are recorded. The form of the histogram in this case is fairly typical of many based on biological measurements; the central intervals contain more observations

---

† This is the first time we have had occasion to use a check but it will certainly not be the last; checking should be an integral part of every analysis. In this connection it is worth emphasizing that, although most of us are brought up to believe that arithmetic accuracy is a thing we can attain by taking sufficient care, this belief is false; we can ensure accuracy only by using independent checks, or, better still, by having someone else duplicate the analysis. It is quite foolish to believe that we can work without making mistakes; the objective is to have enough checks to make sure that every mistake is found.

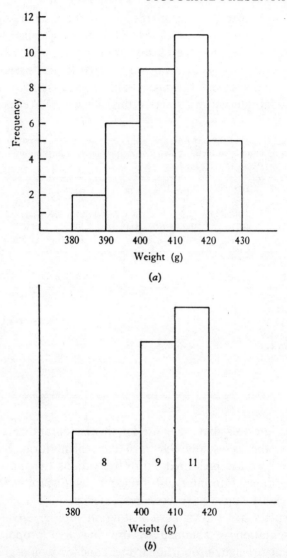

Fig. 2. Histograms of the observations in Tables 1a and 1b; see 2.1.1. In Fig. 2a there was one observation in each of the open classes −379 and 430− and these cannot be plotted. In Fig. 2b the classes were of unequal widths and so the frequencies are marked on the histogram and not on the ordinate: there were respectively 1 and 6 observations in the open classes −379 and 420−.

15

than those near the ends of the range. The interval containing most observations—410 to 419 in this case—is called the *modal* interval. We see that the observations are not evenly spread on the two sides of this interval; we say that the histogram is *skew* and not *symmetric*. The longer 'tail' in this case is to the left of the modal interval; we describe this as *negative* skewness.

Table 1a. *Weights (g) of 35 rats, grouped in intervals of 10 g to obtain frequencies for plotting in a histogram (Fig. 2a). The proportional frequencies are obtained by dividing the frequencies by the total (see 2.1.1.1).*

403	387	416	429	406	421	412
369	394	428	406	389	400	408
424	414	418	399	411	407	391
437	402	426	411	392	407	416
398	416	410	409	419	396	413

Interval		Frequency	Proportional frequency
−379	/	1	0.029
380–389	//	2	0.057
390–399	⫫ /	6	0.171
400–409	⫫ ////	9	0.257
410–419	⫫ ⫫ /	11	0.314
420–429	⫫	5	0.143
430–	/	1	0.029
		35	1.000

The intervals that we chose for this histogram were of equal widths; this is usually the most convenient form. Sometimes however we are presented with observations that have already been grouped into intervals of unequal widths; for example, the rat weights might have been presented as in Table 1b. These frequencies are shown as a histogram in Fig. 2b; a suitable representation is obtained by drawing area proportional to frequency (height proportional to frequency would give a very misleading impression). The vertical axis cannot now be marked with frequencies, so these are shown on the individual blocks.

We can now assess the usefulness of the histogram as a form of presentation. It shows us the range or, if there are open-ended intervals, at least the values covering most of the observations.

The areas of the blocks, which are readily comprehended visually, show us if the observations are uniformly spread or if they are concentrated round a *mode*; and they also give some indication of the skewness or symmetry. The histogram does not tell us how the observations are distributed within the intervals, and the appearance of skewness should be accepted with some caution because it is more liable than the other features to be affected by the choice of intervals.

Table 1*b*. *The rat weights from Table* 1a *grouped in intervals of unequal width; the histogram corresponding to this grouping is in Fig.* 2b.

Interval		Frequency
−379	/	1
380–399	⫫⫫ ///	8
400–409	⫫⫫ ////	9
410–419	⫫⫫ ⫫⫫ /	11
420–	⫫⫫ /	6
		35

2.1.1(c)   Let the computer do the work

The sequence of steps for preparing a histogram described in the last section, whilst not difficult, is nevertheless quite tedious to carry out, especially for a large sample. This should lead us to consider whether or not a computer can be useful, particularly since a single sequence of operations is repeated for every observation.

In practice this means considering whether a suitable program is available on a local computer; the effort involved in writing a new program is almost always too great to be worth contemplating. Usually we shall find that there are several programs available and that we need to choose the one that suits us best. For the present example we will use PLOTALL, a simple graphics package developed at the University of Akron (Seymour and Wiggins, 1981).

The PLOTALL language is designed to be reasonably flexible in the wording of the instructions it will accept. This means that the arrangement of the commands in CTable 1 is not the only

INFORMATION CONTAINED IN ONE SAMPLE

CTable 1. *The PLOTALL commands and data for the preparation of a histogram to illustrate the rat weights in Table 1a.*

```
variable weight
number of observations is 35
read data
403  387  416  429  406  421  412
369  394  428  406  389  400  408
424  414  418  399  411  407  391
437  402  426  411  392  407  416
398  416  410  409  419  396  413
type histogram
ranges 360 370 380 390 400 410 420 430 440
size 6.5 5.0
bar spacing 0
text is -
'360- 370- 380- 390- 400- 410- 420- 430- '
'  369  379  389  399  409  419  429  439'
'
'                    Weight (g)                    '
text height is 0.1625
text location 2.0 1.5
plot weight
stop
```

The interpretation of the commands is discussed in 2.1.1(c); those controlling location and size refer to the PLOTALL output page, which is reproduced to scale but reduced in CFig. 1.

form possible nor necessarily the wording that the individual reader finds most acceptable; but any arrangement must contain the four sections *data description*, *data*, *plot description* and *plot command*.

In CTable 1, the data description consists of the two lines

variable weight
number of observations is 35

whilst the data section contains the command

read data

together with the table of the observations.

The plot description consists of the command

type histogram

18

followed by detailed specification of the layout of the histogram and the text to be used to annotate it. The layout is determined by the commands

ranges
size
bar spacing

The values given for ranges are the divisions of the plotting intervals, as used in Table 1*a*. The size values refer to the horizontal and vertical dimensions (in inches) of the plot area, not including the text. The bar spacing is the gap to be left between successive blocks of the histogram. Since no plot location command is given, the plot is placed in the *default* position which is 2 2 relative to the bottom left-hand corner of the PLOTALL page. The text is controlled by the commands

text is
text height is
text location

Each line of the text is enclosed in single quotes which are not reproduced in the output but all other features of the text layout are copied exactly. Text height (in inches) is chosen to give good legibility and appearance on the plot, remembering that each symbol occupies an area that is square; in the present case five spaces were allowed for the text for each of the eight intervals on the plot so that $5 \times 8 = 40$ symbols were to be spread evenly over the width of 6.5 inches specified in the size command, requiring $6.5/40 = 0.1625$ as the width—and therefore height—of the area for each symbol. Text location, interpreted in the same way as plot location, determines the position of the bottom left-hand corner of the area allocated to the first symbol of the first line of the text.

The plot command contains the single line

plot weight

The resulting histogram is shown in CFig. 1; it contains all the information in the manually drawn histogram in Fig. 2*a*, but the

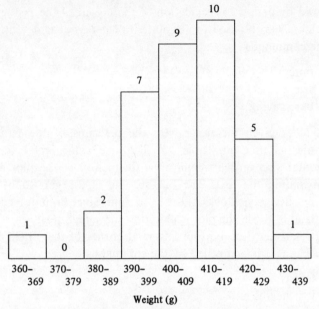

CFig. 1. Histogram of the rat weights in Table 1*a*, drawn in response to the PLOTALL commands in CTable 1.

details of style are different, being determined by the conventions adopted in PLOTALL.

We have thus achieved the objective suggested in the first paragraph above and we should now consider whether the effort was worth while; if this has been our first contact with PLOTALL—and even more if it has been our first contact with any computer package at all—then we may feel some doubts but let us consider just how much effort has been required. Most of the problems will have arisen in learning to give PLOTALL instructions in a form it finds acceptable; these difficulties have now been solved once and for all, at least as far as preparing a histogram is concerned, whilst the labour of presenting the data and commands to the computer has been small and certainly very much less than that involved in preparing the histogram manually. On balance we must conclude that the computer has served us very well and we should be encouraged to consider its use in other analyses.

2.1.1.1 Frequency curve

We will find it useful if we consider how the histogram is modified when the total number of observations is increased; the histograms in Fig. 3 will help us to see what happens. They are based on measurements of yield of wheat grain (g) from 6 inch lengths of row, made as part of a study supported by the Greg Bequest at the University Farm, Cambridge in 1934–5 (unpublished data). Fig. 3a represents 72 measurements on a 36 foot length of one row: Fig. 3b is obtained from 144 measurements on a 72 foot length of the same row, including the values plotted in Fig. 3a: Fig. 3c is based on 288 measurements made on two rows, that used in Fig. 3b and the row next to it: Fig. 3d illustrates 576 measurements made on four rows, those used in Fig. 3c and the two next to them. The succeeding figures are less and less irregular in outline, and appear to approximate to a smooth curve such as the one which has been drawn in by eye on Fig. 3d. At the same time we would expect on intuitive grounds that the larger samples would represent with increasing closeness the population of yields from all the possible 6 inch lengths of row. This consideration, taken with Fig. 3, suggests that the population of yields should be represented by a smooth curve like that in Fig. 3d. Such a curve is in fact the most convenient visual form for illustrating the population of any variable which we can imagine as measurable on a scale so fine that very many different values might be obtained. This kind of variable is called *continuous* because we find that many of its properties agree reasonably well with the mathematical concept of continuity.† The curve representing the population is called the *frequency curve* of the variable.

In 2.1.1 we found that the areas of the blocks in a histogram corresponded to the sample frequencies in the intervals of the range. This property is preserved as the sample size increases and suggests therefore that the area under a frequency curve cut off by a certain interval of the range should correspond to the

---

† We shall have many occasions to borrow concepts from mathematics. We do not need to enter too deeply into the abstractions involved in their precise definitions; what matters is that the mathematical results built upon the definitions should correspond to a good approximation with the real world results which are our concern.

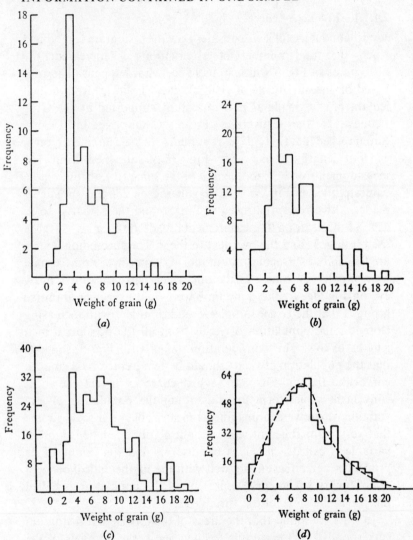

Fig. 3. Histograms of observations of yields of wheat grain, obtained from respectively 72, 144, 288 and 576 six inch lengths of row: see 2.1.1.1. The frequency scales have been changed so that the areas under the four histograms (and the frequency curve) are equal, thus facilitating comparison between the diagrams. There are respectively 0, 1, 2, and 7 observations in the open interval 20–.

'population frequency.' This term however requires some care in interpretation; the number of observations in a population is usually regarded as infinite—we can imagine ourselves going on and on taking measurements under any given sampling procedure—and the number of observations in an interval is therefore some 'fraction of infinity,' which is not an idea that we shall find at all meaningful or helpful. To avoid this difficulty we work with the proportion of the total area cut off by the interval rather than with any direct interpretation in terms of frequency. This proportion is an acceptable concept, and is very easy to appreciate visually. We call it the *probability* of obtaining an observation in the interval. To show that this definition agrees well with our every day understanding of the term, consider first the proportional frequencies given in Table 1a. These are the proportions of the total frequency which correspond to the area in each block, and so, as the sample size increases in the way shown in Fig. 3, each proportional frequency comes to be represented by that proportion of the total area under the frequency curve which is cut off by the appropriate interval. However, proportional frequency for a large sample is about as precise an interpretation of the day to day meaning of probability as we can find in many applications; the representation of the probability of observing a value in a given interval as the proportion of the area under the frequency curve cut off by the interval is thus in satisfactory agreement with common usage.

The 'frequency interpretation' of probability is not however sufficient for all purposes. For example, we often speak of probability as measuring the strength of some belief, and it is hard to see what this has got to do with frequency. This wider sense for the term is sometimes very necessary, and we shall ourselves require it in later chapters; but many simple applications can be covered by the frequency approach, and we shall find this sufficient for the moment.

2.1.1.2 Subgroups in the population

If the sample were selected to take account of subgroups in the population, we might want to know how the differences between subgroups affected the distribution. To examine the differences we could draw a histogram for each subgroup, taking care to use the same scales in all cases; these histograms could be plotted on a single diagram or in adjacent figures. The device of plotting several histograms on top of each other should however be used sparingly; the resulting picture can be very difficult to read. With separate histograms visual comparisons between the subgroups are fairly easy; more sophisticated forms of comparison will be discussed later.

2.2 Average: mean

The histogram gives a useful impression of the sample, and plotting it should be our usual starting point. In most analyses, however, we want more than a visual impression; some kind of numerical summary of the sample is desirable. The simplest – in the sense that most people think of it first – is the sample average, that is the sum of all the observations divided by the number of observations. There are in fact certain limitations on its usefulness as a summary but it can validly be applied in many cases. We will first express its calculation more formally; many statistical quantities are most conveniently expressed with some mathematical symbolism, and the average is a useful first example.

We will represent the quantity that we measure by the symbol x; if we have taken n observations in our sample we will label these with suffices thus $x_1, x_2, x_3, \ldots, x_n$. Here x_1 means the first observation that we took of the variable x; in using formulae involving x_1 we would put in its actual value for our sample, but the use of a general symbol rather than a number enables us to cover all cases with one expression. Similarly x_2 denotes the second observation, and so on; the notation $\ldots, x_n$ enables us to represent a general sample without having to fix the number of observations in advance. We will sometimes want to talk about a general member of the sample – not necessarily the first or the second – and for this purpose we use the symbol x_i, meaning that we can specify that $i = 1$ or 2 or whatever particular value we want.

The sum of all the observations is then

$$x_1 + x_2 + x_3 + \cdots + x_n,$$

the ... again serving as a device to cover all sample sizes. This expression is often abbreviated into the form

$$x_1 + x_2 + x_3 + \cdots + x_n = \sum_{i=1}^{n} x_i,$$

which we read as 'sigma (or sum) from 1 to n of x_i'. The symbol $\sum$ (the Greek capital S) is an instruction to add up, in the same way as 'log' is an instruction to look up in a table of logarithms; the *limits* 1 and n tell us which set of values of the general observation x_i we are to take. They are frequently left out if there is no risk of ambiguity, in which case we write

$$x_1 + x_2 + x_3 + \cdots + x_n = \sum x.$$

We emphasize that these are merely different ways of writing the same thing; their sole justification is to save us trouble – but this is quite a good justification!

The average is now written

$$\frac{\sum x}{n} \quad \text{or} \quad \frac{1}{n} \sum x$$

and is often represented by the symbol $\bar{x}$ (read 'x bar'). We usually call it the *sample mean* rather than the average, because the latter term is sometimes used with a less precise connotation.

2.2.1 Median

The most important limitation on the usefulness of the sample mean can easily be shown by an example. Consider some observations (Table 2) of the numbers of days a group of pigs took to fatten from weaning to bacon; the calculation of the average is shown in the table, and the distribution is plotted on a histogram in Fig. 4a, with the sample mean marked. The distribution is positively skew and the sample mean, regarded as a useful numerical summary – a representative value – for the distribution, is clearly not very satisfactory because it is appreciably removed from the main concentration of frequency. A farmer planning the use of fattening pens would find, if he used

25

the mean fattening period as the basis of his plans, that a substantial proportion of the pens would be nearly empty several days before his calculations led him to expect. The sample mean is indeed liable to be misleading whenever the distribution is noticeably skew and it is often better not to use it in such cases.

The measure of 'average' which is used for a skew sample is the *median*, the central value when the observations are arranged in numerical order. In the present case this is the thirteenth value (there being twelve values less than this and twelve greater) and this is readily found, using the table of frequencies and the observations themselves, to be 116. This value, which is marked on the copy of the histogram in Fig. 4b, is visually more acceptable as a sensible summary of the distribution than is the mean in Fig. 4a. Both quantities leave out a great deal – they tell us nothing about the range of the variable nor the form of the distribution – but as a single number to summarize the histogram the median appears much more reasonable.

When there are an odd number of observations in the sample, as in Table 2, the median is uniquely defined. Let us write n, the sample size, $= 2r + 1$, where r is an integer (any odd number can be expressed in this way); then the $(r + 1)$th observation in order of magnitude has r values below it and r above, and so is the sample median. If n is even ($= 2s$, say) there is no unique central value, and we agree to take the sample median half-way between the sth and $(s + 1)$th observations. In both cases therefore (n odd or even) the median divides the total frequency of the sample into two equal parts; another way of saying this is that the median, when drawn on the histogram, divides its area into two equal parts.

If we imagine the sample to get larger, in the kind of way discussed in 2.1.1.1, this property of bisecting the area will be retained, and it seems natural therefore to define the population median of a continuous variable as that value which bisects the area under the frequency curve. As the areas to the left and to the right of the median are equal, we have equal chances of observing a value less than or greater than the median – remembering that area under the frequency curve corresponds to

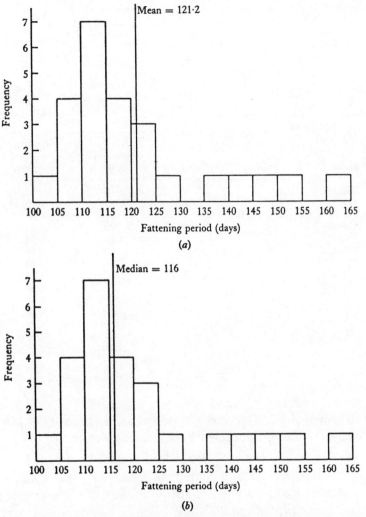

Fig. 4. Histogram of fattening periods of 25 pigs, as detailed in Table 2; see 2.2.1. The sample mean is marked on Fig. 4a and the sample median on Fig. 4b.

probability. As probability is related to proportional frequency, it must be numerically between 0 and 1, and so the two equal chances just mentioned must both be $\frac{1}{2}$, because between them they correspond to the total area under the frequency curve, that

Table 2. *Fattening period (days) of 25 pigs, grouped in intervals of 5 days to obtain frequencies for plotting in a histogram (Fig. 4a). The sample mean is calculated from the individual observations. The cumulative frequencies are discussed in 3.5.1 and used to plot the ogive (Fig. 11).*

105	114	121	117	115
147	119	106	111	142
109	113	163	151	121
114	123	116	109	118
126	115	111	137	106

Interval		Frequency	Cumulative frequency	Cumulative proportional frequency
101–105	/	1	1	0.04
106–110	////	4	5	0.20
111–115	### //	7	12	0.48
116–120	////	4	16	0.64
121–125	///	3	19	0.76
126–130	/	1	20	0.80
131–135		0	20	0.80
136–140	/	1	21	0.84
141–145	/	1	22	0.88
146–150	/	1	23	0.92
151–155	/	1	24	0.96
156–160		0	24	0.96
161–165	/	1	25	1.00
		25		

$$\sum x = 3029, \quad n = 25, \quad \bar{x} = \tfrac{3029}{25} = 121.16.$$

is to probability 1. Thus the area or probability property of the median (which we will denote by x_M) can be written

$$P(x < x_M) = P(x > x_M) = \tfrac{1}{2};$$

i.e. 'the probability that a continuous variable x takes a value less than its median is equal to the probability that it is greater than its median, both probabilities being $\tfrac{1}{2}$.'

2.3 Discrete variables

The restriction of our argument to continuous variables has been mentioned so often above that it is necessary to ask what other kinds of variable can be of interest, and what happens in the non-continuous case to the definitions of the previous sections. The simplest example of a non-continuous variable is a count.

How many brown eggs? How many cultures showing bacterial growth? How many grains in an ear of wheat? A variable which, like a count can take only a restricted set of values is called *discrete*; how do we represent it, and how do we summarize it?

Clearly we can obtain frequencies and proportional frequencies just as easily as in the continuous case; however, we cannot plot them in a histogram nor represent probability by area, because both these devices deal with ranges of values and could not sensibly be used for say, the interval between one brown and two brown eggs in a clutch.† The presentation of frequencies for a discrete variable is shown in Fig. 5, the data (Table 3) being numbers of grains in 20 ears of wheat. We see that frequency is represented by the height of the vertical line drawn at each value of the variable on the horizontal scale. We do not join up the tops of the vertical lines in any way, because we can give no meaning to the probability of, for example, $21\frac{1}{2}$ grains.

The sample mean is calculated exactly as in the continuous case. Denoting the variable by r – the conventional letter for a

Table 3. *The numbers of grains in 20 ears of wheat, with frequencies and proportional frequencies for each value of the count.*

24	21	22	24	26
23	23	21	25	24
22	20	26	24	23
21	23	22	23	25

Count	Frequency	Proportional frequency
20	1	0.05
21	3	0.15
22	3	0.15
23	5	0.25
24	4	0.20
25	2	0.10
26	2	0.10
	20	1.00

$$\sum r = 462, \quad n = 20, \quad \bar{r} = \frac{462}{20} = 23.1.$$

† Sometimes it is convenient to group the values of a discrete variable, and so make a histogram meaningful but from our present point of view the variable then ceases to be truly discrete and becomes a peculiar hybrid between continuous and discrete.

discrete variable – we represent the sample of n observations by $r_1, r_2, r_3, \ldots, r_n$ and define the mean $\bar{r}$ by the equation

$$\bar{r} = \frac{1}{n} \sum r,$$

the sum being taken, as before, over all the members of the sample. For Table 3 this formula gives $\bar{r} = 23.1$.

The median is again defined as the central value of the ordered observations. For the data of Table 3 (20 observations) we take it half-way between the 10th and 11th ordered observations; both these observations are 23 and so the sample median is 23. We note from Fig. 5 that the distribution is reasonably symmetrical; the mean and median are very close together.

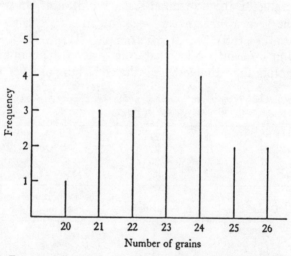

Fig. 5. Frequency diagram for the observations (Table 3) of numbers of grains in an ear of wheat, that is of a discrete variable; see 2.3.

2.3.1 Ranks

A count is one of the commonest forms of discrete variable, but another type which is of some importance arises when the only observation we can make is to arrange the experimental units in order according to some non-measurable characteristic. Thus a housewife may be able to arrange samples of margarine in her order of preference but not to attach any meaningful measure of palatability to each sample. When samples are so ordered we can number them in their order – it does not usually matter whether we take 1 at the 'top' or the 'bottom' of the ordering – and use each sample's number, called its *rank*, as the observation for that sample.

A useful body of statistical methods has been developed for dealing with ranks, and we shall discuss some of them in later sections. Most of them involve very little calculation and it is sometimes convenient to rank observations even when we have measurements available; doing so does waste some information, but often this wastage is quite small and is justified by the saving in calculation.

2.4 Multiple measurements

The discussion in the preceding sections has contained the tacit assumption that only one variable is being studied, that is that only one measurement is made on each unit of experimental material. In many cases, however, we want to observe several different variables on each unit, and we now consider how far the methods developed above are useful in this case. Clearly we can draw a histogram for each variable, and can also find the sample mean and median for each; it is desirable however to have available further methods to enable us to consider all or several of the variables together, because the study of two or more variables and the relationships between them often gives us a better understanding of a problem. In the present section we will outline some of the pictorial methods applicable to multiple measurements; in chapters 4 and 8 we will consider the analytical methods for studying relationships between variables.

2.4.1 Pie and bar charts

In certain cases the means of multiple measurements can usefully be presented in a *pie diagram* or a *bar diagram*. Suppose, for example, that we have observed the composition of cows milk, measuring the proportion of fat, solids-not-fat (SNF) and water. After testing, say, 20 samples of milk, we could average the

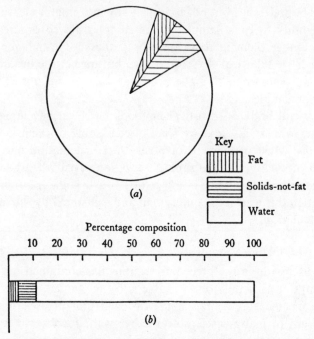

Fig. 6. Pie diagram (Fig. 6a) and bar diagram (Fig. 6b) to illustrate the mean composition of 20 samples of milk; see 2.4.1.

proportions of each constituent – that is average each variable separately – obtaining perhaps sample means of 4 per cent fat, 7 per cent SNF and 89 per cent water. The two forms of diagram for presenting these proportions are shown in Figs. 6a and 6b; both forms show clearly the relative proportions of the three constituents.

Bar diagrams can be used not only to show proportions, as in the last example, but also actual amounts. Table 4 gives the compositions of the growth (in lb/plot) of a lucerne–cocksfoot ley over five two-weekly periods; these values, like those in the milk example, could well be averages, each based on several samples. All the essential information of Table 4 is shown in Fig. 7a; total growth in each period is indicated by the height of the block, whilst the composition of the growth is shown by the differential shading. The reader should notice that this diagram

Table 4. *The yields of lucerne and cocksfoot in a ley, measured in five fortnightly periods. The values are illustrated in Figs. 7a and 7b.*

Period Yields (lb/plot)	1	2	3	4	5
Lucerne	7.1	11.2	13.6	12.5	8.4
Cocksfoot	5.0	4.7	3.7	1.3	1.4
Total	12.1	15.9	17.3	13.8	9.8

is quite different from a histogram; the blocks represent averages of measurements and do not, as in the histogram, correspond to counts.

The data of Table 4 could also be presented in a set of pie diagrams; this is shown in Fig. 7b. The total growth in each period is indicated by the area (*not* the radius) of each diagram. This is a less satisfactory form of presentation than Fig. 7a, because the growths in different periods cannot be compared so easily; on the other hand, changes of proportion are easier to see because they are represented independently of total growth. Thus again we find that no one summary can give a satisfactory representation of all aspects of a sample; furthermore both these methods, being based on averages, give no indication of the variations in growth and composition.

2.4.1(c)   Using PLOTALL to prepare pie charts

A further illustration of the use of the PLOTALL language can be obtained by considering the pie diagrams in Fig. 7b; the instructions for this are shown in CTable 2 and the resulting plots are given in CFig. 2. We note first that the five pie diagrams and the key are introduced by a single data description:

> variables p1 to p6
> number of cases is 2

The 2 *cases* are lucerne and cocksfoot. Similarly, all the diagrams are drawn from a single data section:

> read data
> 7.1   11.2   13.6   12.5   8.4   1
> 5.0   4.7   3.7   1.3   1.4   1

33

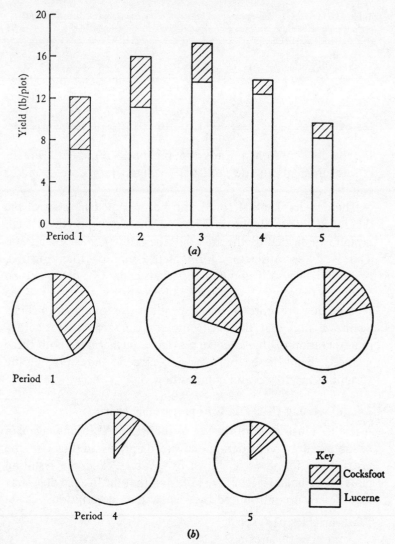

Fig. 7. Bar diagrams (Fig. 7a) and pie diagrams (Fig. 7b) to illustrate changes in yield and in composition of yield in a lucerne–cocksfoot ley; see 2.4.1 and Table 4.

The first five columns give the yields of lucerne and cocksfoot as in Table 4, whilst the last column contains values chosen to give equal segments for the two crops in the key. The six plot

CTable 2. *The PLOTALL commands and data for the preparation of pie diagrams to illustrate the yields of lucerne and cocksfoot set out in Table* 4.

```
variables p1 to p6
number of cases is 2
read data
7.1  11.2  13.6  12.5  8.4  1
5.0   4.7   3.7   1.3  1.4  1
type is pie with option 0
case shade p1 to p6 1 2
radius 0.65
plot location 1.00 5.07
plot p1
subtitle 0.4 4.7 0.15 'Period 1'
radius 0.75
plot location 3.18 4.97
plot p2
subtitle 3.9 4.7 0.15 '2'
radius 0.78
plot location 5.44 4.94
plot p3
subtitle 6.2 4.7 0.15 '3'
radius 0.70
plot location 1.35 2.50
plot p4
subtitle 0.8 2.2 0.15 'Period 4'
radius 0.59
plot location 3.20 2.60
plot p5
subtitle 3.75 2.2 0.15 '5'
radius 0.40
plot location 4.80 2.30
plot p6
subtitle 5.00 3.30 0.15 'Key'
subtitle 5.80 2.85 0.15 'Lucerne'
subtitle 5.80 2.45 0.15 'Cocksfoot'
stop
```

The plot locations and subtitle coordinates are measured in inches from the origin of the PLOTALL page and are therefore in proportion, but not equal, to the coordinates on these printed pages.

descriptions are then in two parts, of which the first is common to all six and so is stated only once:

> type is pie with option 0
> case shade p1 to p6 1 2

Option 0 ensures that no information (i.e. no data values or

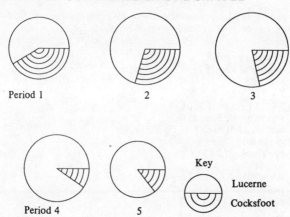

CFig. 2. Pie diagrams to illustrate the lucerne and cocksfoot yields in Table 4, drawn in response to the PLOTALL commands in CTable 2.

labels) is displayed on the plots, whilst the case shading controls how the two cases are indicated on the plots. The second part of the plot description and the plot command differ from plot to plot and so are given separately for each of the five periods: the pattern is sufficiently exemplified by considering that for period 1:

    radius 0.65
    plot location 1.00 5.07
    plot p1
    subtitle 0.4 4.7 0.15 'Period 1'

The radius is chosen so that, as explained in 2.4.1, the area of the pie shall correspond to the total yield in the period. The plot location, specified in inches, gives the coordinates of the bottom left-hand corner of the square area notionally occupied by the pie relative to the bottom left-hand corner of the PLOTALL page; in the same way the subtitle coordinates, which are the first two numbers in the subtitle specification, determine its position. The third number in this specification is the subtitle height (in inches); the determination of the coordinates in relation to this height and to the plot location involves calculations similar to those illustrated in the discussion of text height in 2.1.1(c).

For the key the second part of the plot description and the plot command are constructed in an exactly similar way except that three separate subtitle instructions are necessary to specify the positions of the three separate parts of the subtitle.

The pie diagrams drawn in CFig. 2 by these instructions illustrate the information in Table 4 in just the same way as do those in Fig. 7*b*; as in 2.1.1(c) there are stylistic differences between the two figures because of the conventions used in PLOTALL.

2.4.2 Scatter diagrams

If we wish to examine the relationship between two variables we may conveniently use a *scatter diagram*. Part of the classic observations of Galton (1886) on human heights is given in Table 5 and plotted in Fig. 8; Galton measured the heights of 928 fully grown children and their parents, and related the children's heights to the averages of their parents' heights (the mid-parent heights). It is conventional if there is some sort of 'one-way' relationship between two variables – such as time provides in this case – to plot the *independent* variable on the

Table 5. *Forty-six pairs of observations of mid-parent height* (x) *and adult offspring height* (y), *selected at random from the 928 pairs given by Galton (1886).*†

x	y	x	y	x	y
64.5	63.2	67.5	69.2	68.5	70.2
64.5	64.2	67.5	70.2	69.5	67.2
65.5	72.2	67.5	71.2	69.5	67.2
66.5	65.2	67.5	72.2	69.5	68.2
66.5	66.2	68.5	64.2	69.5	69.2
66.5	67.2	68.5	65.2	69.5	69.2
66.5	70.2	68.5	66.2	69.5	70.2
67.5	65.2	68.5	67.2	69.5	70.2
67.5	66.2	68.5	67.2	69.5	71.2
67.5	66.2	68.5	67.2	69.5	72.2
67.5	66.2	68.5	68.2	69.5	73.2
67.5	67.2	68.5	68.2	70.5	69.2
67.5	67.2	68.5	69.2	71.5	67.2
67.5	68.2	68.5	69.2	71.5	70.2
67.5	69.2	68.5	70.2	71.5	70.2
67.5	69.2				

† The observations were grouped to the nearest inch, but the groups were not centred at whole inches; this is explained in Galton's paper.

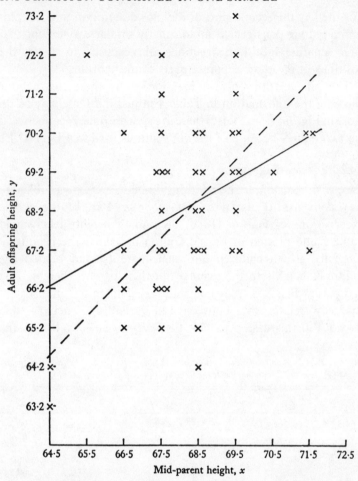

Fig. 8. Scatter diagram illustrating part of Galton's data on mid-parent height (x) and adult offspring height (y); see 2.4.2 and Table 5. The regression line of y on x, fitted by the methods of chapter 8, is shown as a solid line. The relationship which would hold if average offspring height were equal to average parent height—that is if there were no 'regression towards mediocrity'—is shown as a broken line.

horizontal (x) axis and the *dependent* variable on the vertical (y) axis; thus we can readily understand that differences in mid-parent heights may be reflected in differences in children's heights, but we cannot easily conceive the reverse relationship having any meaning. Galton observed that children's heights deviated less from the population average than did their mid-parent heights; he described this by saying that the children's heights had regressed towards the population average, and called the straight line characterizing the average relationship the *regression line* (Fig. 8). This has become the accepted term; we speak of the regression of the dependent variable *on* the independent variable. The analytical methods used in studying regression are discussed in chapter 8.

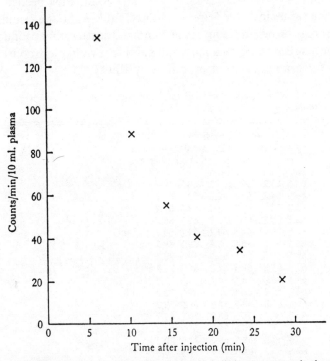

Fig. 9. Scatter diagram illustrating the decay of labelled progesterone in the plasma of a pregnant ewe; see 2.4.2.

A scatter diagram need not, of course, always suggest a linear relationship between the two variables plotted. If we inject labelled progesterone into a pregnant ewe and measure the radio-activity of the blood plasma over a period after the injection we may find a scatter like that in Fig. 9. These points approximate to a smooth curve of the type known as *exponential*; it occurs, for example, when the rate of decay of a substance is proportional to the excess of its concentration over a certain threshold value. The fitting of such a curve, when the measurements contain some random variation, can be a very elaborate statistical exercise.

2.4.2(c)   Using SPSS-X to plot the diagram

The X version of the SPSS package is presented from a statistical viewpoint by Norusis (1983, 1985). The commands needed to plot the values in Table 5 are set out in CTable 3 and the resulting diagram is shown in CFig. 3; the command scattergram produces a plot that is printed on a lineprinter and so is rather less polished than the graphics output produced by PLOTALL.†

CTable 3. *The SPSS-X commands and data for the preparation of a scatter diagram to illustrate the observations in Table 5.*

```
data list free / mph aoh
variable labels mph  ' Mid-parent height'
                aoh  'Adult offspring height'
begin data
64.5 63.2   67.5 69.2   68.5 70.2
64.5 64.2   67.5 70.2   69.5 67.2
65.5 72.2   67.5 71.2   69.5 67.2
66.5 65.2   67.5 72.2   69.5 68.2
66.5 66.2   68.5 64.2   69.5 69.2
66.5 67.2   68.5 65.2   69.5 69.2
66.5 70.2   68.5 66.2   69.5 70.2
67.5 65.2   68.5 67.2   69.5 70.2
67.5 66.2   68.5 67.2   69.5 71.2
67.5 66.2   68.5 67.2   69.5 72.2
67.5 66.2   68.5 68.2   69.5 73.2
67.5 67.2   68.5 68.2   70.5 69.2
67.5 67.2   68.5 69.2   71.5 67.2
67.5 68.2   68.5 69.2   71.5 70.2
67.5 69.2   68.5 70.2   71.5 70.2
67.5 69.2
end data
scattergram aoh (60,75) mph (63,73)
option 4
execute
```

† A scatter diagram can be drawn using PLOTALL but does not indicate multiple points, which occur in these data; two possible methods of indicating them are shown in Fig. 8 and CFig. 3.

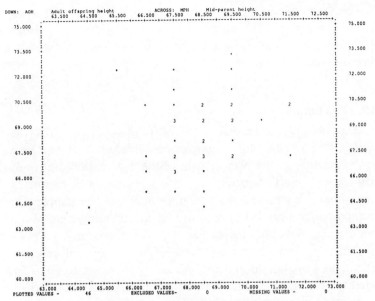

CFig. 3. Scatter diagram of the observations in Table 5, drawn in response to the SPSS-X commands in CTable 3.

The first three lines of the command sequence specify the variables that are to be analysed and the form of data list in which they are presented.† The variable names are very brief because only eight characters are allowed but the specification of variable labels allows a fuller description. The data are in freefield-format and so it is possible to include more than one case on each line.

The two lines of the command sequence that control the plotting specify that variable aoh will be used for the y-axis – because it is the first variable named – and also determine the range of values that will be represented on each axis; these ranges are chosen bearing in mind that scattergram gives 10 divisions of the vertical axis and 20 of the horizontal. Option 4 suppresses the division of the plotting area into nine parts by vertical and horizontal grid lines.

† File handle commands which are necessary in many implementations of SPSS-X are not used on the Cambridge University mainframe because files are accessed by using their identifiers in the local command language.

The diagram in CFig. 3 is only a part of the output; as in most languages the full output gives certain information about the computer installation and the resources used, as well as repeating the commands.

2.5 Estimation

The mean and the median were introduced in 2.2, 2.2.1, as representative values for a sample; we agreed that neither gave any indication of the overall form of the distribution, but both seemed, on intuitive grounds, worth considering as summary quantities. We also saw that when we imagined the sample size to increase we were led to consider the mean and median of the population. We explained earlier (1.2) that the overall purpose of statistical methods was to make the sample give information about the population, and it is necessary therefore to consider what information the sample mean and median give about their population equivalents. We regard the sample values as *estimates* of the population values: another way of expressing the question of the previous sentence is to ask if they are good estimates. This terminology however at once raises a fresh question, what do we mean by a good estimate? To answer this we must introduce a new concept and we will do this by an example.

Consider again the fattening period of bacon pigs; we gave a sample of 25 observations of this variable in Table 2. Suppose we now measured this period for a further sample of 25 pigs and calculated the mean and median for the new sample. We would then have two estimated values for the population mean, one from each sample, and also two estimates for the population median. What would we expect to find if we compared the two estimates of the mean with each other? Would the two values be the same? Would the two sample medians be equal? We need very little experience of randomly varying material to know that the answers to these last two questions would usually be no; only in very rare cases will the values for two samples be exactly equal. This is not in fact very surprising; the individual values in the second sample are most unlikely to repeat those in the first, so there is no reason why the mean and median should repeat. This does not in any way imply that the sample mean and median are poor estimates of the population values; all we have found is an example of a very general result, that any quantity which we calculate from a sample of randomly varying material will itself exhibit a certain amount of random variation. A quantity calculated from a sample is called a *statistic*, and a more formal statement of this result is that a sample statistic is not in general constant from one sample to another, but has a *sampling distribution*. This last term is slightly misleading, because all the distributions we have considered arise from sampling; the name is however usually reserved for the distribution from sample to sample of a statistic calculated for each sample.

The concept of a sampling distribution is the essential new idea in specifying what makes a good estimate. Using it, we can say first that any single quantity used as an estimate is practically never 'right', in the sense that the sample value is practically never equal to the population value. Even if the sampling distribution covers the population value – and we certainly hope it does – the chance that the sample value will equal the population value is, like the chance that it equals any specified value, extremely small.† The single value of the estimate is in fact of very little use and the best we can hope for is that it is 'fairly near' the population value. It is more helpful if we can give a range which we are 'fairly sure' will enclose the population value; the distinction can be expressed by saying that a *point estimate* (i.e. a single value) is not nearly as valuable as an *interval estimate* (i.e. a range of values). We will take the first steps to make this argument more precise when we consider in the next section the interval estimate of the population median; the discussion of the interval estimate of the population mean cannot be given until 6.4.

---

†This is a simple property of any continuous distribution and most sampling distributions are more or less continuous. The area under the frequency curve corresponding to any single value of the variable is zero; we have not explored the exact meaning of zero area, but it must clearly correspond, at the very least, to a most unlikely event.

2.5.1 Interval estimation

We must begin by explaining what is meant by saying that we are 'fairly sure' that the interval given by our estimation procedure covers the population median. As before we imagine that we have taken n observations $x_1, x_2, x_3, \ldots, x_n$ from a population whose median M is not known to us. We want to give a rule for finding, from the n observations, two numbers, which we will call M_L and M_U, with the property that we are 'fairly sure' M lies in the interval between them, that is that $M_L < M < M_U$; 'fairly sure' is taken to mean that, if we use this procedure on a large number of samples, we shall be right 'most of the time,' and this expression in turn is conventionally interpreted as '19 times out of 20'. Several aspects of this interpretation need to be explained.

(*a*) 19 times out of 20, equivalent to 95 times out of 100, corresponds to a probability of 0.95 (measured on the usual scale from 0 to 1) or of 95 per cent (measured on a scale from 0 to 100). Thus our requirement can be expressed more formally as:

$$P(M_L < M < M_U) = 0.95.$$

The limits M_L and M_U are known as *95 per cent confidence limits* for the median.

(*b*) If our limits enclose M 19 times out of 20, they will fail to enclose it once in 20 times. Thus if we make a practice of using this interval estimate for the population median, our procedure will give us a wrong answer once in 20 times, that is it will, with this frequency, give limits which do not enclose the population value. This frequency will work out only in the very long run; it is no use thinking the rule will give a correct answer the first 19 times we use it and a wrong answer next time – indeed we can never know in any given case whether the answer is right or wrong. This is the price that we must pay for working with a variable that exhibits random variation; we cannot find a (useful) procedure which always gives the right answer.

(*c*) The choice of 0.95 as the critical probability is no more than a convention, but the practical success that has followed much research and development work in which this probability has been used is evidence of a sort that the convention is not unreasonable for many applications. Sometimes, however, we wish to be more sure than this and to impose a stricter requirement on the limits. Two other probabilities which are conventionally used are 99 per cent (right 99 times out of 100, wrong once in 100 times) and 99.9 per cent (right 999 times out of 1000, wrong once in 1000 times); as before these frequencies must be interpreted only in the very long run. If we use strict probabilities we find, of course, that the intervals to which they lead are wider than those given by less stringent requirements; to be more sure that the interval is correct, we have to take it wider.

2.5.1.1 Interval estimate for the median of one population

The problem of finding confidence limits for the median was solved approximately by Thompson (1936). His method was to arrange the n observations of the sample in order of magnitude, and to use as limits the kth observation from each end of the run, choosing k so that the probability that the limits enclosed the median was greater than 0.95, but as near to it as possible – it is not usually possible to obtain a probability of exactly 0.95 because k is necessarily a whole number. Nair (1940) calculated the value of k for $n = 6, 7, \ldots, 81$, giving values appropriate to 95 per cent and 99 per cent confidence; Nair's table is reproduced as Table A2.

We will illustrate the use of this procedure by considering again the observations on fattening period in Table 2; here $n = 25$. The observations rearranged in ascending order are: 105, 106, 106, 109, 109, 111, 111, 113, 114, 114, 115, 115, 116, 117, 118, 119, 121, 121, 123, 126, 137, 142, 147, 151, 163. The median, as noted above, is 116. For $n = 25$, Table A2 gives $k = 8$ for approximately 95 per cent confidence and $k = 6$ for approximately 99 per cent confidence. The eighth observations from each end are 113 and 121; the sixth are 111 and 126. The exact confidence levels, given in the table, are 95.68 per cent (or 0.9568) and 99.60 per cent (or 0.9960). Thus we can write

$$P(113 < M < 121) = 0.9568$$

and $\qquad P(111 < M < 126) = 0.9960.$

The sample median, 116, is the best we can do if we are asked to give a single value for the population median; it is almost certainly wrong, and gives us no indication of how wrong it may be. The limits on the other hand define a range which we can say, with known confidence, will enclose the population median: we should give limits to an estimate on every possible occasion.

2.6(c)   Working with a computer

The three sections in this chapter illustrating the use of a computer package have contained only the sequences of commands needed for the particular analyses under consideration. There has been no discussion of the two stages necessary before inputting the commands, which are those of establishing contact with the computer and then telling it which package you wish to use. These are essential stages but the details of the procedures involved vary between installations and cannot usefully be discussed in a book concerned with statistical analyses rather than with computation; whether you are using a micro or a mainframe you will need to consult the local professionals and the relevant manuals.

One matter of procedure which you are almost certain to need to consider is the choice between the two methods of instructing a computer known as *interactive* (or *on-line*) and *batch mode*. The choice does not exist in all packages at all installations but is nevertheless of very wide importance. Working on-line means that the computer responds to each command as it receives it from your keyboard, either executing the command or storing it for use later in the computation. In batch mode on the other hand the computer receives and stores all the commands needed for an analysis and executes them together at a convenient later moment. The selection of this moment may be entirely up to you or may be determined largely or completely by the computer; here again you need to be aware of the practices used at your local installation.

If this choice is available you will wish to balance the advantages of the two methods. Working on-line has all the flexibility and directness of using pen and paper and a desk calculator; you can see the result of each step as you take it and can use this result in your consideration of what to do next. On the other hand, mistakes may be more troublesome than with pen and paper; you may find for example that you have inadvertently altered or removed results that you need to use again at a later stage of the analysis and that your error cannot be corrected simply by using the edit facilities which the package may contain. Working in batch mode means that you can eliminate many errors, both typographical and other, before attempting to submit your analysis; however, most mainframe computers use a queueing system in processing jobs in batch mode so that there may be some delay before your analysis is carried out. In some circumstances this possibility may actually be an advantage; thus if you have a long series of analyses you may be able to use less of your allocation of computer resources by instructing the computer to run your program at night or at a cheap time of day.

Given that your local installation is powerful enough for queueing delays to be in most cases only a few seconds, the balance of advantage is usually with working in batch mode.

2.6.1(c) Inputting data and other uses of files

In the previous section we saw that one of the advantages of working in batch mode is the opportunity this gives to detect and correct errors in commands and data before these are submitted to the computer; this consideration is especially relevant when preparing large amounts of data. In this circumstance it is often useful to develop still further the concept of preparing the program in advance by preparing separately the commands and the data; the easiest way of doing this is to set up separate *files* for each part of the program. In some installations this is done within the package but in others it can be undertaken within the local command language; this may well prove to be an easier method of working but it will require that at the start of the package commands we give a statement of which existing local files are to be used in the analysis.

In the same way it may well be convenient to place the output from our analysis in a file which is readable in the command language as well as in the statistical package. To do so may make available to us a wider range of output devices – printers, plotters and so on – than is available using only commands within the package.

A further category of files occurs when using SPSS-X; these files are the *system files* of that language which contain commands, data and computations arranged in the particular format it uses. The characteristics of this language mean that its system files often need to be stored and handled somewhat separately from files in the local command language; but they are important and their correct use is an integral part of working in SPSS-X.

All these topics can be mentioned here only in very general terms because the details depend on the local installation but it is worth a lot of effort to get to know the local procedures and possibilities. To do so is an essential part of learning to make efficient use of the great assistance that a computer can offer.

2.7 Exercises

Exercises 1–8 require the plotting of histograms from sets of observations, as is described in 2.1.1. Before you begin to extract the frequencies in each example, try to assess the form of the distribution from looking over the individual observations, and see also how well you can estimate the mean and median. You may be suprised at how difficult it is to make these assessments from the complete table of observations.

1 Measurements of concentration of spermatozoa were made on 45 samples of bull semen; plot the results, given below in units of 10^8 spermatozoa/ml, in a suitable histogram.

13.8	12.3	9.7	8.6	10.2	12.8	14.1	12.6	12.7
9.0	14.2	13.3	14.8	14.3	13.9	11.0	9.7	14.3
9.5	12.3	11.3	13.9	15.1	11.0	9.6	13.2	12.0
11.8	7.5	12.5	12.1	11.7	10.5	12.2	11.0	12.2
14.2	10.4	15.9	12.0	13.8	9.3	12.9	12.4	10.4

2 The amount of DNA in 95 rat liver cells was scored on stained sections; plot the results, given below, in a suitable histogram. Can you suggest an explanation for the trimodal form of the histogram?

3.4	13.2	6.7	1.4	1.3	3.8	3.9	2.9	13.2
3.9	2.7	4.4	3.6	4.4	2.4	3.6	3.1	7.5
2.9	7.8	2.7	3.9	3.3	1.7	2.0	4.4	3.3
0.7	3.9	1.6	5.6	3.0	3.4	1.4	3.5	2.8
1.4	1.9	2.3	2.9	2.8	1.5	4.1	5.9	3.1
8.7	2.8	3.8	13.0	3.0	3.0	4.1	2.8	3.8
1.5	2.2	4.3	2.4	1.1	2.8	3.7	14.1	2.6
3.5	4.6	4.4	3.8	6.7	2.6	1.2	9.9	3.1
5.4	2.9	3.5	4.3	2.6	7.2	3.2	3.1	2.4
2.4	3.2	6.7	2.1	3.1	3.5	2.3	3.0	3.5
3.4	5.1	2.9	2.8	7.6				

3 The weight gains (kg) of 40 beef steers in their first fattening period are given below: plot these values in a suitable histogram.

49.5	58.7	46.6	62.5	49.5	59.1	56.1	53.4	49.8
46.4	49.7	50.4	55.9	55.5	56.0	47.1	50.6	47.3
57.0	53.2	58.5	54.0	48.9	48.7	53.4	53.7	54.5
55.1	53.3	53.5	50.3	49.7	47.2	49.7	50.9	45.8
64.6	52.7	50.3	57.8					

4 The numbers of eggs laid by 60 hens in a two-week period are given below; plot these values in a suitable diagram.

3	7	8	9	7	9	4	8	10	8	10	10
10	7	11	7	9	5	11	9	9	9	8	8
5	12	10	8	6	8	11	9	9	7	8	4
9	10	9	6	6	9	8	11	8	9	9	9
9	8	10	10	7	9	8	12	7	9	9	9

5 The numbers of seeds in 65 seed cases from a new variety of sweet pea were as follows; plot these values in a suitable diagram.

5	4	8	8	5	3	3	5	11	5	7	7
9	4	6	13	8	4	14	8	2	5	16	8
14	7	8	3	8	9	10	6	12	5	5	5
6	5	16	2	8	6	10	9	7	6	6	2

11	11	8	9	9	8	11	5	7	15	5	2
5	11	4	8	3							

6 In a hospital investigation, the numbers of hours which 35 patients slept after a certain form of anaesthetic were recorded. Plot the results, given below, in a suitable histogram.

3.1	0.9	17.3	10.3	7.6	11.0	7.1	6.6	4.0
5.4	3.1	1.4	5.2	3.7	13.0	7.6	7.0	4.5
2.3	13.2	7.7	11.8	11.9	3.0	6.6	2.9	9.7
7.8	3.4	8.3	4.6	7.2	1.1	3.5	3.9	

7 As part of the validation of a new assay procedure, 25 independent determinations of relative potency against a standard were made on a single batch of a drug. Plot the results, given below, in a suitable histogram.

2.41	2.34	2.90	2.75	2.66	2.86	2.29	1.83	2.08
3.22	2.37	2.27	2.38	2.51	2.22	2.97	3.08	2.57
2.85	3.04	3.22	2.88	1.88	3.04	2.20		

8 In a growth study involving 40 schoolgirls the following bone lengths (cm) were obtained: plot these values in a suitable diagram.

17.6	14.1	13.9	13.1	16.8	14.7	13.2	15.7	13.5
15.4	14.3	13.9	13.2	13.6	15.3	13.2	16.6	14.7
18.1	18.4	15.6	15.5	15.4	14.7	17.6	14.4	19.8
13.6	14.7	14.7	13.9	17.8	18.2	13.8	13.1	13.5
13.6	14.9	14.7	14.1					

Exercises 9–13 are based on numbers 1, 3 and 6–8 above: in each case the histogram is to be drawn, but the frequencies are given for intervals which are not all of the same width. The method of drawing the histogram in this circumstance is shown in Fig. 2b.

9 The observations in exercise 1 led to the following table; plot the histogram.

Concentrations	7.0–8.9	9.0–10.9	11.0–11.9	12.0–12.9	13.0–13.9	14.0–15.9
Frequency	2	10	6	13	6	8

10 The observations in exercise 3 led to the following table; plot ths histogram.

Weight gain	44.0–47.9	48.0–49.9	50.0–51.9	52.0–53.9	54.0–57.9	58.0–65.9
Frequency	6	8	5	7	9	5

11 The observations in exercise 6 led to the following table; plot the histogram.

Hours of sleep	0–1.9	2.0–3.9	4.0–5.9	6.0–7.9	8.0–11.9	12.0–17.9
Frequency	3	9	5	9	6	3

12 The observations in exercise 7 led to the following table; plot the histogram.

Relative potency	1.80–2.19	2.20–2.39	2.40–2.79	2.80–2.99	3.00–3.39
Frequency	3	7	5	5	5

13 The observations in exercise 8 led to the following table; plot the histogram.

Bone length	13.0–13.4	13.5–13.9	14.0–14.4	14.5–14.9	15.0–15.9
Frequency	5	9	4	7	6

Bone length	16.0–16.9	17.0–19.9
Frequency	2	7

To gain some experience of sample to sample variations, try to devise a measurement which you can make on a large number of experimental units – plants, animals, culture dishes or whatever is readily available. Try to choose a measurement which has some practical meaning in your work rather than select a 'statistical' trial such as tossing pennies or rolling a die. If you cannot find any measurement which you can make a large number of times, you can use Table A22, which contains 1500 observations from one population; but it is much better to use your own measurements. Exercises 14–16 are based on your sets of observations.

14 Take a series of samples of 30 observations and plot the histogram for each; note both the similarities and the differences between the histograms.

15 Repeat exercise 14 using samples of 60 observations: compare the results with those in that exercise.

16 In the manner of Fig. 3 draw histograms for samples of 60, 120, 180, 240, 300, ... observations; note how the irregularities gradually become less evident as the sample size increases.

Exercises 17–19 require the pictorial presentation of multiple measurements; the methods used in 2.4.1 are appropriate.

17 The numbers of 6 types of cells in unit volume of each of 15 samples are given below; present these data to show as clearly as possible the average composition of the samples and the differences between samples.

	Cell type					
Sample	A	B	C	D	E	F
1	4	25	30	157	66	7
2	5	29	21	148	68	16
3	2	25	29	171	66	23
4	5	24	26	131	44	17
5	5	27	31	154	56	16
6	6	28	28	144	55	12
7	5	32	32	168	55	14
8	2	39	24	164	53	20
9	4	16	32	160	61	19
10	7	30	29	160	52	17
11	5	27	26	166	60	21
12	4	32	25	146	58	9
13	7	39	29	158	69	19
14	2	33	32	158	76	6
15	8	27	26	161	52	8

18 The numbers of 7 species of gulls in each of 12 colonies are given below; present these observations to show as clearly as possible the average composition of the colonies and the differences between colonies.

	Gull species						
Colony	A	B	C	D	E	F	G
1	1437	219	91	464	21	176	401
2	1430	186	41	368	17	77	200
3	1615	364	62	461	14	84	270
4	1340	152	41	360	20	150	512
5	1631	168	43	282	14	71	234
6	2047	133	43	412	19	73	290
7	1136	160	66	343	21	70	528
8	1622	156	55	370	18	72	1211
9	1228	325	47	301	15	103	326
10	1749	161	43	377	15	83	222
11	2209	237	40	323	28	77	232
12	1911	132	46	452	15	75	395

19 As part of an investigation of land use, samples of farms were taken from 3 districts, and the acreages in 5 main types of crop were noted on each farm. Present the observations, given below, as clearly as possible.

			Crop		
Farm	Permanent grass	Ley	Cereals	Roots	Woodland
District *A*					
1	190	46	162	103	10
2	50	89	248	13	103
3	137	10	175	30	0
4	282	95	270	105	0
5	230	15	226	27	7
6	68	49	459	45	13
7	33	86	313	92	11
District *B*					
8	16	35	436	60	0
9	82	39	152	9	6
10	93	98	481	0	26
11	70	70	247	11	0
12	81	186	167	104	10
13	160	23	444	0	53
14	150	148	195	23	10
15	31	178	183	0	0
District *C*					
16	70	19	343	8	7
17	252	266	245	14	11
18	21	43	138	33	3
19	261	27	291	31	107
20	23	20	334	82	22
21	284	36	248	27	85
22	458	17	378	12	84
23	7	14	445	10	0
24	11	21	344	153	0
25	34	49	265	24	9

Exercises 20–3 involve plotting scatter diagrams, as in 2.4.2. Just as in the earlier exercises (1–8) it was instructive to try to assess from the table of results the form of the distribution for the single variable, so in the present exercises it is worthwhile to try to assess the form of the relationship between the two variables. Once again you may find it is not very easy to do this from the table of observations, but one important feature of the diagram which is even more difficult to assess from the table is how far the width

of the scatter is the same all the way along the diagram. These considerations emphasize the importance of drawing diagrams and not trying to rely on 'running your eye over the figures'.

20 As part of a growth study on calves a certain body measurement (cm) was taken at birth (x) and at one year of age (y) on 35 heifer calves; plot the observations, given below, in a scatter diagram.

x	y	x	y	x	y
22.2	28.5	19.3	28.3	24.0	38.5
19.0	22.4	22.3	34.2	22.9	30.0
21.3	27.1	25.7	37.2	18.2	23.7
20.5	29.0	23.2	31.6	21.2	28.4
18.5	27.7	21.5	33.0	24.7	39.8
19.0	29.1	25.2	36.3	18.9	28.6
24.8	31.5	25.0	37.6	23.2	34.9
19.1	23.6	19.7	31.7	19.9	34.1
25.1	38.3	25.0	36.6	23.1	33.2
24.7	39.3	20.3	30.1	18.9	28.7
22.1	34.3	19.3	29.4	19.5	32.5
20.8	35.0	24.0	39.9		

21 The following observations of height (x, cm) and weight (y, kg) were taken on a group of 50 overweight children; plot the observations in a scatter diagram.

x	y	x	y	x	y
123.5	56.6	129.2	62.1	144.7	68.1
148.2	84.3	122.7	56.0	128.7	58.4
139.6	70.2	144.3	71.9	124.4	51.3
121.5	47.2	123.8	48.7	135.8	62.6
144.3	71.3	131.0	61.5	128.7	56.7
133.6	71.4	118.9	48.6	142.2	66.9
127.8	48.7	145.7	77.7	148.5	52.3
124.7	52.7	145.2	65.7	138.8	63.0
139.0	67.9	127.8	56.0	136.3	65.5
136.0	61.3	135.0	57.1	124.5	61.4
133.0	53.5	147.4	81.3	134.7	55.9
146.2	82.7	137.3	55.2	119.8	47.9
139.5	62.0	129.4	54.6	140.1	66.2
135.3	62.3	128.6	57.9	131.6	48.7
129.4	52.4	141.4	59.4	143.5	77.8
148.8	61.5	138.1	58.8	137.9	50.2
143.1	73.6	125.6	54.3		

22 A psychologist, investigating response of sea anemones to a standard stimulus, measured the duration of response (y, sec) and the size of the anemone (x, conventional scale); plot the following observations, obtained on 21 individuals, in a scatter diagram.

x	y	x	y	x	y
0.2	1.1	6.2	6.3	6.9	6.0
3.9	5.2	0.3	2.2	4.7	6.0
5.8	6.0	4.2	5.3	2.5	3.8
7.8	6.2	0.9	2.2	1.6	3.3
4.9	5.9	3.5	5.0	8.2	6.3
8.6	6.6	1.8	3.1	5.3	5.7
3.1	4.5	3.7	5.1	1.9	4.8

23 A mycologist measured the response of a certain mould to 10 levels of manganese in the culture medium, obtaining the following growths of mould (y, μg). Plot the observations in a scatter diagram, using the logarithm of manganese concentration as independent variable (x).

Concentration (p.p.m.)	1	2	4	8	16
Growth (μg)	9.8	14.3	18.7	21.1	22.4
Concentration (p.p.m.)	32	64	128	256	512
Growth (μg)	22.8	21.8	20.7	19.7	18.4

Exercise 24 *also concerns the relationships between variables, but is more difficult as more than two variables are involved; not only does this require us to plot more than one diagram, but we may also wish to consider whether the relationship between two variables is in any way influenced by a third variable. This last point is difficult to investigate by pictorial methods, requiring* multiple regression *techniques, which are beyond the scope of this book.*

24 Thirty sides of bacon were cut across and measurements made on the cut face: A and B were measured on the lean and C and H on the fat, all in mm. Draw scatter diagrams to investigate the relationships between these variables.

Side	A	B	C	H		Side	A	B	C	H
1	76	41	18	21		16	76	47	20	34
2	83	44	16	29		17	83	50	20	35
3	90	49	18	35		18	95	55	19	38
4	64	39	21	24		19	82	45	18	34
5	74	47	21	34		20	94	55	17	42
6	97	51	17	43		21	104	50	14	43
7	68	47	22	32		22	90	54	18	42
8	81	41	20	27		23	91	53	19	39
9	97	49	15	36		24	110	59	15	48
10	79	55	20	32		25	98	51	16	35
11	94	46	17	35		26	90	65	19	49
12	96	58	17	46		27	107	66	14	53
13	89	40	15	27		28	93	54	18	39
14	90	55	18	41		29	111	59	13	49
15	106	51	14	38		30	118	66	13	54

Exercises 25–32 use the observations given in exercises 1–8 above. In each case obtain point estimates of the population mean, using the sample mean $\bar{x}$ (see 2.2), and the population median, using the sample median (see 2.2.1). In each case compare these two estimates, and try to detect how the difference between them is related to the skewness of the distribution: for a positively skew distribution the mean is greater than the median, but sample estimates of skewness, mean and median do not always show this. Obtain also confidence limits for the population median (see 2.5.1.1); remember that we have not yet considered how to obtain confidence limits for the population mean.

25 Use the observations in exercise 1 in the manner just described.

26 Use the observations in exercise 2 in the manner just described.

27 Use the observations in exercise 3 in the manner just described.

28 Use the observations in exercise 4 in the manner just described.

29 Use the observations in exercise 5 in the manner just described.

30 Use the observations in exercise 6 in the manner just described.

31 Use the observations in exercise 7 in the manner just described.

32 Use the observations in exercise 8 in the manner just described.

Exercises 33–6 are concerned with estimating means and medians using the grouped observations given in exercises 10–13 above. Grouping observations wastes information, so that estimates obtained after grouping are less satisfactory than those obtained from individual observations; nevertheless it is useful to have methods available for use with grouped observations as occasionally these are all that are available. The methods are illustrated on the values given in exercise 9.

Denote by $x_0, x_1, \ldots$ the dividing values between the groups, by $f_1, f_2, \ldots$ the numbers of observations in the groups (f_1 in x_0 to x_1, etc.) and by $F_1, F_2, \ldots$ the total number of observations in and below each group (F_1 less than or equal to x_1, F_2 less than or equal to x_2, etc.). Then for exercise 9 we have

$$
\begin{aligned}
x_0 &= 6.95 \\
x_1 &= 8.95 & f_1 &= 2 & F_1 &= 2 \\
x_2 &= 10.95 & f_2 &= 10 & F_2 &= 12 \\
x_3 &= 11.95 & f_3 &= 6 & F_3 &= 18 \\
x_4 &= 12.95 & f_4 &= 13 & F_4 &= 31 \\
x_5 &= 13.95 & f_5 &= 6 & F_5 &= 37 \\
x_6 &= 15.95 & f_6 &= 8 & F_6 &= 45
\end{aligned}
$$

Note how the values of x_i are defined; thus $x_1 = 8.95$ means that values less than 8.95, recorded as 8.9 or less, appear in the first group, as we are given, whilst values greater than 8.95, recorded as 9.0 or more, appear in the second group.

In estimating the mean we need the central values of the groups, which we may denote by $X_1, X_2, \ldots$ Thus for exercise 9

$$X_1 = 7.95, \qquad X_2 = 9.95, \qquad X_3 = 11.45,$$
$$X_4 = 12.45, \qquad X_5 = 13.45, \qquad X_6 = 14.95.$$

Then the point estimate of the population mean is

$$\frac{f_1 X_1 + f_2 X_2 + \cdots}{f_1 + f_2 \cdots} = \frac{\sum fX}{\sum f}.$$

In the present case this gives

$$\frac{2 \times 7.95 + 10 \times 9.95 + 6 \times 11.45 + 13 \times 12.45 + 6 \times 13.45 + 8 \times 14.95}{2 + 10 + 6 + 13 + 6 + 8}$$
$$= (15.9 + 99.5 + 68.7 + 161.85 + 80.7 + 119.6)/45$$
$$= 546.25/45 = 12.14$$

In estimating the median we need first the value of $F_{0.5}$, the cumulative frequency which divides the sample into two equal parts in the manner described in 2.2.1: for the observations of exercise 9 we have $F_{0.5} = 23$. We then note the limits x_i, x_{i+1} of the group in which $F_{0.5}$ occurs, noting also the corresponding cumulative frequencies F_i, F_{i+1}; for the present case $x_i = 11.95$, $x_{i+1} = 12.95$, $F_i = 18$, $F_{i+1} = 31$. $F_{0.5}$ will, of course, satisfy $F_i \leqslant F_{0.5} \leqslant F_{i+1}$. Then the point estimate of the median is

$$x_i \frac{F_{i+1} - F_{0.5}}{F_{i+1} - F_i} + x_{i+1} \frac{F_{0.5} - F_i}{F_{i+1} - F_i}$$

In the present case this gives

$$11.95 \times \frac{31 - 23}{31 - 18} + 12.95 \times \frac{23 - 18}{31 - 18} = \frac{11.95 \times 8 + 12.95 \times 5}{13}$$
$$= \frac{95.6 + 64.75}{13}$$
$$= \frac{160.35}{13} = 12.33.$$

Both these estimation procedures are based on the approximation that the observations in each group are uniformly distributed. This is clearly no more than an approximation but no other assumption is any better.

The method for the median can be extended to provide an estimate of the value of the variable corresponding to any chosen cumulative frequency. Thus if we wanted to estimate the confidence limits of the median we would first refer to Table A2, which shows that for $n = 45$ and approximately 95 per cent confidence we have $k = 16$, and would then determine the cumulative frequencies corresponding to the 16th observation from each end of the ordered sample, namely $F = 16$ and $F = 30$. We would then use these values in place of $F_{0.5}$, defining x_i, x_{i+1}, F_i and F_{i+1} for each chosen value of F. Such a procedure can however give only confidence limits with a much lower confidence level than the nominal value; the exact value of the confidence level would depend on the grouping used and would not be easy to determine.

The complexity of the calculations may perhaps help to convince the reader that grouping observations is a practice to be used sparingly.

33 Use the observations in exercise 10 in the manner just described.

34 Use the observations in exercise 11 in the manner just described.

35 Use the observations in exercise 12 in the manner just described.

36 Use the observations in exercise 13 in the manner just described.

Exercises 37 and 38 use the observations obtained in exercises 14 and 15 to study the sample-to-sample variations of the point estimates of mean and median, and of the confidence limits for the median. The 'confidence' property of these limits requires that they should enclose the true median in a known proportion of samples in the very long run; it is not however possible to make any useful statement about confidence in the short run. This point is illustrated in Fig. A2 which shows, for two distributions with known medians, the confidence limits obtained for the median in 100 samples each containing 17 observations. We see from Table A2 that the 'approximately 95 per cent confidence' limits for $n = 17$ have in *fact 95.10 per cent confidence, but the limits in the first set of 100 samples actually enclose the median in 98 cases, whilst the limits in the second set enclose the median in 95 cases.*

37 Using the observations you obtained in exercise 14, calculate point estimates of the mean and median for each set of 30 observations; examine how the values of these estimates vary between the sets. Obtain also for each set the confidence limits for the median, and note how these limits vary between the sets.

38 Repeat exercise 37, using the sets of 60 observations you obtained in exercise 15. Try to compare the variations of these estimates between sets with the sampling distributions obtained in exercise 37 for sets of 30 observations.

3 COMPARING SEVERAL SAMPLES

3.1 The problem

The methods described in chapter 2 enable us to summarize some of the information contained in one sample. Further kinds of summary for use in this case will be given later, but we turn first to consider how the methods already set out need to be modified and extended when we have several samples and wish to investigate the possibility that they come from different populations. We remind ourselves that all the questions that really matter will concern the populations; it is of no interest to know whether the samples differ because, whether they do or not, the next samples we take will almost certainly do so even if the populations are identical. The fact of random variation means, as before, that we are interested in the samples only for what they tell about the populations.

3.1.1 Pictorial presentation

The first kind of summary that we used for one sample was a histogram. When we have several samples we can draw a histogram for each, as we did for subsamples in 2.1.1.2. Similarly if we have multiple measurements we can, in the manner of 2.4.1, draw a pie diagram or a bar chart for each sample. We shall usually find however that, although these methods are useful first summaries, they do not enable us to see very clearly the differences between the samples; in particular they give no indication whether these differences are reflections of genuine population differences or whether they can be regarded simply as artefacts produced by random variation. Such random differences will of course occur even when the populations are identical; this is no more than saying that samples from a single population will not all be the same.

3.2 Interval estimate for a difference of medians

Since pictorial methods are not sufficient for evaluating differences between populations, we consider how the various numerical summaries, which we used for a single sample in chapter 2, can be applied to the case of many samples. The median and mean can be obtained for each sample, but these by themselves are no more helpful than were histograms or pie diagrams for each sample, because we do not know how far the apparent differences are real. As in the case of pictorial representations, so in the case of medians and means, differences between samples will occur even if the populations are identical.

We need therefore to try to extend the idea of confidence interval, which was introduced for one sample to allow for the random variation unavoidable in an estimate and might be hoped to extend to deal with random variation in two or more estimates. Thus, taking first the simple case of two samples, we could consider the difference of sample medians as a point estimate of the difference of population medians and could seek to replace the point estimate by a suitable interval estimate.

A solution to this problem is given by Lehmann (1963), and is somewhat similar to the one-sample procedure used in 2.5.1.1. Suppose we have n_1 observations $x_1, x_2, \ldots, x_{n_1}$ of the variable x and n_2 observations $y_1, y_2, \ldots, y_{n_2}$ of the variable y. Suppose further that we are justified in assuming that the distributions in the two populations are of the same shape and size but cover different ranges; the distributions in Fig. 10a satisfy this condition, but those in Figs. 10b and 10c do not. We say that the distributions differ in *location*; any quantity which specifies the location, as the median does in the present discussion, is called a *location parameter*. The assumption that two distributions differ only in location is frequently justified in practice, because x and y will often represent the same quantity (greenness, fattening period, and so on) measured under two sets of conditions which are sufficiently similar not to change the variability of the experimental material.

We calculate first all the differences that can be obtained by taking an observation in the x-sample from one in the y-sample;

there are $n_1 n_2$ of these differences, and we shall usually find that some are positive and some negative. We then arrange these differences in order of increasing magnitude, taking account of their signs when doing so. Thus suppose $n_1 = n_2 = 6$ and that the

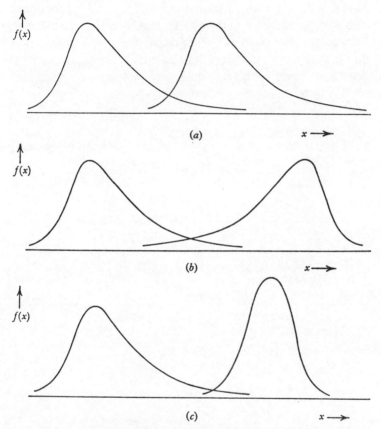

Fig. 10. Frequency functions differing only in location (Fig. 10a), or in location and shape (Figs. 10b and 10c); see 3.2.

values of x are 5, 4, 7, 10, 6, 14 and those of y are 11, 9, 12, 13, 8, 15. The calculation of the differences between the x's and the y's and their rearrangement into order is shown in Table 6.

The confidence limits for the difference between the population medians are two values in this sequence of sample differences, the proper values being chosen with the help of

Table A3. The quantity U given in this table is known as the *Wilcoxon statistic*; the method of calculating it is explained in 3.2.1. To choose the appropriate value of U we need to specify the sample sizes (n_1 and n_2) and the level of confidence required. With $n_1 = n_2 = 6$ and with (approximately) 95 per cent confidence, Table A3 gives U as 5 and we apply this by taking the fifth value from each end in the ordered sequence of differences, namely -2 and 8 (Table 6). This procedure enables us to say, with 95 per cent confidence, that the difference of population medians, $M_y - M_x$, lies between -2 and 8.

Table 6. *The 36 differences formed by subtracting each x-value from each y-value (see 3.2). In the second part of the table the differences are arranged in ascending order, taking account of sign.*

The 36 differences		y_1 11	y_2 9	y_3 12	y_4 13	y_5 8	y_6 15
x_1	5	6	4	7	8	3	10
x_2	4	7	5	8	9	4	11
x_3	7	4	2	5	6	1	8
x_4	10	1	-1	2	3	-2	5
x_5	6	5	3	6	7	2	9
x_6	14	-3	-5	-2	-1	-6	1

The differences arranged in order

-6	-5	-3	-2	-2	-1	-1	1	1
1	2	2	2	3	3	3	4	4
4	5	5	5	5	6	6	6	7
7	7	8	8	8	9	9	10	11

3.2.1* The distribution of U

The method of calculating the distribution, and hence the confidence values, of U will be illustrated for the case $n_1 = n_2 = 4$; larger values of n increase the length of the calculation beyond what is desirable in an example. We consider first the definition of the Wilcoxon statistic and then the calculation and use of its distribution.

We first rearrange the four observations x_1, x_2, x_3, x_4 of x with the four observations y_1, y_2, y_3, y_4 of y in a single sequence in ascending order of magnitude. If the samples were 1, 5, 3, 8 and 4, 10, 7, 6 this procedure would give

x_1	x_3	y_1	x_2	y_4	y_3	x_4	y_2
1	3	4	5	6	7	8	10.

We now count, for each value of x in turn, the number of y-values which precede it in the sequence; the total of these counts we denote by U_1. In the example x_1 is preceded by 0 y-values, x_2 by 1, x_3 by 0 and x_4 by 3; thus for these samples $U_1 = 4$. We obtain similarly U_2 for the y observations; y_1 is preceded by 2 x-values, y_2 by 4, y_3 by 3 and y_4 by 3, so that $U_2 = 12$. The smaller of U_1 and U_2 is the Wilcoxon statistic, so that in this case $U = 4$.

The relationship between this procedure and the ordering of differences discussed in 3.2 is that U_1 is the number of negative differences and U_2 is the number of positive differences. Thus for the present samples the ordered sequence of differences – a y-value minus an x-value – is

$$-4, -2, -1, -1, 1, 1, 2, 2, 3, 3, 4, 5, 5, 6, 7, 9;$$

four of these are negative $(U_1 = 4)$ and twelve are positive $(U_2 = 12)$. It is not difficult to see that this relationship holds in general.

The distribution of U is obtained by considering all the possible arrangements of $x_1, \ldots, x_4$ and $y_1, \ldots, y_4$ in a single sequence, each arrangement giving a value of U. The number of these arrangements is 70, this being the number of different ways in which $x_1, \ldots, x_4$ can be allocated to 4 places in the sequence of 8 values. The arrangements are all given, with their values of U_1, U_2 and U, in Table 7.

To apply this set of values to the problem of finding confidence limits for the difference of population medians, we consider first the distribution when the medians are equal; we shall find that this extends easily to the general case. When the medians are equal the populations are identical – we assumed in 3.2 that they were the same except in location – so that the 70 possible

Table 7. *The calculation of all possible values of U_1, U_2 and U for $n_1 = n_2 = 4$.†*

1	2	3	4	5	6	7	8	U_1	U_2	U		1	2	3	4	5	6	7	8	U_1	U_2	U
×	×	×	×					0	16	0		×	×	×	×					4	12	4
×	×	×		×				1	15	1		×	×	×		×				5	11	5
×	×	×			×			2	14	2		×	×	×			×			6	10	6
×	×	×				×		3	13	3		×	×	×				×		7	9	7
×	×	×					×	4	12	4		×	×		×	×				6	10	6
×	×		×	×				2	14	2		×	×		×		×			7	9	7
×	×		×		×			3	13	3		×	×		×			×		8	8	8
×	×		×			×		4	12	4		×	×			×	×			8	8	8
×	×		×				×	5	11	5		×	×			×		×		9	7	7
×	×			×	×			4	12	4		×	×				×	×		10	6	6
×	×			×		×		5	11	5		×		×	×	×				7	9	7
×	×			×			×	6	10	6		×		×	×		×			8	8	8
×	×				×	×		6	10	6		×		×	×			×		9	7	7
×	×				×		×	7	9	7		×		×		×	×			9	7	7
×	×					×	×	8	8	8		×		×		×		×		10	6	6
×		×	×	×				3	13	3		×		×			×	×		11	5	5
×		×	×		×			4	12	4		×			×	×	×			10	6	6
×		×	×			×		5	11	5		×			×	×		×		11	5	5
×		×	×				×	6	10	6		×			×		×	×		12	4	4
×		×		×	×			5	11	5		×				×	×	×		13	3	3
×		×		×		×		6	10	6			×	×	×	×				8	8	8
×		×		×			×	7	9	7			×	×	×		×			9	7	7
×		×			×	×		7	9	7			×	×	×			×		10	6	6
×		×			×		×	8	8	8			×	×		×	×			10	6	6
×		×				×	×	9	7	7			×	×		×		×		11	5	5
×			×	×	×			6	10	6			×	×			×	×		12	4	4
×			×	×		×		7	9	7			×		×	×	×			11	5	5
×			×	×			×	8	8	8			×		×	×		×		12	4	4
×			×		×	×		8	8	8			×		×		×	×		13	3	3
×			×		×		×	9	7	7			×			×	×	×		14	2	2
×			×			×	×	10	6	6				×	×	×	×			12	4	4
×				×	×	×		9	7	7				×	×	×		×		13	3	3
×				×	×		×	10	6	6				×	×		×	×		14	2	2
×				×		×	×	11	5	5				×		×	×	×		15	1	1
×					×	×	×	12	4	4					×	×	×	×		16	0	0

† The positions occupied by members of the first sample in the ranking $1, 2, \ldots, 8$ are indicated by × and the positions occupied by the second sample are left blank. Count U_1 is the total of the numbers of second sample members preceding the first sample members; U_2 is similarly defined, interchanging the samples. U is the smaller of U_1, U_2. The lines across the table help to show the frequency pattern.

arrangements are all equally likely. The distribution of U is therefore obtained simply by counting the frequencies with which the various values occur; there is no need to weight the values with their probabilities because they are all equally probable.

The distribution in the case of our example ($n_1 = n_2 = 4$) is given in Table 8; we see that in only 5.7 per cent of trials would U be less than or equal to 1, that is in only this proportion of trials would we find as few as one negative value or as few as one positive value in the sequence of ordered $(y - x)$ differences. Remembering that the true difference of medians has been taken as zero, so that negative values correspond to sample values below the true difference and positive values correspond to sample values above the true difference, we can say that in only 5.7 per cent of trials would we find either as few as one sample difference less than or equal to the true difference or as few as one sample difference greater than or equal to the true difference. We thus take as 5.7 per cent confidence limits for the true difference of medians, the first value from each end of the ordered sequence of sample differences.

Table 8. *The distribution of U for $n_1 = n_2 = 4$, summarized from Table 7; the last column contains the probabilities used in the Mann–Whitney test.*

U	Frequency	Cumulative frequency	Cumulative percentage frequency
0	2	2	2.9
1	2	4	5.7
2	4	8	11.4
3	6	14	20.0
4	10	24	34.3
5	10	34	48.6
6	14	48	68.6
7	14	62	88.6
8	8	70	100.0
	70		

Since 5.7 per cent is greater than the conventional confidence level of 5 per cent, we find that we cannot for $n_1 = n_2 = 4$ obtain confidence limits at a level less than or equal to 5 per cent because to do so we would have to take terms in the ordered sequence which were less than one from each end, which is clearly impossible! The above form of calculation is too lengthy to give in full for $n_1 = n_2 = 6$, which we used in 3.2, because we would have to list 924 (rather than 70) possible arrangements; if this is

done we find that in only 4.2 per cent of arrangements is U less than or equal to 5, the value used in 3.2, so that the fifth differences from each end of the ordered sequence can be used as 4.2 per cent confidence limits for the true difference of medians. We are not in general able to give limits with exactly 5 per cent confidence because U is a discrete variable.

3.2.2 Significance test for a difference of medians

The limits obtained in 3.2 provide a useful and meaningful statement about the difference between the population medians, but it is sometimes possible to draw useful conclusions by considering only the question 'can the difference of population medians be regarded as zero?' If, for example, we were comparing a new cheap drug with a well established but expensive treatment, we might feel that the crucial question was whether the new drug was as effective as the old; we would therefore wish to concentrate our analysis on examining the reasonableness of the hypothesis that the drugs were equivalent.† On the other hand, if we were evaluating two forms of the new drug, we would want as complete a statement as possible about the difference between them, which the confidence limits would provide but an analysis limited to investigating the one possibility of zero difference would not. The contrast between these examples shows that the choice of analysis may be affected by our aims in the investigation, and is therefore to that extent a subjective choice. In other words, the statistical analysis gives objective answers to the questions we choose to ask but does not necessarily help us in the choice of questions; this very general limitation is of real importance, and should be constantly in the mind of anyone using statistical methods.

An analytical procedure which concentrates on assessing the reasonableness of regarding two (or more) populations as the same in some respect is called a *significance test*, and the particular hypothesis of equality which is studied is called the *null hypothesis*. In the present case the null hypothesis is that the population medians are equal. This was the assumption on which the

† Such an approach can be criticized on the grounds that we ought to consider the seriousness of a wrong conclusion. This aspect of the problem is considered in chapter 5.

distribution of U was calculated in 3.2.1, so that this distribution can be used as the basis of the significance test.† To decide whether the hypothesis of zero population difference is compatible with the sample values, we obtain the probability of such values occurring when the populations really are the same; if this probability is not too small we can regard the agreement between the null hypothesis and the samples as reasonable and *accept* the null hypothesis. If on the other hand this probability is small we reject the hypothesis. By convention, as in the calculation of confidence limits, we regard a probability greater than 5 per cent as reasonable, that is we accept the null hypothesis if, when we assume it to be true, we find a probability greater than 0.05 of values such as those in our sample. Stricter probabilities (1 or 0.1 per cent) are used for the same reasons as in calculating confidence limits.

The distribution of U when the population medians were equal showed that for $n_1 = n_2 = 4$ values of U less than or equal to 1 occurred in 5.7 per cent of trials whilst the value 0 occurred in 2.9 per cent of trials. Thus if we found $U = 0$ in our samples we would reject the null hypothesis, because such a value is (conventionally) unlikely when the medians are equal, but if we found $U = 1$ we would accept the hypothesis because such a value could occur (conventionally) reasonably often.‡ A significance test performed in this way will from time to time be wrong, in just the same way as are confidence limits, because there will be occasions when the null hypothesis is true but we nevertheless reject it because, by chance, a small value of U has occurred. All that we can hope to do is to determine how often, in the very long run, this kind of error will occur.§

† Readers who omitted 3.2.1 can follow the next stages of the argument by accepting the distribution without working through its derivation.

‡ A form of wording which should never be used is 'the probability that the null hypothesis is true is (say) 2.9 per cent'. On the frequency approach to probability this phrase has no meaning whatever, because we cannot talk about a distribution of hypotheses or about the probability of any particular hypothesis being true. Some other definitions of probability (see, for example, 5.2) do give meaning to these statements, but they are not permissible when we use the frequency definition.

§ This error, rejecting the null hypothesis when it is true, is referred to as the *Type I error*. There is also the possibility of accepting the hypothesis when it is false; this is known as the *Type II error*. We do not at this stage consider how often it occurs.

We note that if we accept the null hypothesis we are not entitled to say that we have proved the populations to be the same in that respect. We could well imagine many other hypotheses which would agree reasonably well with the sample values, and we cannot claim that any one of these hypotheses is 'right' in any useful sense. The significance test approach should be used only when the null hypothesis is meaningful in the context of our investigation; forgetting this caveat, as far too often happens, can lead to conclusions that are of no relevance to the object of the analysis.

3.2.2.1 The Mann–Whitney procedure

The simplified calculations which suffice for the significance test will be illustrated on the samples used in 3.2. We first rearrange each sample in ascending order as follows:

x-sample	4	5	6	7	10	14
y-sample	8	9	11	12	13	15

We now assign to each observation its rank in the joint ordering of the two samples together, as follows:

x-sample	1	2	3	4	7	11
y-sample	5	6	8	9	10	12

We now total the ranks for each sample, obtaining

$$R_1 = 28, \quad R_2 = 50;$$

we note that $\quad R_1 + R_2 = \tfrac{1}{2}(n_1 + n_2)(n_1 + n_2 + 1),$

which is a useful check on the arithmetic. We now calculate U_1 and U_2 by the formulae:

$$U_1 = n_1 n_2 + \tfrac{1}{2}n_2(n_2 + 1) - R_2,$$
$$U_2 = n_1 n_2 + \tfrac{1}{2}n_1(n_1 + 1) - R_1,$$

giving $U_1 = 7$, $U_2 = 29$; in this case the appropriate check is $U_1 + U_2 = n_1 n_2$. The significance test is performed by noting whether either U_1 or U_2 is less than or equal to the tabulated value U, which was first used in 3.2 in obtaining confidence limits for the difference of population medians. In the present case the tabulated value is 5 and as neither U_1 nor U_2 is less than nor equal to this value we conclude that the difference between the sample medians is not significant so that, as explained in 3.2.2, 'we can reasonably regard the populations as the same in this respect' (i.e. in their medians).

It is easy to verify that the values of U_1 and U_2 obtained above are the same as those which would be obtained by the counting procedure used in 3.2.1 and that, as in that section, they correspond to the numbers of negative and positive differences in the sequence of differences $(y - x)$. Thus the present procedure represents a worthwhile saving of calculation when, as in the case of the significance test, all we need is the values of U_1 and U_2; these are of course not sufficient for the determination of confidence limits and so the shortened method is no help when these are required. The simplified procedure is due to Mann and Whitney (1947) and the test is known as the *Mann–Whitney U test*.

3.2.2.2 Ties

In the example in the previous section, the twelve observations were distinct. If this is not the case the ranking procedure requires further definition. Consider the following two samples:

x-values	5	4	8	11	6	14
y-values	11	9	11	13	8	15

Re-arranged in order within samples these give:

x-values	4	5	6	8	11	14
y-values	8	9	11	11	13	15

The ranking procedure begins as previously, the x-values 4, 5 and 6, being 1, 2 and 3. The two values 8 constitute a *tie* and are given the same rank, calculated as the average of the ranks corresponding to their positions in the single ordered sequence. These positions are 4 and 5, and so each value 8 is ranked $4\frac{1}{2}$. The next value (9) is ranked according to its position in the sequence, that is 6. The three 11's occupy positions 7, 8 and 9 and are therefore given rank 8, the average of these three positions. The remaining values, 13, 14 and 15, are ranked by their positions in the sequence, namely 10, 11 and 12. The final ranks are thus:

x-values	1	2	3	$4\frac{1}{2}$	8	11
y-values	$4\frac{1}{2}$	6	8	8	10	12

The quantities R_1, R_2, U_1, U_2 are calculated as before, using the same checks, and significance is tested in the usual way (a correction is sometimes used in the significance test, but it is necessary only when the ties are very long, in which case the value of any comparison of the samples is open to question).

3.2.2.3 Large sample test

The table of critical values of U (Table A3) refers only to the case where neither n_1 nor n_2 is greater than 20. For larger samples we obtain U_1 as previously and then calculate

$$\frac{|U_1 - \frac{1}{2}n_1 n_2|}{\sqrt{\{\frac{1}{12}n_1 n_2(n_1 + n_2 + 1)\}}}.$$

(The sign | | indicates that we take the positive difference between U_1 and $\frac{1}{2}n_1 n_2$; we call this the *modulus* of $U_1 - \frac{1}{2}n_1 n_2$.) This statistic follows approximately the distribution of a standardized normal deviate, which is discussed in 6.2; critical values for the significance test are 1.960 (5 per cent significance), 2.576 (1 per cent significance) and 3.291 (0.1 per cent significance).

3.3 More than two samples – the Kruskal–Wallis test

If we have samples from more than two populations we can use the procedure of 3.2 to obtain confidence limits for the difference of any two of the population medians, but it is sometimes convenient to extend the significance test to include all the populations in a single analysis. Thus if several forms of a new drug are being compared, it is sensible to have an overall test to investigate whether there are any differences between the forms. The required procedure is known as the *Kruskal–Wallis analysis of variance of ranks*.†

Suppose then that we have samples from k populations

---

† The term analysis of variance (anovar) belongs properly to the comparison of means, and elementary examples of the method are discussed in chapter 7. The present analysis has certain similarities to anovar and that term is therefore used descriptively rather than literally.

containing $n_1, n_2, \ldots, n_k$ observations; we denote the total number of observations by n, that is $n = \sum_{i=1}^{k} n_i$. As in the case of the Mann–Whitney analysis we assume that the populations differ only in location; the null hypothesis of the test is that the population medians are also equal, so that the populations are identical. In the example worked in Table 9 we have $k = 3$, $n_1 = 5$, $n_2 = 4$, $n_3 = 4$, $n = 13$.

We begin the calculation by replacing each observation by its rank in the ordering of all the samples together; ties are dealt with as in 3.2.2.2. We then total the ranks for each sample, denoting the totals by $R_1, R_2, \ldots, R_k$; we have the check that

$$R_1 + R_2 + \cdots + R_k = \sum_{i=1}^{k} R_i = \sum R = \tfrac{1}{2}n(n+1).$$

For each sample we calculate R_i^2/n_i, the square of the total of the ranks divided by the sample size, and then sum these quantities over the samples. This calculation is denoted by

$$\frac{R_1^2}{n_1} + \frac{R_2^2}{n_2} + \cdots + \frac{R_k^2}{n_k} = \sum_{i=1}^{k} \frac{R_i^2}{n_i} = \sum \frac{R^2}{n}.$$

The test statistic, H, is now calculated from the formula

$$H = \frac{12}{n(n+1)} \sum_{i=1}^{k} \frac{R_i^2}{n_i} - 3(n+1)$$

which in the present example gives $H = 5.118$. The table to be used in deciding the significance of H depends on the number and size of the samples. If $k \leqslant 3$ and $n_i \leqslant 5$ (that is if we have three or fewer samples, all containing five or fewer observations), as in the present case, we use Table A4, entering the table at the position specified by the sample sizes and locating the calculated value of H among the tabulated values. In our case the relevant tabulated values are 5.6176 ($P = 0.050$) and 4.6187 ($P = 0.100$); the sample value of H lies between these values and so the probability relevant to our test lies between the tabulated probabilities, and is thus greater than 0.05 (5 per cent). We are therefore entitled, with the conventional degree of confidence, to accept the null hypothesis of the test, that is to believe that the medians of the three populations are the same.

Table 9. *An example of the calculations for the Kruskal–Wallis test.*
The sample values are given first and these are then replaced by their ranks
which are used in calculating H.

Sample	Observed values							
A	15	8	12	14	8			
B	4	9	3	5				
C	3	6	10	8				

	Ranks					n_i	R_i	R_i^2/n_i
A	13	7	11	12	7	5	50	500
B	3	9	$1\frac{1}{2}$	4		4	$17\frac{1}{2}$	76.5625
C	$1\frac{1}{2}$	5	10	7		4	$23\frac{1}{2}$	138.0625
						$\overline{13}$	$\overline{91}$	$\overline{714.625}$

$$H = \frac{12}{13 \times 14} \times 714.625 - 3 \times 14$$

$$= 47.118 - 42 = 5.118.$$

If the values of k or of some or all of the n_i are too large for Table A4 to be used, we refer to the χ^2 distribution given in Table A5, obtaining an approximate test by entering the table in the row $v = k - 1$ and locating the sample value of H between tabulated values of χ^2 in this row. The probabilities heading the appropriate columns of the table then enclose the probability associated with our sample, in the same way as did those found for smaller samples from Table A4.

3.3.1 Using observed ranks

In the examples of the Mann–Whitney and Kruskal–Wallis procedures we began with observations representing the result of some measuring processes, but at once replaced these measurements with appropriate ranks. The test can therefore be applied even when the observations are not measurements but only an ordering (ranking) of the experimental material. For example, in many subjective assessments (taste, smell and so on) it is often difficult to obtain an agreed and reliable scoring system but it may be relatively easy for each judge to arrange the samples in a meaningful order.

We note that, although significance test procedures can be applied to observed ranks, confidence limits cannot usefully be given, as they would have no simple interpretation when not referring to a measured quantity.

3.4 The two-factor case

In the introductory discussion of experimental design in 1.3 and 1.3.1, we saw that any prior knowledge of naturally occurring groupings within the experimental material should be used in setting up the experiment, one convenient way of doing this being the use of a randomized block layout. If such knowledge is used in the design it must also be incorporated in the analysis; the methods of 3.2 and 3.3, whilst entirely appropriate to the completely randomized layout, are not suitable when blocks have been used. In this case the required significance tests for differences of location are due to Wilcoxon (two samples) and Friedmann (more than two samples).

3.4.1 Two samples – Wilcoxon's test

The *Wilcoxon signed rank test* is used when we wish to compare the medians of two populations, which differ (if at all) only in location, and from which observations have been taken in pairs, one from each population. For example, in comparing the colour fastness of two dyes we might have available ten samples of cloth and could then usefully proceed by dividing each sample in half and applying each dye to one half – allocating dyes to halves at random. We would then obtain ten pairs of values of colour fastness. Another form of pairing that is particularly valuable in much biological work is the use of identical twins, one of the two treatments being allocated at random to each twin of a pair, with a fresh randomization for each pair.

In the colour fastness example we might obtain scores such as those in Table 10. To perform the test we first calculate the differences between samples within pairs; zero differences contribute no information to the comparison and so are ignored. In the present case the number of effective comparisons (n) is thus 9. We shall usually find that some of the differences are positive and some negative, but in the next step we ignore these signs and rank only the numerical values of the differences; ties

75

are dealt with as usual (3.2.2.2). We now sum the ranks of the positive differences, obtaining R_+, and, separately, those of the negative differences, obtaining R_-; as check we have

$$R_+ + R_- = \tfrac{1}{2}n(n+1).$$

The test statistic is the smaller of R_+ and R_-, denoted by T, and the level of significance associated with it is found from Table A6. For $n = 9$ we find that if T is less than or equal to 6 (2) we should reject at the 5 per cent (1 per cent) level of significance the null hypothesis that the medians are equal. In the present case ($T = 5$) we conclude therefore that the two dyes are not the same in colour fastness.

Table 10. *An example of the calculations for the Wilcoxon signed rank test.*

Cloth sample	1	2	3	4	5	6	7	8	9	10
Score for dye A	4	4	5	1	3	3	5	6	6	3
Score for dye B	7	6	9	5	6	3	4	7	4	7
Difference ($B - A$)	3	2	4	4	3	0	-1	1	-2	4
Rank of $|B - A|$	$5\frac{1}{2}$	$3\frac{1}{2}$	8	8	$5\frac{1}{2}$		$1\frac{1}{2}$	$1\frac{1}{2}$	$3\frac{1}{2}$	8
R_+	$5\frac{1}{2}$	$+3\frac{1}{2}$	$+8$	$+8$	$+5\frac{1}{2}$			$+1\frac{1}{2}$		$+8 = 40$
R_-							$1\frac{1}{2}$		$+3\frac{1}{2}$	$= 5$

$$R_+ + R_- = 45 = \tfrac{1}{2}n(n+1), \quad (n = 9), \quad T = 5.$$

Confidence limits for the median difference can be obtained by applying the procedure of 2.5.1.1 to the sample of ten differences (there is no reason to exclude the zero difference when obtaining the confidence limits). For $n = 10$ in Table A2 we find $k = 2$ which leads to confidence limits $-1, 4$, with 97.86 per cent confidence. These limits enclose zero, but this does not contradict the significance level. Even an apparent contradiction would not, however, be a cause for despondency because the reliability of all these procedures is specified only in the very long run; apparent anomalies arise not infrequently when several different procedures are used together.

When the sample size (n) exceeds 25 we cannot use Table A6. Instead we calculate a standardized deviate z from the formula

$$z = \frac{|T - \tfrac{1}{4}n(n+1)|}{\sqrt{\{\tfrac{1}{24}n(n+1)(2n+1)\}}};$$

we reject the null hypothesis at the 5 per cent (1 per cent, 0.1 per cent) level of significance if z is greater than or equal to 1.960 (2.576, 3.291).

3.4.1.1* The distribution of T

The distribution of the statistic T is easy to calculate for small samples, and we will consider the case $n = 6$ as a useful illustration. As in 3.2.1 we consider the distribution under the null hypothesis, that is when the medians of the two populations are equal; since we also assume that the populations differ at most in location, the null hypothesis is equivalent to supposing the the populations are identical. In this case any difference has an equal chance of being positive or negative and hence the ranks $1, 2, \ldots, n$ are equally likely to be associated with the positive or negative rank total. Thus in the case $n = 6$ (larger values of n require rather tedious enumeration) the 64 ($= 2^6$) combinations of signs listed in Table 11 are all equally likely to occur. The distribution of T, and the associated cumulative probabilities, are therefore obtained directly from the frequencies without any weighting. We find that there are probabilities of 3.1 per cent of obtaining $T = 0$ and of 6.3 per cent of obtaining $T = 0$ or 1 when in fact the population distributions are identical. Thus if we observed $T = 1$ we would (conventionally) regard the null hypothesis as reasonable because a value of T as small as this would occur by chance in more than 5 per cent of trials when the hypothesis was true. On the other hand if we observed $T = 0$ we would have (conventionally) to reject the null hypothesis because a value as small as this would occur in less than 5 per cent of trials when the hypothesis was true.

3.4.1.2 The sign test

Wilcoxon's test requires that the variable used shall be measurable (i.e. not a rank) and that the populations can safely be assumed to differ only in location, but another test is available when these conditions are too restrictive. This is the *sign test*, which requires only that we be able to say, for each pair of experimental units, which treatment has given the 'greater' or

Table 11. *Calculating the distribution of T for n = 6.†*

1	2	3	4	5	6	R_+	R_-	T		T	Frequency	Cumulative frequency	Cumulative percentage frequency
+	+	+	+	+	+	21	0	0		0	2	2	3.1
−	+	+	+	+	+	20	1	1		1	2	4	6.2
+	−	+	+	+	+	19	2	2		2	2	6	9.4
+	+	−	+	+	+	18	3	3		3	4	10	15.6
+	+	+	−	+	+	17	4	4		4	4	14	21.9
+	+	+	+	−	+	16	5	5		5	6	20	31.3
+	+	+	+	+	−	15	6	6		6	8	28	43.7
−	−	+	+	+	+	18	3	3		7	8	36	56.3
−	+	−	+	+	+	17	4	4		8	8	44	68.7
−	+	+	−	+	+	16	5	5		9	10	54	84.4
−	+	+	+	−	+	15	6	6		10	10	64	100.0
−	+	+	+	+	−	14	7	7			64		
+	−	−	+	+	+	16	5	5					
+	−	+	−	+	+	15	6	6					
+	−	+	+	−	+	14	7	7					
+	−	+	+	+	−	13	8	8					
+	+	−	−	+	+	14	7	7					
+	+	−	+	−	+	13	8	8					
+	+	−	+	+	−	12	9	9					
+	+	+	−	−	+	12	9	9					
+	+	+	−	+	−	11	10	10					
+	+	+	+	−	−	10	11	10					
−	−	−	+	+	+	15	6	6					
−	−	+	−	+	+	14	7	7					
−	−	+	+	−	+	13	8	8					
−	−	+	+	+	−	12	9	9					
−	+	−	−	+	+	13	8	8					
−	+	−	+	−	+	12	9	9					
−	+	−	+	+	−	11	10	10					
−	+	+	−	−	+	11	10	10					
−	+	+	−	+	−	10	11	10					
−	+	+	+	−	−	9	12	9					
+	−	−	−	+	+	12	9	9					
+	−	−	+	−	+	11	10	10					
+	−	−	+	+	−	10	11	10					
+	−	+	−	−	+	10	11	10					
+	−	+	−	+	−	9	12	9					
+	−	+	+	−	−	8	13	8					
+	+	−	−	−	+	9	12	9					
+	+	−	−	+	−	8	13	8					
+	+	−	+	−	−	7	14	7					
+	+	+	−	−	−	6	15	6					

† The left-hand part of the table lists the possible arrangements of + and − signs which contain 6, 5, 4 or 3 +'s; the arrangements containing 2, 1 or 0 +'s are obtained

'better' response, defining these terms in any way which is meaningful for our investigation. Suppose that when we have made this judgement for each of n pairs we find that one of the treatments has been judged greater in r pairs and the other in $(n - r)$. The test statistic is the smaller of r and $n - r$; denote it by R. We enter Table A7 in the row determined by n. If R is less than or equal to the value in the 5 per cent (1 per cent) column we reject at the 5 per cent (1 per cent) level of significance the null hypothesis that the two populations are the same in the character we have studied.

The probabilities used in preparing this table are easy to calculate if we note that the null hypothesis implies that judgements in favour of either treatment are equally likely (probability $= \frac{1}{2}$). Thus the distribution of the score for one treatment in n pairs is the same as the distribution of the number of heads in n throws of a fair coin; this is a special case of the *binomial distribution*, which is discussed briefly in sections 9.2.1 and 9.2.1.1. The relevant formula is that the chance of s heads in n throws is

$$\frac{n(n-1)\cdots(n-s+1)}{1\cdot 2\cdot 3\cdots s}\left(\frac{1}{2}\right)^n,$$

so that the required probability, that of either population receiving a sample score less than or equal to R when the populations are the same, is twice the sum of this term over all values of s less than or equal to R, that is twice

$\left(\frac{1}{2}\right)^n$ (these terms given by $s = 0$

$+ n\left(\frac{1}{2}\right)^n$ $s = 1$

$+ \dfrac{n(n-1)}{2}\left(\frac{1}{2}\right)^n$ $s = 2$

$+ \cdots\cdots$ $\cdots\cdots$

$+ \dfrac{n(n-1)\cdots(n-R+1)}{1\cdot 2\cdots R}\left(\frac{1}{2}\right)^n$ $s = R$).

by interchanging $+$'s and $-$'s in arrangements with 4, 5 or 6 $+$'s (i.e. above the double line). The values of R_+, R_- and T are given for each arrangement; the values of T for the omitted arrangements are the same as for those above the double line. The single dividing lines serve only to show the pattern of the arrangements.

The right-hand part of the table summarizes the distribution of T, including the probability used in the Wilcoxon signed rank test.

We have then

$$\text{Probability} = 2 \sum_{s=0}^{R} \frac{n(n-1)\cdots(n-s+1)}{1 \cdot 2 \cdots s} \left(\frac{1}{2}\right)^n ;$$

the coefficient 2 is necessary because either population might receive the smaller sample score and so summation over one sample is not sufficient.

For $n = 15$ the *binomial coefficients*

$$\frac{n(n-1)\cdots(n-s+1)}{1 \cdot 2 \cdots s} \quad \text{are as follows:}$$

s	0	1	2	3	4	5	6	7
coeff.	1	15	105	455	1365	3003	5005	6435
s	8	9	10	11	12	13	14	15
coeff.	6435	5005	3003	1365	455	105	15	1

The totals of the first $(R + 1)$ of these for $R = 0, 1, 2, \ldots, 7$ are given in Table 12 and the corresponding probabilities are easily calculated, using $(\frac{1}{2})^{15} = 1/32\,768$.

Probabilities for $n > 25$ are not given in Table A7. In this case we calculate $(n - 2R - 1)/\sqrt{n}$ and obtain the probability by using this value as a standardized deviate (1.960 for 5 per cent significance, 2.756 for 1 per cent and 3.291 for 0.1 per cent).

3.4.2 More than two samples – Friedmann's test

If we have n treatments to compare (in 3.4.1 we had $n = 2$), in a randomized block layout with m blocks we use *Friedmann's test*, which, like most of the earlier tests, is designed to compare location (medians) for populations which are the same in other respects. In the example in Table 13 we have $n = 4$ and $m = 5$. We first replace each observation by its rank within its block – this procedure is noticeably easier than that used in the Kruskal–Wallis test, where all the observations were ranked in a single sequence. The ranks for each treatment are totalled; we note the check that the sum of these totals is

$$\sum_{i=1}^{n} R_i = R = \tfrac{1}{2}mn(n + 1),$$

where R_i is the total of the ranks for treatment i ($i = 1, 2, \ldots, n$).

Table 12. *The calculation of probabilities for use with the sign test when* $n = 15$.†

	$\sum_{s=0}^{r} \dfrac{n(n-1)\ldots(n-s+1)}{1 \cdot 2 \ldots s}$	
r	with $n = 15$	Percentage probability
0	1	0
1	16	0.1
2	121	0.7
3	576	3.5
4	1941	11.8
5	4944	30.2
6	9949	60.7
7	16384	100.0

† The percentage probability corresponding to r is calculated as

$$200 \sum_{s=0}^{r} \frac{n(n-1)\ldots(n-s+1)}{1 \cdot 2 \ldots s}\left(\frac{1}{2}\right)^{n}.$$

The value for $r = 0$ is more accurately 0.006.

Table 13. *An example of the calculations for Friedmann's test. The actual observations are given first and are then replaced by their ranks within blocks.*

	Observations Treatment				Ranks Treatment			
Block	A	B	C	D	A	B	C	D
1	6	8	11	12	1	2	3	4
2	2	5	10	10	1	2	$3\frac{1}{2}$	$3\frac{1}{2}$
3	5	5	9	11	$1\frac{1}{2}$	$1\frac{1}{2}$	3	4
4	1	5	8	9	1	2	3	4
5	3	7	8	7	1	$2\frac{1}{2}$	4	$2\frac{1}{2}$
Totals					$5\frac{1}{2}$	10	$16\frac{1}{2}$	18
					R_A	R_B	R_C	R_D

$$R_A + R_B + R_C + R_D = R = 50 = \tfrac{1}{2} \times 5 \times 4 \times 5$$

$$S = 5.5^2 + 10^2 + 16.5^2 + 18^2 - \frac{50^2}{4}$$

$$= 30.25 + 100 + 272.25 + 324 - 625$$

$$= 101.5.$$

The test statistic is

$$S = R_1^2 + R_2^2 + \cdots + R_n^2 - \frac{R^2}{n}$$

$$= \sum_{i=1}^{n} R_i^2 - \frac{R^2}{n}.$$

The significance level associated with this is found from Table A8, where for $n = 4$, $m = 5$ we find that for S greater than or equal to 83 (105) we reject the null hypothesis at the 1 per cent (0.1 per cent) level. Our result ($S = 101.5$) is between these values, and so we reject at the 1 per cent level of significance the null hypothesis that the populations have the same median.

Table A8 deals only with the cases $n = 3$, $m = 3, 4, \ldots, 10$; $n = 4$, $m = 2, 3, \ldots, 6$; $n = 5$, $m = 3$. For larger values of m and n we calculate $12S/mn(n + 1)$ and determine its approximate significance from the χ^2 table (Table A5) with $n - 1$ degrees of freedom.

3.5 A more general comparison of two samples – the Kolmogorov–Smirnov two-sample test

All the significance tests discussed in the previous sections have aimed at detecting differences of location when the populations may be assumed to be the same in all other respects. This assumption is valid in many practical problems, but sometimes we have reason to doubt it and so look for a more general test of differences between the populations. If, for example, we have measured the proportions of ears of two varieties of wheat which have been affected with rust, we may conclude that any difference of location is bound to imply a difference of form, simply because the range of values of a proportion is bounded; we could not therefore use the Mann–Whitney test to compare the two varieties. In this case (two samples) we can use the *Kolmogorov–Smirnov two-sample test*, which is sensitive to any difference, either of form or location, between the populations.

To carry out this test, we first arrange each sample in order of magnitude. If we are working with a continuous variable, we divide the scale into intervals, using the same division for both samples. If we have fairly large samples the division of the scale should be chosen so that each interval contains no more than 2 or 3 observations; coarser division reduces the sensitivity of the test. Thus if, in the comparison between varieties, we have taken samples from 12 plots of each variety, obtaining the proportions infected given in Table 14, we could conveniently divide the proportion scale (0–100 per cent) into 5 per cent intervals; the ordered values are shown in the table. We now count for each sample the cumulative frequencies up to each division of the scale; these cumulative frequencies tell us how many observations we have less than or equal to each interval's upper bound. We now take for each interval the modulus of the difference between samples of the cumulative frequencies, and use as test statistic the largest value of this modulus; for the data in Table 14 this

Table 14. *An example of the calculations for the Kolmogorov–Smirnov two-sample test.*

Sample values of proportions of ears infected:

Variety *A*	43	62	81	69	73	55	64	18	89	67	59	61
Variety *B*	24	31	19	47	35	29	18	13	43	17	65	28

	Rearranged values		Cumulative frequencies		
Interval	*A*	*B*	*A*	*B*	|Difference|
0–4			0	0	0
5–9			0	0	0
10–14		13	0	1	1
15–19	18	19, 18, 17	1	4	3
20–24		24	1	5	4
25–29		29, 28	1	7	6
30–34		31	1	8	7
35–39		35	1	9	8
40–44	43	43	2	10	8
45–49		47	2	11	9
50–54			2	11	9
55–59	55, 59		4	11	7
60–64	62, 64, 61		7	11	4
65–69	69, 67	65	9	12	3
70–74	73		10	12	2
75–79			10	12	2
80–84	81		11	12	1
85–89	89		12	12	0
90–94			12	12	0
95–100			12	12	0

value is 9. The significance of this value is determined from Table A9; here for $n = 12$ we find that maximum differences of 6 and 8 respectively indicate significance at the 5 per cent level and 1 per cent level (Table A9 gives significance values for use only when the two samples have the same size). In the present case we thus have significance at least at the 1 per cent level, and so we conclude that the two varieties cannot reasonably be regarded as having the same distribution of susceptibility to rust infection.

If the samples each contain more than 40 observations we have an approximate test applicable even when the samples have different sizes, n_1 and n_2. To use this, we begin as before but when the cumulative frequencies have been obtained we divide each by its appropriate sample size (n_1 or n_2) to obtain the cumulative proportional frequencies on a 0 to 1 scale; we use as test statistic the largest numerical difference between these cumulative proportional frequencies. To determine significance we calculate

$$\sqrt{\{(n_1 + n_2)/n_1 n_2\}},$$

and judge our largest difference to be 5 per cent (1 per cent, 0.1 per cent) significant if it exceeds 1.36 (1.63, 1.95) times this quantity.

3.5.1 Ogive

The cumulative proportional frequencies introduced in 3.5 lead to a pictorial summary of the sample which is sometimes useful as an alternative to the histogram. If we plot with the scale of the variable on the horizontal axis and the frequency scale (0 to 1) on the vertical axis, we obtain a step-function known as the *ogive*. The ogive for the data of Table 2 is shown in Fig. 11 and may be compared with the histogram in Fig. 4. The median of the sample is readily obtained from the ogive as the value on the horizontal axis corresponding to 0.5 on the vertical axis; similarly, we can at once obtain the *quartiles*, values of the variable which divide the sample into quarters, by considering the horizontal values corresponding to 0.25 and 0.75 on the vertical axis. For most purposes however the ogive is not as convenient a summary as the histogram, because details of the form of the distribution (skewness for example) are less easy to see in the ogive.

3.5.1.1* Distribution function

We are nevertheless led to an important theoretical tool if we consider how the ogive changes as the sample size increases (compare the introduction of the frequency curve in 2.1.1.1). Once again we may expect to obtain a smooth curve, and the functional expression relating this to the observed variable x is known as the *distribution function* of x; it is conventionally denoted by $F(x)$. We have already seen that $F(x_M) = 0.5$ where x_M is the

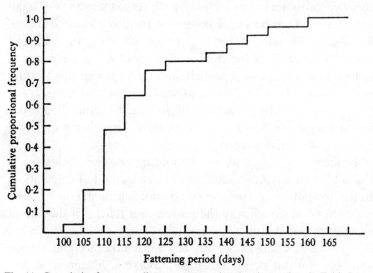

Fig. 11. Cumulative frequency diagram representing the observations in Table 2; see 3.5.1 and compare Fig. 4.

median, and more generally $F(x_0)$ measures the proportion of the population for which our variable is less than or equal to x_0. This function is very useful in theoretical work, but we shall have little occasion to refer to it.

We note that if the frequency curve is functionally expressed by the *frequency function*, conventionally denoted by $f(x)$, then

$$f(x) = \frac{dF(x)}{dx}.$$

85

The element of area under the frequency curve, which we saw was taken as the pictorial representation of probability, is denoted (in the usual calculus notation) by $f(x)\,dx$, and this may alternatively be written $dF(x)$.

3.6 One- or two-sided tests

The purpose of this section is to explain why, in several of the tests considered in this chapter, we have included both ends of a distribution when calculating probabilities. Thus in 3.2.1 U was defined as the smaller of U_1 and U_2 and so, by using the probability associated with U being less than or equal to a specified value, we in fact included both very large and very small values of U_1 (and hence, of course, of U_2 also, since the sum of U_1 and U_2 is constant for a given sample size). In 3.4.1 and 3.4.1.1 we took T as the smaller of R_+ and R_- and so, by the same argument, included probabilities for large and small values of R_+ or R_-. In 3.4.1.2 we used similarly the smaller of r and $(n-r)$ to obtain the probability of 'so large a sample difference,' introducing a coefficient 2 so as to allow for either population receiving the smaller score.

In these three cases we are comparing two populations, and this makes easier the explanation of our method of calculation. In the two-population problem we do not in general know, before we start a test, which way the populations differ – if they do at all. Either population may have the larger median (if we are comparing locations) and our test must take account of our ignorance on this point. When we observe a difference, we use the appropriate significance test procedure to decide whether this must be regarded as evidence of a real difference, or whether we can still accept it as a chance deviation occurring when the null hypothesis is true. When assessing the probability on which to base this decision, we remember that, were the null hypothesis true – and the basis of our calculation is to assume that it is – deviations in either sense could occur, and so we include the probabilities corresponding to both senses. Although only one sense can occur in one sample, we have to regard this as a representative not just of the chance deviations in this sense, but of all the deviations, in either sense, that might occur when the null hypothesis was true.

If however we do have sufficient prior knowledge of the populations to be sure that sample deviations in one sense can be due only to chance and cannot be indicative of a real population difference, then the probabilities used above are wrong. For example if we have added an extra source of lysine to a diet for chicks, we may be very confident from our previous knowledge of such additives that the new diet is at least as good as the old, but we are not so sure whether it is actually better. In this case if we observe an improvement with the new diet we will test to see if it must be regarded as significant, but if we find that the new diet has done worse we will be confident that this is a chance effect and will not wish to perform a significance test. We say that we require a *one-sided* significance test in contrast to the *two-sided* procedures used in the absence of prior knowledge.

The choice between one- and two-sided procedures depends, therefore, on our assessment of our prior knowledge. The author's view is that one-sided tests are justified only rather rarely but that we ought always to consider their possible relevance.

3.6.1 One-sided tests require smaller differences for significance

In all the two-sample tests so far discussed, the distributions from which the probabilities were calculated (U_1 leading to U, R_+ leading to T, r leading to the sign test) were symmetric, and so one-sided tests are obtained by dividing the tabular probabilities by 2. Thus in the sign test with $n = 15, r = 3$ we found a probability of 3.5 per cent, appropriate to the two-sided test. The corresponding probability for a one-sided test is thus 1.75 per cent; the observation $r = 3$ is thus more convincingly significant in a one-sided test than in a two-sided – always provided that $r = 3$ is in that end of the distribution which our prior knowledge led us to expect!

This relationship between one- and two-sided procedures can be justified heuristically as follows. In a one-sided test, we have some prior knowledge to incorporate and so we need less sample evidence to convince us; thus a result that fails to convince us in a two-sided test may be enough to be judged significant in a one-sided procedure. Alternatively, if we look for differences which will be judged just significant, we find that we require smaller values in a one-sided test than in a two; the smaller difference is enough to convince us when we take account of our prior knowledge.

3.7(c)   The SPSS-X command npar tests

All the significance tests described in this chapter can be carried out by using appropriate subcommands with the SPSS-X command npar tests. The structures of the subcommands are shown in CTable 4, where the tests are listed in the order in

CTable 4. *The subcommands to be used with the SPSS-X command npar tests in carrying out the significance tests considered in chapter 3; the subcommands are interpreted in* 3.7(c).

Test	Subcommand
Mann-Whitney	M-W - *varlist* by *var(value 1, value 2)*
Kruskal-Wallis	K-W - *varlist* by *var(value 1, value 2)*
Wilcoxon signed rank	Wilcoxon - *varlist/*
sign	sign test - *varlist/*
Friedman	Friedman - *varlist/*
Kolmogorov-Smirnov 2 sample	K-S - *varlist* by *var(value 1, value 2)*

which they were introduced above. We see that the subcommands for the tests used with a single-factor design (the Mann–Whitney, the Kruskal–Wallis and the Kolmogorov–Smirnov two-sample) are of a structure different from those for the tests used with a two-factor design (the Wilcoxon signed rank, the sign and the Friedman); within these two groups there are, moreover, further differences of structure which are not evident in the condensed versions given in the table. Commands, data and output for the six tests are shown in CTables 5 to 10; in all six cases the data have been submitted *in-stream* rather than in a separate file.

In the three single-factor tests, the subcommand, after identifying the test, specifies the variable to be analysed – according to the name used in the data list command – and the treatment groups to be compared. In CTable 5 – the Mann–Whitney test – the statistic U is the smaller of U_1 and U_2 as defined in 3.2.2.1 and so can be compared with the value of U from Table A3; the statistic W, the smaller of R_1 and R_2, is used in another, equivalent, version of the procedure which is known as the *Wilcoxon rank sum test*. The use of Table A3 allows only that the significance test probability be located relative to the 5 per cent significance level, but the SPSS-X calculation does more than this. When, as here, the sum of the two sample sizes

CTable 5. *SPSS-X commands, data and output for the Mann–Whitney test on the data in 3.2.*

Commands and data

    data list free / group variable
    begin data
    1 5   1 4   1 7   1 10   1 6   1 14
    2 11   2 9   2 12   2 13   2 8   2 15
    end data
    npar tests M-W = variable by group (1,2)

Output

- - - - MANN-WHITNEY U - WILCOXON RANK SUM W TEST

        VARIABLE
    BY GROUP

        MEAN RANK   CASES

          4.67   6   GROUP = 1.00
          8.33   6   GROUP = 2.00
                          - -
                          12   TOTAL

		EXACT	CORRECTED FOR TIES	
U	W	2 - TAILED P	Z	2 - TAILED P
7.0	28.0	0.0931	-1.7614	0.0782

---

is less than 30, the calculation gives the exact significance level and in all cases the output includes the level estimated with the use of the normal approximation mentioned in 3.2.2.3. In the present example, the exact level is 9.31 per cent, whilst the normal approximation gives 7.82 per cent.

In the Kruskal–Wallis test, shown in CTable 6, the structure of the subcommand used is very similar to that appropriate to the Mann–Whitney test, but we notice that the identification of the treatment groups to be compared although appearing similar is interpreted differently in the two tests; under M–W the notation group (1, 2) means that groups 1 and 2 are compared but under K–W the notation sample (1, 3) means that all the samples with labels from 1 to 3 are included – the samples are labelled 1, 2

CTable 6. *SPSS-X commands, data and output for the Kruskal–Wallis test on the observed values in Table 9.*

Commands and data

```
data list free / sample variable
begin data
1 15  1 8  1 12  1 14  1 8
2 4  2 9  2 3  2 5
3 3  3 6  3 10  3 8
end data
npar tests K-W - variable by sample (1,3)
```

Output

```
- - - - - - -   KRUSKAL-WALLIS 1-WAY ANOVA
```

VARIABLE
BY SAMPLE

MEAN RANK	CASES		
10.00	5	SAMPLE =	1
4.38	4	SAMPLE =	2
5.88	4	SAMPLE =	3
	- -		
	13	TOTAL	

			CORRECTED FOR TIES	
CASES	CHI-SQUARE	SIGNIFICANCE	CHI-SQUARE	SIGNIFICANCE
13	5.1181	0.0774	5.1894	0.0747

and 3 and not *A*, *B* and *C* as in Table 9 because in this procedure, as in many others, SPSS-X will not accept *alpha-numeric* labelling. The value of the test statistic *H* is given and tested as χ^2 for all sample sizes; its probability level in CTable 6, 7.74 per cent, exceeds 5 per cent – as we found in 3.3. The values given as 'corrected for ties' are derived from a modification in the formula for *H* to take account of the numbers and lengths of the ties in the data; like the correction mentioned at the end of 3.2.2.2 this modification has little effect unless there is a considerable amount of tying, in which case the value of the analysis is somewhat limited.

In the three two-factor tests the groups to be compared are regarded in SPSS-X as defining separate variables so that in CTables 7 and 8 the scores for dyes *A* and *B* from Table 10 have to be taken to define variables which make up the varlist in the subcommand. Similarly in CTable 9 for the Friedman test

CTable 7. *SPSS-X commands, data and output for the Wilcoxon signed rank test applied to the scores in Table* 10.

Commands and data

```
data list free / A B
begin data
47 46 59 15 36 33 54 67 64 37
end data
npar tests Wilcoxon = A B /
```

Output

`- - - - - - -` WILCOXON MATCHED-PAIRS SIGNED-RANKS TEST

```
          A
WITH  B
```

MEAN RANK	CASES	
2.50	2	- RANKS (B LT A)
5.71	7	+ RANKS (B GT A)
	1	TIES (B EQ A)
	--	
	10	TOTAL

Z = -2.0732   2-TAILED P = .0382

COMPARING SEVERAL SAMPLES

CTable 8. *SPSS-X commands, data and output for the sign test applied to the scores in Table* 10.

==

Commands and data

```
data list free / A B
begin data
47  46  59  15  36  33  54  67  64  37
end data
npar tests sign - A B /
```

Output

- - - - - - - -SIGN TEST

```
        A
WITH  B
```

```
            CASES

            2   - DIFFS (B LT A)
            7   + DIFFS (B GT A)     (BINOMIAL)
            1     TIES               2-TAILED P -      .1797
            --
           10     TOTAL
```

==

CTable 9. *SPSS-X commands, data and output for Friedmann's test on the data in Table* 13.

==

Commands and data

```
data list free / A B C D
begin data
6 8 11 12  2 5 10 10  5 5 9 11  1 5 8 9  3 7 8 7
end data
npar tests Friedman - A B C D /
```

Output

- - - - - - FRIEDMAN TWO-WAY ANOVA

MEAN RANK	VARIABLE
1.10	A
2.00	B
3.30	C
3.60	D

CASES	CHI-SQUARE	D.F.	SIGNIFICANCE
5	12.1800	3	.0068

==

CTable 10. *SPSS-X commands, data and output for the Kolmogorov–Smirnov two-sample test applied to the proportions in Table 14.*

Commands and data

```
data list free / variety pei
variable labels pei 'Proportion of ears infected'
value labels variety 1 'Variety A' 2 'Variety B'
begin data
1 43   1 62   1 81   1 69   1 73   1 55   1 64   1 18   1 89
1 67   1 59   1 61   2 24   2 31   2 19   2 47   2 35   2 29
2 18   2 13   2 43   2 17   2 65   2 28
end data
npar tests K-S - pei by variety (1,2)
```

Output

- - - - - - - - -KOLMOGOROV-SMIRNOV 2-SAMPLE TEST

     PEI   Proportion of ears infected
BY   VARIETY

      CASES

     12   VARIETY - 1.00   Variety A
     12   VARIETY - 2.00   Variety B
    --
     24   TOTAL

WARNING - DUE TO SMALL SAMPLE SIZE, PROBABILITY TABLES SHOULD
BE CONSULTED.

MOST EXTREME DIFFERENCES				
ABSOLUTE	POSITIVE	NEGATIVE	K-S Z	2-TAILED P
0.75000	0.00000	-0.75000	1.837	0.002

analysis on the data in Table 13, the four treatments to be compared are taken to define the variables included in the varlist. For the Wilcoxon test in CTable 7, the test statistic T is not given but only the standardized normal deviate derived from it, with its probability level – note that in calculating the deviate by the formula in 3.4.1 the value of n is the number of non-zero differences. For the sign test in CTable 8 the probability level is calculated from the binomial distribution because the number of non-zero differences is not greater than 25 and so the normal approximation is not satisfactory. For the Friedman test in

CTable 9 the test statistic S is not given but rather the value of χ^2 calculated from it by the formula in 3.4.2.

For the Kolmogorov–Smirnov test in CTable 10, the structure of the commands is exactly the same as those for the other single-factor two-sample test considered, the Mann–Whitney test, but since the variables in Table 14 are specified more fully than those in 3.2 it is appropriate to use variable labels and value labels commands. In the output of this test the extreme differences quoted are calculated as described in the last paragraph of 3.5, using cumulative frequencies divided by sample size. In this test, as in CTables 7 and 9, the test statistic is presented in its normalized form, using the calculation described in 3.5, but a warning is given that because the sample sizes are small exact probability tables should be consulted.

3.8 Exercises

The purpose of exercises 1–6 is to suggest examples in which you can try to discover how far sampling under different conditions affects the scatter of a distribution. The methods developed in 3.2 require that the two distributions to be compared should differ only in location and not in shape: whether this condition holds in any given case can be decided either by examining the data or by drawing on past experience of the variable under investigation. The following exercises suggest some simple cases where you can take measurements on two groups; you should be able to suggest other examples from your own work. In each case plot the histograms for the two groups separately: it is a good idea to draw them on detail paper (semi-transparent graph paper) as this makes the visual comparison much easier.

Such comparison is a useful first step, but in later sections we will consider simple analytical methods to make comparisons more objective; it is very difficult, without a lot of experience, to judge whether the apparent differences between two histograms are of real importance or are just chance effects.

1 Select a species of plant which is readily available and define a convenient quantity to count or measure on each plant. Make this observation on two groups of 20 plants grown either under different conditions or at different sites.

2 Take the measurement of exercise 1 on groups of 40 plants, and again on groups of 100 plants.

3 Take the measurement of exercise 1 on two groups of plants at different stages of maturity – you will often find that this affects the shape and width of a distribution.

4 Apply the procedures of exercises 1–3 to a suitable animal subject. In addition to groups of differing ages consider also the two sexes.

5 Apply the procedures of exercises 1–3 to a suitable laboratory measurement.

6 Look again at the histograms you drew for exercises 14 and 15 of 2.6; these will illustrate how much variation of shape and width you can get between samples from a single population.

Exercises 7–9 contain sets of observations which can be used to obtain confidence limits for the difference between the population medians, in the manner described in 3.2.

7 Twenty-nine oedematous patients were divided at random into two groups and treated with one or other of two diuretics, one (group *A*) a well-tested drug and the other (group *B*) a new preparation. The following values refer to the volume of urine (litres) excreted by the patients in the 24-hour period after treatment.

A	1.27	2.12	1.15	2.01	1.83	2.18	2.14	1.41
	1.15	1.11	1.56	1.30	2.69	1.77		
B	1.89	2.32	2.27	1.93	2.10	1.97	1.89	3.63
	2.26	2.66	1.29	1.80	2.31	2.93	1.86	

8 In a test for two treatments for angina pectoris, 32 patients were divided at random into two groups and given a three-week course of one or other of two drugs. One of the measurements

95

taken to compare these drugs was related to the amount of effort the patient could make in a standard exercise test; the values observed were as follows:

P	1072	1001	990	1017	982	1026	957	1072
	915	1045	1068	993	1010	1014	1014	997
Q	758	1012	967	997	817	880	823	875
	869	880	844	884	944	951	824	1002

9 A nurseryman, wishing to compare two varieties of cyclamen, counted the number of flowers produced in the flowering season by each of 28 plants, 13 of variety V and 15 of variety W. The numbers he obtained were as follows:

V	22	23	23	28	23	29	22	28	34	25	26	33	24	
W	32	43	33	44	31	27	30	29	31	32	30	35	27	29	34

10* Calculate the distribution of U for $n_1 = n_2 = 5$, using the method applied in 3.2.1. (There are 252 possible arrangements.)

11* Calculate the distribution of U for $n_1 = 5$, $n_2 = 4$, using the method applied in 3.2.1. (There are 126 possible arrangements; notice that the same distribution is obtained if we interchange n_1 and n_2, taking $n_1 = 4$, $n_2 = 5$.)

Exercises 12–15 require the use of the Mann–Whitney significance test for the null hypothesis that the medians of the two groups are the same. The calculation is described in 3.2.2.1. In the first three of these examples the confidence limits for the difference of medians are already available, so that the significance test does not provide any new information; if however we had been, in these particular problems, satisfied that all that was needed was a conclusion concerning the null hypothesis, then the significance test would have required a simpler calculation than that needed to obtain the confidence limits.

12 Using the observations in exercise 7, test the null hypothesis that the population medians for the two treatment groups were equal.

13 Using the observations in exercise 8, test the null hypothesis that the population medians for the two treatment groups were equal.

14 Using the observations in exercise 9, test the null hypothesis that the population medians for the two varieties were equal.

15 The following observations of density were obtained from 25 samples of wood from each of two varieties of tree. Test the null hypothesis that the median densities for the two populations were the same. (Note that n_1 and n_2 are greater than 20, so that the large sample test, given in 3.2.2.3, is required.)

L	0.76	0.74	0.75	0.72	0.73	0.71	0.73	0.73
	0.77	0.74	0.73	0.78	0.75	0.73	0.75	0.76
	0.75	0.69	0.73	0.74	0.74	0.75	0.76	0.72
	0.77							
M	0.68	0.67	0.72	0.69	0.66	0.71	0.69	0.71
	0.71	0.71	0.70	0.72	0.71	0.72	0.71	0.66
	0.70	0.72	0.68	0.70	0.69	0.69	0.73	0.71
	0.70							

Exercises 16–20 require comparisons between more than two groups. The null hypothesis that all the medians are equal may be tested by the Kruskal–Wallis test (see 3.3). This requires the condition that the groups should differ only in location and it is therefore desirable to begin the analysis by drawing a histogram for each group; in many cases however – as in some of these examples – the groups are too small for the procedure to be helpful, and we can therefore only assume the distributions to be the same and treat the results obtained with appropriate caution. Confidence limits for the difference between any two medians can be calculated by the method used in 3.2.

16 Ten measurements of the butter-fat percentage of the milk of each of four cows gave the following values:

A	B	C	D
3.43	3.63	3.43	3.48
3.46	3.64	3.48	3.54
3.47	3.57	3.39	3.43
3.41	3.55	3.43	3.38
3.41	3.66	3.45	3.49
3.32	3.70	3.37	3.49
3.37	3.58	3.42	3.48
3.46	3.64	3.39	3.53
3.48	3.58	3.43	3.43
3.45	3.66	3.52	3.53

17 Fifteen women who had agreed to take part in a test of two low-calorie breads were allocated at random to 3 groups of 5; two groups (A and B) were given the test breads and the third group C was used as control. The women were all asked to use a specified dietary regime, the groups differing only in the bread used; weight losses (kg) over the first seven-day period were as follows:

A	B	C
1.66	1.89	3.77
2.24	2.23	0.78
2.97	3.28	1.61
1.82	3.62	2.04
1.57	2.57	3.56

18 In an experiment to compare the yield of 5 varieties of mushroom spawn, 6 plots were grown with each variety and the yield (kg) recorded over a standard period; the results were as follows:

P	Q	R	S	T
38.2	42.9	18.2	50.4	63.3
31.7	64.3	19.7	65.0	66.7
28.1	36.9	41.7	48.5	60.4
29.1	35.9	26.2	51.0	57.4
33.4	41.0	21.0	52.9	65.6
25.2	38.8	52.2	59.1	61.2

19 In a survey of the period for which frozen foods are kept in retail deep-freezes, 3 types of foodstuff (*F*, *G* and *H*) were examined: 8 samples of each were traced in retail shops. Some of these shops provided samples of more than 1 of the 3 types, but this occurrence was not sufficiently common to justify a two-factor analysis, so that the Kruskal–Wallis test is used as the best available. The observations below are the age (in days of storage in the retail deep-freeze) of the 24 samples.

F	G	H
27	42	47
24	42	48
34	53	52
32	57	47
31	44	59
20	63	63
24	47	47
32	46	69

20 In a nutrition experiment with laboratory rats one of the characters recorded was the liver weight expressed as a percentage of body weight. The values of this percentage for the rats in the four treatment groups are as follows:

A	B	C	D
3.42	3.17	3.34	3.64
3.96	3.63	3.72	3.93
3.87	3.38	3.81	3.77
4.19	3.47	3.66	4.18
3.58	3.39	3.55	4.21
3.76	3.41	3.51	3.88
3.84	3.55		3.96
	3.44		3.91

Exercises 21–4 involve the comparison of two groups in paired-sample designs, for which the Wilcoxon signed rank test (see 3.4.1) is appropriate. Note that in exercise 21 the problem is the same as in exercise 7 and that, further, the 'experimental material' – 29 patients – is the same, but that since a two-factor

design is used rather than a single factor, a different method of analysis is appropriate.

21 Twenty-nine oedematous patients (as in exercise 7) were used in the comparison of two more diuretics. Each patient was given one drug (chosen at random from the two under test) and then, after an interval of 6 days, the other; the doctor in charge of the investigation was prepared to assume that the order of using the two drugs could be ignored. The observations obtained (volume of urine (litres) excreted in 24 hours) were as follows:

Patient	Treatment A	Treatment B	Patient	Treatment A	Treatment B
1	1.66	2.24	16	0.85	1.64
2	2.01	2.18	17	1.06	1.73
3	1.84	2.40	18	0.93	2.06
4	0.62	1.30	19	1.83	2.72
5	2.25	2.57	20	0.94	2.01
6	1.17	1.87	21	0.47	1.80
7	1.20	1.38	22	0.82	1.24
8	1.04	1.58	23	1.28	1.79
9	2.50	2.79	24	0.58	2.43
10	2.39	3.16	25	0.95	1.76
11	1.04	1.53	26	1.85	2.91
12	1.55	2.19	27	0.89	1.69
13	3.90	4.61	28	0.92	0.97
14	2.11	2.67	29	2.19	3.05
15	1.76	1.56			

22 (Compare exercise 9.) A nurseryman, wishing to compare two varieties of cyclamen, obtained the co-operation of 20 housewives. He gave each of them two plants, one of each variety, and asked them to keep the two plants together, treating them similarly as far as possible, and to record the number of flowers produced by each plant during the flowering season. The results he obtained were as follows:

Household	Variety		Household	Variety	
	V	W		V	W
1	24	31	11	25	32
2	25	27	12	19	25
3	27	30	13	28	41
4	26	28	14	31	35
5	22	28	15	23	34
6	21	28	16	26	29
7	23	27	17	24	29
8	23	28	18	27	41
9	23	30	19	18	26
10	25	28	20	28	26

23 An ecologist, interested in comparing two species of primrose in a certain type of habitat, selected 18 sites at which neither species was present, and in 1953 planted 10 plants of each species at each site. He recorded the number of plants in each colony year by year, obtaining the following counts in 1972:

Site	Species		Site	Species	
	S	T		S	T
1	187	226	10	51	89
2	17	57	11	153	172
3	5	48	12	22	45
4	65	109	13	87	98
5	18	55	14	32	47
6	181	222	15	34	61
7	28	68	16	49	68
8	41	76	17	18	44
9	44	92	18	26	44

24 A biochemist, wishing to compare two methods of assaying the lysine content of chick feeds, undertook this comparison as an extra factor in an experiment to compare the effect of two storage temperatures on the feed. He took therefore 15 samples of feed, divided each in half and stored one part at each of the two temperatures, finally comparing the two parts by both assay methods. The slope-ratios obtained in the assays were as follows:

Feed sample	Assay method 1	Assay method 2	Feed sample	Assay method 1	Assay method 2
1	2.25	2.31	9	1.88	2.21
2	2.16	1.95	10	2.29	2.29
3	1.48	1.52	11	1.83	1.95
4	2.34	2.53	12	2.02	2.00
5	2.00	2.08	13	1.55	1.54
6	1.54	1.62	14	2.15	2.00
7	2.08	2.28	15	1.74	1.75
8	2.03	2.17			

In exercises 25–30 the sign test (given in 3.4.1.2) is used for paired-sample comparisons. Exercises 25–8 use the same observations as 21–4, reminding us that if the Wilcoxon test can be used then so can the sign test, although this latter test makes slightly less efficient use of the observations. Exercises 29 and 30, however, deal with situations in which no meaningful measurement is available, so that the Wilcoxon test cannot be used, but nevertheless the significance test is of valid interest.

25 Use the sign test for the data of exercise 21.

26 Use the sign test for the data of exercise 22.

27 Use the sign test for the data of exercise 23.

28 Use the sign test for the data of exercise 24.

29 Obtain the co-operation of other members of your class, or other workers in your laboratory, to compare the quality of, say, two brands of instant coffee – or two makes of canned beer, or the service at two local supermarkets, or whatever comparison is meaningful to the group you work with.

Different members of the group will base their preference on different features of the subjects compared, but nevertheless the overall comparison can be meaningful and have practical applications: the same comment applies to the next exercise.

30 Conduct an opinion survey amongst your friends as to whether a dog or a cat is the better domestic pet.

Exercises 31–5 involve two-factor comparisons with more than two levels for the interesting factor, and so require the use of Friedmann's test, described in 3.4.2.

31 The histologist referred to in exercise 1.4(8) used a two-factor design to compare 4 differential stains on each of 20 semen samples. The percentages of stained cells that he observed were as follows:

Semen sample	Differential stain			
	A	B	C	D
1	82.1	65.5	68.4	62.9
2	71.9	61.3	54.5	64.8
3	35.9	21.7	20.6	24.3
4	36.7	30.3	14.7	28.7
5	23.6	14.9	11.1	18.6
6	24.5	17.3	19.4	16.9
7	30.6	26.8	29.8	30.7
8	40.1	28.5	34.0	45.7
9	30.7	33.3	22.1	27.7
10	41.8	27.4	23.0	39.1
11	20.7	22.0	17.9	23.8
12	30.1	24.9	23.2	19.1
13	70.8	66.8	65.7	72.6
14	43.9	32.9	30.0	34.3
15	33.6	28.8	25.8	19.9
16	17.8	16.4	12.0	25.0
17	48.6	36.7	33.6	44.0
18	36.8	30.5	21.2	33.9
19	38.7	31.6	37.3	33.1
20	40.6	35.3	28.8	31.9

32 The experiment designed in exercise 1.4(6) to compare 4 diets for fattening pigs led to the following observations of weight gain (kg) for the individual pigs:

Block	Diet			
	P	Q	R	S
1	72.0	69.8	71.4	70.9
2	72.0	73.7	73.9	74.1
3	69.4	70.2	70.4	71.9
4	69.3	70.2	69.9	71.5
5	71.0	68.6	70.9	71.1
6	69.0	69.8	69.8	70.1
7	70.6	70.5	70.1	71.2
8	73.4	74.1	70.3	72.2

33 The experiment designed in exercise 1.4(5) to compare the effects of various plant hormones on growth, gave the following observations of dry matter weight gain (g per individual seedling):

			Row of pots on the staging		
Treatment	1	2	3	4	5
A	5.08	7.20	6.27	8.96	3.79
B	4.11	6.86	7.04	8.43	4.62
C	5.23	7.96	7.65	10.01	4.27
D	5.00	7.68	6.33	9.74	4.40
E	7.98	10.43	9.50	13.46	7.43
F	7.74	10.84	10.44	12.99	8.13
G	10.99	12.90	12.32	15.17	10.24
H	11.56	13.14	12.67	16.16	10.85
I	10.08	13.54	13.37	16.69	10.48
J	10.90	12.74	13.00	15.93	10.39

34 In an experiment on the effect of different proteins on the growth of young chicks, cages of 3 chicks were fed one of 16 diets, the variable observed being the total weight gain of the 3 chicks over the period 10 to 20 days of age. There were 6 replicates of all the treatments, the 96 cages being arranged in two batteries with the 16 cages of a replicate as close together as possible and the treatments allocated randomly to cages within replicates. The weight gains (g) were as follows:

			Replicates			
Diets	1	2	3	4	5	6
A	125	95	121	92	80	87
B	201	169	152	174	141	128
C	251	216	209	231	226	230
D	332	323	310	317	320	291
E	224	170	176	193	163	153
F	294	290	268	279	274	267
G	206	187	172	180	170	147
H	298	237	281	291	267	184
J	116	101	103	146	94	80
K	135	137	129	138	121	131
L	171	160	156	207	171	144
M	262	277	233	249	213	221
N	165	155	135	165	145	124

Diets	Replicates					
	1	2	3	4	5	6
P	222	196	184	200	164	167
Q	180	156	187	187	162	157
R	247	264	211	247	222	229

35 In an experiment validating a laboratory technique in a large hospital, 3 technicians made a certain measurement, each technician repeating the measurement with 5 different instruments: all 15 observations were made on subsamples from a single sample of material. The values obtained were as follows:

Technician	Instrument				
	1	2	3	4	5
1	5.2	11.7	2.1	4.7	10.6
2	7.9	12.0	6.4	5.1	10.8
3	4.1	6.2	3.8	3.2	9.2

Note that this example differs from exercises 31–4 in that both factors in the design are of interest to us; it is therefore necessary to use Friedmann's test twice, once to test the null hypothesis that technician medians do not differ ($n = 3$, $m = 5$) and again to compare instrument medians ($n = 5$, $m = 3$).

Exercises 36–41 require the use of the Kolmogorov–Smirnov two-sample test, discussed in 3.5. As explained in that section, this is a general test for difference between two populations, and does not require the assumption that the forms of the two distributions are the same. We may expect that a test which does not require any conditions on the populations will be on average less sensitive than a test that does impose a condition, and therefore in exercises 36–9, which use the same data as exercises 12–15, we may not get the same results as were given by the Mann–Whitney test.

36 Carry out the Kolmogorov–Smirnov test on the data of exercise 12.

37 Carry out the Kolmogorov–Smirnov test on the data of exercise 13.

38 Carry out the Kolmogorov–Smirnov test on the data of exercise 14.

39 Carry out the Kolmogorov–Smirnov test on the data of exercise 15.

40 As part of an investigation comparing two processes in the manufacture of instant tea, an industrial taste panel scored 28 samples processed by method *A* and 30 samples processed by method *B*. Working on a 0–25 scale, the taste scores awarded were as follows:

Process *A*	13	13	11	9	10	9	5	13	9	11
	11	9	10	11	11	10	9	9	14	12
	12	12	9	13	10	13	12	13		
Process *B*	24	17	20	17	21	18	19	18	22	17
	22	19	22	20	24	19	21	21	21	20
	21	22	17	19	21	21	21	20	21	22

41 A commercial chick hatchery tested 25 samples of 20 eggs from each of two suppliers: the numbers of live chicks were as follows:

Suppliers *S*	19	16	19	20	17	17	18	20	17	19
	17	17	18	19	18	17	19	19	18	17
	16	19	18	18	14					
Suppliers *T*	16	14	15	12	12	15	13	12	15	13
	16	11	13	13	9	14	14	16	17	10
	11	15	17	15	11					

In 3.6 we considered the distribution between one- and two-sided tests; this distinction is relevant to some of the preceding exercises, so we now consider this topic.

42 Consider the tests carried out in exercises 12, 14, 21, 23 and 29, deciding in each case whether you think that a one-sided or a two-sided test is appropriate.

43 Consider the tests carried out in exercises 13, 15, 22, 24 and 30, deciding in each case whether you think that a one-sided or a two-sided test is appropriate.

4 ASSOCIATION

4.1 The idea of association

In the methods discussed in earlier chapters we have shown how to summarize one sample and how to compare two or more samples, but in every case we have supposed that we are dealing with only one variable. This is a reasonable starting point, but many new and interesting problems require attention if we observe more than one facet of our material. Thus in our earliest discussion of green grass, we could well have considered how the greenness was related to, say, the botanical composition of the sward; in considering the fattening period of bacon pigs (2.2.1) we might reasonably have asked how this period was related to each animal's weight at the start. In both these cases, and in many others, we should be interested in the association between two or more variables; the purpose of this chapter is to describe some of the methods available for the analysis of association.

We shall consider three main groups of methods. The first (4.2) is used when the observations are counts (frequencies); the second and third apply to measured variables, the second (4.3) being a simple test of correlation and the third (chapter 8) giving more precise information about the form of the relationship.

4.2 An example of a 2 × 2 contingency table

Suppose we wish to test the usefulness of a measles vaccine in a group of 200 children. We allocate the children, at random, into two groups of 100; the children in one group receive the vaccine, the children in the other (control) group receive only the inert diluent in which the vaccine was administered. After a suitable period, we count the numbers of children from each group who have contracted measles; perhaps in the vaccinated group we find 10 have done so, and in the control group 34. These results can conveniently be displayed (Table 15a) in a *contingency table*, which shows the association between the two classifications of the children according to treatment and to infection. These classifications both have 2 subclasses and the table is therefore described as a 2 × 2 contingency table.

Table 15a. 2 × 2 *contingency table showing possible observed frequencies obtained in a trial of measles vaccine.*

Treatment	Result of exposure		Total
	Infected	Non-infected	
Vaccinated	10	90	100
Control	34	66	100
Total	44	156	200

Table 15b. *The expected frequencies corresponding to the observed frequencies in Table 15a, derived under the null hypothesis that the classifications by treatment and result of exposure are not associated.*

Treatment	Result of exposure		Total
	Infected	Non-infected	
Vaccinated	22	78	100
Control	22	78	100
Total	44	156	200

Since the purpose of the experiment was to determine whether the vaccine had any effect on the proportion of children who contracted the disease, we wish therefore to decide whether the proportion 10:90 differs from 34:66 by a larger amount than we would accept as due to chance. We thus require a significance test for the null hypothesis that the population proportions in the two rows of the table are the same. A suitable approximate test can be obtained using the χ^2 distribution, which we have already encountered in the large sample forms of the Kruskal–Wallis and Friedmann tests. To apply the distribution in the present case, we need to determine the numbers that the null hypothesis leads us to expect in the cells of the table, and to compare these *expected frequencies* with the *observed frequencies* which are the data.

In calculating the expected frequencies we would like to know what is the (common) population proportion between the columns; we do not know this *a priori* and so the best that we can do is to estimate it from the observations. The sensible way to do this would seem to be to use the overall proportion between columns in the total sample of 200 children – since we have assumed by the null hypothesis that the proportions in the two rows are the same, this common proportion clearly also applies to the margin of the table giving the totals. We thus estimate the population proportions as 44/200 infected and 156/200 non-infected; these lead us to expected frequencies in both rows of

$$22 = \frac{44}{200} \times 100 \quad \text{infected}$$

and
$$78 = \frac{156}{200} \times 100 \quad \text{non-infected}.$$

(The values for the two rows are the same because the total numbers are equal.) We are able therefore to prepare the table of expected frequencies given in Table 15*b*; we note that our argument has had the effect of making the margins the same in the two tables.

To compare the observed and expected frequencies we calculate for each cell of the table (but not the margins) the value of the expression

$$\frac{(\text{observed frequency} - \text{expected frequency})^2}{\text{expected frequency}}$$

and total these quantities over the (four) cells of the table to obtain the χ^2 statistic. This gives

$$\chi^2 = \frac{(10 - 22)^2}{22} + \frac{(90 - 78)^2}{78} + \frac{(34 - 22)^2}{22} + \frac{(66 - 78)^2}{78}$$
$$= \frac{(-12)^2}{22} + \frac{12^2}{78} + \frac{12^2}{22} + \frac{(-12)^2}{78}$$
$$= 6.545 + 1.846 + 6.545 + 1.846$$
$$= 16.782.$$

To determine the probability level appropriate to this value of χ^2 we enter Table A5 with 1 degree of freedom; this number is determined from the general rule for contingency tables

degrees of freedom = number of rows minus one

× number of columns minus one.

The probability that we require is that for a one-sided test (3.6) because we can be reasonably sure that the vaccine will not increase the chance of infection. Table A5 gives probabilities for two-sided tests, and so we shall need to halve the probability obtained from the table. Our value of χ^2 (16.782) is larger than that (10.83) given for a (two-sided) probability of 0.001 (0.1 per cent); the appropriate (one-sided) probability is thus less than one half of 0.1 per cent, i.e. 0.05 per cent. This means that there is less than 0.05 per cent (i.e. less than 1 in 2000) chance of so large a value of χ^2 occurring when the population proportions for the two rows are really equal, so that we (conventionally) reject the null hypothesis and conclude that the vaccine has improved the children's resistance to the infection.

4.2.1 The general 2 × 2 table

Denoting the two classifications in a 2 × 2 contingency table by A and B, and the observed frequencies by a, b, c, d (Table 16), we obtain by the argument above expected frequencies

$$\frac{a+c}{N} \times (a+b) \quad \text{for the } a \text{ cell},$$

$$\frac{b+d}{N} \times (a+b) \quad \text{for the } b \text{ cell},$$

$$\frac{a+c}{N} \times (c+d) \quad \text{for the } c \text{ cell}$$

and $$\frac{b+d}{N} \times (c+d) \quad \text{for the } d \text{ cell}.$$

The contribution of the a cell to the χ^2 statistic is thus

$$\frac{\left\{ a - \dfrac{(a+c)(a+b)}{N} \right\}^2}{\dfrac{(a+c)(a+b)}{N}},$$

Table 16a. *2 × 2 contingency table with general observed frequencies.*

	Classification B		
	a	b	a + b
Classification A	c	d	c + d
	a + c	b + d	a + b + c + d = N

Table 16b. *The expected frequencies corresponding to the observed frequencies in Table 16a, derived under the null hypothesis that the row and column classifications are not associated.*

	Classification B		
	$\dfrac{(a+c)(a+b)}{N}$	$\dfrac{(b+d)(a+b)}{N}$	a + b
Classification A	$\dfrac{(a+c)(c+d)}{N}$	$\dfrac{(b+d)(c+d)}{N}$	c + d
	a + c	b + d	N

which may be shown to reduce to

$$\frac{(ad - bc)^2}{N(a + c)(a + b)}.$$

Similar expressions are obtained for the other cells of the table, and adding these and simplifying further leads to the result

$$\chi_1^2 = \frac{(ad - bc)^2 N}{(a + c)(b + d)(a + b)(c + d)}.$$

The 1 as suffix in χ_1^2 indicates that the statistic has 1 degree of freedom. We note that the four factors of the denominator are the four marginal totals of the tables of observed and expected frequencies. Using this formula it is possible to calculate the value of χ^2 for the 2 × 2 table without explicitly obtaining the expected frequencies; thus with the data of Table 15a we have

$$\chi_1^2 = \frac{(10 \times 66 - 90 \times 34)^2 \times 200}{44 \times 156 \times 100 \times 100}$$

$$= \frac{2400 \times 2400 \times 200}{44 \times 156 \times 100 \times 100} = \frac{7200}{429} = 16.783.$$

This is the same value as was obtained in 4.2, but is here the result of a much shorter calculation.

4.2.1.1 The significance test is symmetrical in rows and columns

The discussion in 4.2 and 4.2.1 set out a test for comparing the proportions in the two rows of a 2×2 contingency table, the null hypothesis being that the row population proportions were equal. We consider now what test is relevant to the null hypothesis that the column population proportions are equal. By the same argument as previously we estimate the common proportion from the marginal column, and so we obtain expected frequencies

$$\frac{a+b}{N} \times (a+c) \quad \text{for the } a \text{ cell (Table 16)},$$

$$\frac{a+b}{N} \times (b+d) \quad \text{for the } b \text{ cell},$$

$$\frac{c+d}{N} \times (a+c) \quad \text{for the } c \text{ cell}$$

and $\quad \dfrac{c+d}{N} \times (b+d) \quad$ for the d cell.

We see at once that these are exactly the same as the expected frequencies used in 4.2.1 to compare the rows of the table. Since the expected frequencies (and, of course, the observed frequencies) are the same in the two types of comparisons, it follows that the χ^2 statistic must be the same. In other words the same test applies whether we set out to compare row populations or column populations; the test can therefore truly be regarded as a test of association.

In the example discussed there was no question as to which null hypothesis was sensible and relevant; the time sequence of the experiment made it clear that we could reasonably ask whether vaccination (row) affected infection (column), whilst the reverse question had no meaning. In some cases, however, we do not have this degree of clarity, and the interpretation of the test becomes more difficult. Consider for example the (invented) observations of Table 17. Here we suppose that a group of children have been classified on the results of an intelligence test and on an assessment of their parents' income group – how this latter character is determined need not concern us here! Let us suppose that both classifications are into two subclasses, that for intelligence into quotients above and below 100, and that for income into values above and below £12 000 per annum. The frequencies in Table 17 give $\chi_1^2 = 9.255$ which, in a two-sided test (as we have no prior idea which way an association would occur), gives a probability rather less than 0.01 (1 per cent, or 1 in 100). Thus there is significant evidence of an association between the two classifications, but what is not clear is what this implies in practical terms. Consider, for example, the following interpretations, which might appeal to investigators of differing political convictions.

Table 17. *An invented example to illustrate one of the difficulties of interpreting the results of statistical tests; the example is discussed in 4.2.1.1.*

Child's intelligence quotient	Parents' income group		Total
	⩾ £12000 p.a.	< £12000 p.a.	
⩾ 100	19	8	27
< 100	11	24	35
Total	30	32	62

$$\chi_1^2 = \frac{(19 \times 24 - 8 \times 11)^2 \times 62}{30 \times 32 \times 27 \times 35}$$

$$= \frac{368 \times 368 \times 62}{30 \times 32 \times 27 \times 35} = 9.255.$$

(*a*) A conservative might argue that, since intelligence is at least partially inherited, the row classification tells us something about the parents' intelligence and the comparison between the rows shows how the more intelligent parents – potentially the more valuable members of the community? – have been the more adequately rewarded in terms of this world's goods. This, he might argue, is clearly desirable; society should encourage its more valuable members, The table shows, in fact, just one aspect of the beneficial workings of a free enterprise economy.

(b) A socialist's reaction might be rather different. Comparing the columns, he considers the second, the children of the downtrodden poor; he notes how their development, judged by intelligence, is held back and contrasts this with the favoured situation of the children of bourgeois parents (column 1). How inequitable this is, he thinks, how wasteful of the potentialities of so many members of the new generation; but is not this very typical of the evils of a free enterprise economy.

Both interpretations use the same data, and, by the argument above, the same test; but the conclusions appear diametrically opposed. Each investigator may be tempted to say 'statistics prove . . .' and advance his argument; clearly both cannot be right!

In fact neither is right, not because his analysis is faulty, but because he has not grasped the place of statistical techniques in the inductive process. The different conclusions that the investigators drew were inherent in the different questions they asked, and these in turn were inherent in their prior beliefs. Statistical techniques help us to give objective answers to our questions, but they provide very little guidance as to what questions are sensible; the choice of questions is the real art of the investigator. Statistical methods are in many cases an indispensable tool, but any statement that begins 'statistics prove . . .' is necessarily and unavoidably a lie.

4.2.2 The general $p \times q$ table

In the general case of a $p \times q$ table (p subclasses in the row (A) classification and q in the column (B) classification) it is convenient to extend our earlier use of a suffix as label for our observations by using two suffices, one to specify row and the other to specify column. Thus O_{ij} can conveniently denote the observed frequency in row i (subclass i of the A classification) and column j (subclass j of the B classification). This leads to the notation of Table 18; it is convenient further to denote the row totals by $R_1, R_2, \ldots, R_p$ and the column totals by $C_1, C_2, \ldots, C_q$. We note that a check on the additions is given by totalling both margins; the sums should be the total number of observations, N, so that

$$\sum_{i=1}^{p} R_i = \sum_{j=1}^{q} C_j = N.$$

Table 18. *The general notation for the observed frequencies forming a $p \times q$ contingency table; the calculation of the expected frequencies is discussed in 4.2.2.*

		Classification B					
		1	2	3	$\cdots$	q	
	1	O_{11}	O_{12}	O_{13}	$\cdots$	O_{1q}	R_1
	2	O_{21}	O_{22}	O_{23}	$\cdots$	O_{2q}	R_2
Classification A	3	O_{31}	O_{32}	O_{33}	$\cdots$	O_{3q}	R_3
	$\vdots$	$\vdots$	$\vdots$	$\vdots$		$\vdots$	$\vdots$
	p	O_{p1}	O_{p2}	P_{p3}	$\cdots$	O_{pq}	R_p
		C_1	C_2	C_3	$\cdots$	C_q	N

The purpose of the test is to detect any difference between rows in the population proportions from column to column, or alternatively any difference between columns in the population proportions from row to row; as in the discussion of the 2×2 table in 4.2.1.1, these forms of difference are tested together. An argument very similar to that of 4.2 suggests that suitable expected frequencies can be calculated from the marginal totals: to estimate the frequencies a row at a time we would use the proportions given by the column totals, so that for row i we should get

$$\frac{C_1}{N} \times R_i, \quad \frac{C_2}{N} \times R_i, \quad \ldots, \quad \frac{C_q}{N} \times R_i,$$

whilst to estimate a column at a time we would use the proportions given by the row totals, so that for column j we should get

$$\frac{R_1}{N} \times C_j, \quad \frac{R_2}{N} \times C_j, \quad \ldots, \quad \frac{R_p}{N} \times C_j.$$

Clearly, as we expect, these two approaches give the same table of expected frequencies, that corresponding to the ij position (row i and column j) being

$$E_{ij} = \frac{R_i C_j}{N}.$$

As a useful check, we note that the marginal totals for the table of E_{ij}'s should be the same as those for the observed frequencies. We now calculate

$$\frac{(O_{ij} - E_{ij})^2}{E_{ij}}$$

for each of the pq cells in the table and sum over all these values to obtain the χ^2 statistic with $(p-1)(q-1)$ degrees of freedom. Thus

$$\chi^2_{(p-1)(q-1)} = \sum_{i=1}^{p} \sum_{j=1}^{q} \frac{(O_{ij} - E_{ij})^2}{E_{ij}}.$$

Except in a few cases, this general formula does not permit any worthwhile simplification.

4.2.3 Conditions for validity of the test

The mathematical analysis on which the contingency table tests are based requires two conditions for its validity, first that the sampling circumstances are, in a certain sense, uniform, and secondly that the frequencies are not too small.

The uniformity condition can most easily be explained by reference to a small contingency table such as that discussed in 4.2. Considering, for example, the children in the treated group, we require first that there are within this group no subgroups which affect the chance of infection and second that the children can be regarded as *independent* in the statistical sense of the term, that is that the fact of one particular child becoming infected does not influence the chance for some other. Whether we can be sure that these conditions are satisfied depends on our prior knowledge of the population and the disease. For example, if we suspected that measles was more common in boys than in girls we would have to make our groups of one sex only, perhaps treating the two sexes as separate experiments. The independence requirement would probably lead us not to include brothers or sisters in the trial; presumably if one child contracted measles its brother would have an above average degree of exposure and so might not have the average chance of infection. Clearly it is not easy to be confident about these conditions, because we can never be sure that our prior knowledge is adequate; this is an unavoidable hazard in these, and indeed in many other, statistical procedures.

The requirement that the frequencies be not too small is usually considered to be met if all the expected frequencies exceed 5 (lower values than this are quoted by some authors). If we find in a contingency table that this condition is not met it is usual to combine adjacent rows or columns – thus adding the observed frequencies for the combined rows (columns) and, separately, the corresponding expected frequencies – so as to make the new expected frequencies sufficiently large. No rules are available to tell us how to group the rows or columns; the only requirements are that the newly formed rows/columns are biologically meaningful and that the expected frequencies of the final table should be large enough. The degrees of freedom of the test must, of course, be calculated from the final size of the table after the combination of rows and/or columns is complete.

The case of the 2 × 2 table requires special consideration; if we tried to apply the general procedure of combining rows or columns we should find no table left at all. In this case an empirical device known as *Yates's correction* is used. The effect of this is to replace the formula of 4.2.1 by the following expression

$$\chi_1^2 = \frac{(|ad - bc| - \frac{1}{2}N)^2 N}{(a + c)(b + d)(a + b)(c + d)}.$$

The notation here is as in 4.2.1; the modulus $|ad - bc|$ was explained in 3.2.2.3.

4.2.4(c) Contingency table tests using Minitab

The arithmetic involved in testing for association in a large contingency table is substantial and repetitive and encourages us to look for a suitable computer program. The analyses can be carried out using SPSS-X but we will take the opportunity to illustrate another package, Minitab (Ryan, Joiner and

Ryan, 1980); these authors describe Minitab as:

a worksheet of rows and columns, in which data are stored; and a collection of about 150 commands, which operate on the data stored in the worksheet.

CTable 11. *Minitab commands, data and output for the contingency table analysis on the observations given in Table 15a.*

Commands and data

```
read c1 c2
10  90
34  66
chis c1 c2
stop
```

Output

```
-- read c1 c2
COLUMN      C1       C2
COUNT        2        2
ROW
  1        10.      90.
  2        34.      66.

-- chis c1 c2
```

EXPECTED FREQUENCIES ARE PRINTED BELOW OBSERVED
    FREQUENCIES

	I	C1	I	C2	I	TOTALS
	I		I		I	
1	I	10	I	90	I	100
	I	22.0	I	78.0	I	
2	I	34	I	66	I	100
	I	22.0	I	78.0	I	
TOTALS	I	44	I	156	I	200

TOTAL CHI SQUARE -

$$6.55 + 1.85 +$$
$$6.55 + 1.85 +$$

$$- 16.78$$

DEGREES OF FREEDOM - (2-1) X (2-1) -   1

-- stop

The application of Minitab to the 2×2 contingency table in Table 15a is shown in CTable 11; the calculation of χ^2 and its degrees of freedom is shown but the associated probability level has to be determined from the appropriate table.

To illustrate the use of Minitab on a larger contingency table we will consider exercise 4 of 4.5. The commands and data are set out in CTable 12; the output is similar in layout to that for a 2×2 contingency table in CTable 11. The command

$$\text{let } c6 = c2 + c3 + c4 + c5$$

shows the use of one of the column manipulations possible in Minitab; others will occur in later sections.

CTable 12. *Minitab commands and data for the contingency table analyses required for exercise 4 in 4.5.*

```
read c1 - c5
100  146  149  146  147
71   32   28   37   33
29   17   16   17   18
let c6 = c2 + c3 + c4 + c5
chis c1 - c5
chis c2 - c5
chis c1 c6
stop
```

4.3 Associated measurements

In the contingency tests just discussed each individual had two properties which could be specified by classification, the number of classes in each classification being fairly small; thus each child in the measles investigation (4.2) had a place in the treatment classification (treated or control) and a place in the infection classification (infected or non-infected). In many cases we are able to do more than merely classify our individuals; we can make measurements relevant to the factors we are investigating, and can therefore examine the relationships between the numerical values of the two variables. This more specific form of analysis usually gives worthwhile results with far fewer observations than are needed for a contingency table; we can of course base a contingency table on measurements, using (grouped) values of each variable as the classes of the two classifications, but this requires a number of observations at each value. Moreover a contingency table tests for any form of association, whereas with measured variables we are usually interested only in a fairly simple form of relationship; by limiting ourselves to a test to detect such simple forms we have less need to be prodigal with our observations.

In the present section we take only a first step in the utilization of measurements, considering only the ranks associated with them. This procedure has the advantage that we do not need to make precise assumptions about the distributions of the variable, but on the other hand it means that we can examine the relationship only in very general terms; a more exact method of analysis, in which the relationship is made more specific, is given in chapter 8.

4.3.1 Kendall's coefficient of rank correlation, τ

The measure we shall use of the association between the ranks of two variables is *Kendall's rank correlation coefficient*; we will introduce it by illustrating its calculation. Suppose that for the first 10 of the pigs used in compiling Table 2 we know not only the days to bacon weight (as in that table) but also their starting weight; we will denote this weight by x and number of days by y. The pairs of observations are given in Table 19a. The first step is to rearrange the pairs in order of one of the variables; we will order in x, but so long as each y was kept with its x it would not alter the answer to order in y. This step gives the sequence in Table 19b. In the second step we work on the variable which we did not order, which in this case is y. Starting with left-hand pair (39, 142) we count, in the y row, the number (P) of values which exceed the value in the left-hand pair and the number (Q) of values which are below the left-hand value; this gives $P = 1$, $Q = 8$. We now repeat this process starting with the second pair from the left, including in the count only pairs to the right of this; we obtain $P = 0$, $Q = 8$. This procedure is continued across the row, in every case counting only values to the right of the starting pair. For each pair we then calculate $(P - Q)$ and total the values of this quantity to give the test statistic S; in the present example this gives $S = -37$. The rank correlation coefficient, τ, is defined by the formula

$$\tau = \frac{2S}{n(n-1)},$$

which in this case give $\tau = -0.822$, n being the number of pairs of observations. To test whether this is a significant correlation, that is to test the null hypothesis that the orderings of x and y are independent, we refer the value of S to the appropriate column ($n = 10$) of Table A10; we find the probability associated with $S = 37$ to be given as 0.04 per cent for a two-sided test, but a one-sided test is required in this example because a pig which starts heavier cannot be at a disadvantage compared with a lighter animal, and so we need only to test whether it requires a shorter fattening period. The appropriate probability is thus 0.02 per cent, one half of the tabulated value; this probability is that of obtaining so large (negatively) a value of S when the population rankings of the two variables are unrelated. It is (conventionally) very small, and so we reject the null hypothesis and conclude that higher starting weight is associated with a shorter fattening period.

For values of n greater than 10 we cannot use Table A10 to test significance: we obtain S as before, then calculate

$$\frac{S\sqrt{18}}{\sqrt{[n(n-1)(2n+5)]}},$$

and determine significance by treating this quantity as approximately a standardized normal deviate (5 per cent significance if it exceeds 1.960, 1 per cent if it exceeds 2.576 and 0.1 per cent if it exceeds 3.291).

Table 19a. *The weaning weight (lb) and fattening period (days) for the first 10 pigs from Table 2.*

Pig number	1	2	3	4	5	6	7	8	9	10
Weaning weight (lb) (x)	59	56	46	50	48	41	49	57	52	39
Days to bacon weight (y)	105	114	121	117	115	147	119	106	111	142

Table 19b. *The calculation of Kendall's coefficient of rank correlation for the observations in Table 19a; the procedure is described in 4.3.1.*

x	39	41	46	48	49	50	52	56	57	59
y	142	147	121	115	119	117	111	114	106	105
P	1	0	0	2	0	0	1	0	0	
Q	8	8	7	4	5	4	2	2	1	
$P - Q$	-7	-8	-7	-2	-5	-4	-1	-2	-1	

$$S = \sum (P - Q) = -37, \qquad n = 10,$$

$$\tau = \frac{2S}{n(n-1)} = \frac{-2 \times 37}{10 \times 9} = -\frac{74}{90} = -0.822.$$

4.3.1.1 The interpretation of τ

The sign of S, and hence of τ, indicates the sense of the relationship between the two variables. When, as in the example of 4.3.1, S is negative we have an inverse relationship, one variable becoming, on average, larger as the other decreases. When S is positive the relationship is direct and the variables increase together.

The precise form of the relationship between x and y cannot, however, be inferred from the value of τ because this is calculated from ranks and not from the actual values of the variables. For example, the three sets of observations plotted in Fig. 12 all give $\tau = 1.0$ (the largest possible value, corresponding to perfect agreement between the rankings of x and y), yet they differ considerably in form. Again, the three sets plotted in Fig. 13 all give $\tau = 0.5$, but visually they convey very different impressions. These examples show that, whilst the correlation coefficient is a useful measure of closeness of association, it is not by itself a complete summary of the observations.

The same comment applies with even more force to the significance level of the coefficient. This level depends on the sample size n and the value of τ; with large samples quite small values of τ can be significant, whereas with smaller samples larger values of τ are needed. The three sets of observations plotted in Fig. 14 all give a rank correlation just significant at the 5 per cent level, but our judgement is that they differ appreciably in importance.

The method of calculation when there are ties is illustrated in Table 20. The procedure is basically as before with the restriction that if, at any stage, the current starting pair is tied with a pair to its right (either x tied with x or y with y or both) then this second pair is excluded from the counts for P and Q. Thus with

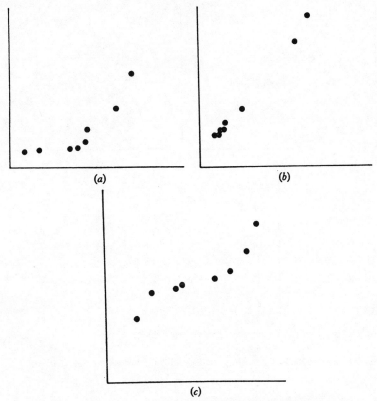

Fig. 12. Scatter diagrams for which the Kendall coefficient of rank correlation, τ, is 1.0; see 4.3.1.1.

the first two pairs of this table (39, 142; 41, 147) the count proceeds as before, but when we come to the third pair (46, 121) we omit the fourth pair (46, 115) from the counts because these pairs are tied in x. The fourth pair does not have any ties to its right and so counts are straightforward. The counts are concluded in this way and S is calculated as before. The calculation of τ, however,

123

is rather more complex, because it is necessary to change the denominator $n(n-1)$ to allow for the ties. Suppose that the ties in x are of lengths $\xi_1, \xi_2, \ldots$; in the example of Table 20 $\xi_1 = 2$

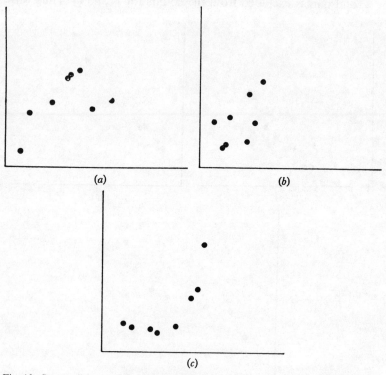

(a)   (b)   (c)

Fig. 13. Scatter diagrams for which the Kendall coefficient of rank correlation, τ, is 0.5; see 4.3.1.1.

(two 46's) and $\xi_2 = 2$ (two 50's). Define similarly $\eta_1, \eta_2, \ldots$ for the ties in y; in the example this gives $\eta_1 = 3$ (three 111's). Calculate now

$$X = \tfrac{1}{2}[\xi_1(\xi_1 - 1) + \xi_2(\xi_2 - 1) + \cdots]$$
$$= \tfrac{1}{2} \sum \xi(\xi - 1)$$

and

$$Y = \tfrac{1}{2}[\eta_1(\eta_1 - 1) + \eta_2(\eta_2 - 1) + \cdots]$$
$$= \tfrac{1}{2} \sum \eta(\eta - 1);$$

in the example
$$X = \tfrac{1}{2}(2 \times 1 + 2 \times 1) = 2,$$
$$Y = \tfrac{1}{2} \times 3 \times 2 = 3.$$

Then τ is calculated as
$$\tau = S/\sqrt{\{\tfrac{1}{2}n(n-1) - X\}} \sqrt{\{\tfrac{1}{2}n(n-1) - Y\}}.$$

[If there are no ties, so that $X = Y = 0$, this expression reduces to
$$\tau = S/\sqrt{\{\tfrac{1}{2}n(n-1)\}} \sqrt{\{\tfrac{1}{2}n(n-1)\}} = 2S/n(n-1),$$
as before.] In the example we have

$$\tau = \frac{-34}{\sqrt{\{(45-2)(45-3)\}}}$$
$$= \frac{-34}{\sqrt{(43 \times 42)}}$$
$$= \frac{-34}{42.50}$$
$$= -0.800.$$

The theory underlying this correction for ties, together with an explanation of the (rare) cases where it is not required and of the modified significance test procedure, is too elaborate to discuss here, but can be found in Kendall (1970). It is worth

Table 20. *The calculation of Kendall's coefficient of rank correlation when there are ties among the observations.*†

x	39	41	46	46	49	50	50	56	57	59
y	142	147	121	115	119	117	111	111	111	105
P	1	0	0	2	0	0	0	0	0	
Q	8	8	6	4	5	3	1	1	1	
$P - Q$	−7	−8	−6	−2	−5	−3	−1	−1	−1	

$$S = \sum (P - Q) = -34,$$
$$X = \tfrac{1}{2}\sum \xi(\xi - 1) = \tfrac{1}{2}(2 \times 1 + 2 \times 1) = 2,$$
$$Y = \tfrac{1}{2}\sum \eta(\eta - 1) = \tfrac{1}{2} \times 3 \times 2 = 3,$$
$$\tau = \frac{S}{\sqrt{\{\tfrac{1}{2}n(n-1) - X\}} \sqrt{\{\tfrac{1}{2}n(n-1) - Y\}}} = \frac{-34}{\sqrt{(45-2)} \sqrt{(45-3)}}$$
$$= \frac{-34}{\sqrt{(43 \times 42)}} = \frac{-34}{42.50} = 0.800.$$

† The pairs of (x, y) values have been ordered on x, and the counts of P and Q are therefore made on y. The procedure for these counts and the definitions of ξ and η are given in 4.3.1.1.

noting, however, that in many cases the correction has relatively little effect, and no great error is introduced by omitting it; thus in the example if we use the uncorrected formula we obtain $\tau = -0.756$. This value is numerically less than that given by the full formula, and it is easy to see that this will always be the case; thus if we omit the correction we tend to underestimate the correlation – which is probably less serious than overestimating it.

As in the case of the Mann–Whitney and Kruskal–Wallis tests (3.3.1), we can calculate S and τ if we observe only the ranks of the x and y variables and not actual measurements.

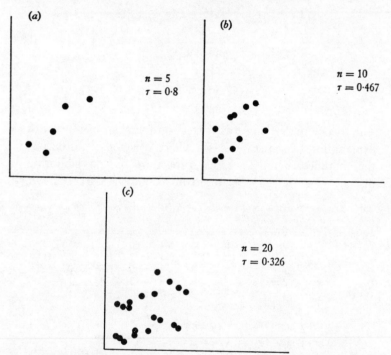

Fig. 14. Scatter diagrams for which the Kendall coefficient of rank correlation, τ, is just significant at the 5 per cent level; see 4.3.1.1.

4.3.1.2*   The distribution of S

We will calculate the distribution of S under the null hypothesis when $n = 4$; the calculation is simple and is a useful illustration of the meaning of the probability used in the significance test.

The null hypothesis that the rankings of x and y are independent means that, when the pairs are ordered on x, the 24 possible orderings of the ranks of y are equally likely. Half these orderings are listed in Table 21 and the value of S is

Table 21. *The distribution of S when $n = 4$ and the rankings of x and y are not associated.*†

Rank ordering of y

				P	S
1	2	3	4	6	6
1	2	4	3	5	4
1	3	2	4	5	4
1	3	4	2	4	2
1	4	2	3	4	2
1	4	3	2	3	0
2	1	3	4	5	4
2	1	4	3	4	2
2	3	1	4	4	2
2	4	1	3	3	0
3	1	2	4	4	2
3	2	1	4	3	0

When there are no ties among the n pairs
$P + Q = \frac{1}{2}n(n-1)$ so that
$$S = P - Q$$
$$= P - \{\tfrac{1}{2}n(n-1) - P\}$$
$$= 2P - \tfrac{1}{2}n(n-1).$$
In the present case $n = 4$ so that
$$S = 2P - 6.$$

Distribution of $|S|$

$|S|$	Frequency	Cumulative frequency	Percentage cumulative frequency
6	2	2	8.3
4	6	8	33.3
2	10	18	75.0
0	6	24	100.0
	24		

† Half the possible y rankings are shown; the remainder are the reverse of these and give values of S with opposite sign (see 4.3.1.2).

calculated for each. The remaining 12 orderings are obtained by reversing the 12 given, and the values of S for these remaining 12 are therefore obtained by changing the signs of the values for the first 12. The distribution of S is then as shown, and the distribution of $|S|$, which is required for a two-sided test, is obtained easily. The cumulative frequencies and probabilities for this latter distribution are given and show that, for example, there is 33 per cent probability of obtaining a value of S numerically as large as 4.

None of the probabilities is conventionally small, so that with

127

$n = 4$ no value of S can give significance in a two-sided test. However, for larger values of n the calculation, though tedious, follows exactly the same lines as in this example and we find we can get S extreme enough to justify rejection of the null hypothesis.

4.3.2(c)   Using SPSS-X to obtain Kendall's τ

The SPSS-X command crosstabs is used to obtain counts of observations classified in a two-factor table; a range of associated statistic n commands provides several measures of association, amongst them τ_b, which differs from the coefficient defined in 4.3.1 only in containing a correction for ties. The commands, data and output in CTable 13 show the use of the procedure on the values recorded in Table 19a; option 13 is used to prevent

CTable 13. *The SPSS-X commands, data and output for the calculation of Kendall's τ_b on the observations given in Table 19a.*

```
Commands and data

        data list free / x y
        variable labels x 'Weaning weight (lb)'
                        y 'Days to bacon weight'
        begin data
        59 105  56 114  46 121  50 117  48 115
        41 147  49 119  57 106  52 111  39 142
        end data
        crosstabs tables = x by y
        option 13
        statistic 6
        execute

Output

- - - - - - - - - - - - - - - -  STATISTICS FOR - - - - - - - - - - - - -
   X   Weaning weight (lb)            BY  Y    Days to bacon weight
NUMBER OF OBSERVATIONS =   10
- - - - - - - - - - - - - - - - - - - - - - - - - - - - - - - - - - - - - -
        STATISTIC              VALUE           SIGNIFICANCE
        - - - - - - - - -      - - - - - -     - - - - - - - - - - - - -


KENDALL'S TAU  B            -0.82222

NUMBER OF MISSING
            OBSERVATIONS  =      0
```

the outputting of the two-factor table of counts – which would of course consist of a large number of blanks with ten 1's scattered amongst them – and statistic 6 calls for τ_b. Since Table 19*a* contains no ties, the value of τ_b is the same as the value calculated in Table 19*b*; the probability of this value is not given and has to be determined from the relevant table.

4.4 Interpreting an association

In the preceding sections we have given two methods, each relevant to a wide variety of circumstances, for detecting the presence of an association between two factors. The example invented in 4.2.1.1 showed that the interpretation of an association is not always easy, but we have not so far discussed the general logic of such interpretation; this is the subject of the present section.

If we observe an association between two factors, A and B, there are three main forms of interpretation available to us:

(*a*) change in A causes change in B;

(*b*) change in B causes change in A;

(*c*) explanations (*a*) and (*b*) are both false, changes in A and B being both effects of changes in some other factor C.

The mere existence of the association does not give us any evidence as to which of these three interpretations is correct; this is the essential point, and it is very often overlooked. The only way we can choose between the interpretations is to bring in prior knowledge or belief; the danger is that we do this without realizing it, and, in consequence, without declaring it.

The sociologists we imagined to be analysing the example of 4.2.1.1 used different prior beliefs: the conservative believed that intelligence affected social position, whilst the socialist thought that economic accidents affected the child's development. Because of their difference in prior beliefs they were led unavoidably to different conclusions: our prior ideas are a fundamental part of every analysis we ever carry out, and are bound to have immense influence on our conclusions. We may well think that the prior beliefs of both workers are false, and that some explanation of type (*c*) is desirable; but we should not confuse what we think with known facts. Without further evidence, the choice of interpretation (c) is just another form of prior belief.

4.5 Exercises

Exercises 1–14 involve the use of the χ^2-test for contingency tables, as described in 4.2 et sqq. In each case where the test is found to be significant, consider the interpretation of the association, especially the outside information which you bring into use in the interpretation. How far might another statistician disagree with your judgement of what information is relevant?

1 An anthropologist studying ethnic types in the British Isles obtained the following observations in a certain district:

19 men had a 'long' head and red hair;
46 men had a 'long' head but did not have red hair;
8 men had a 'short' head and red hair;
73 men had a 'short' head but did not have red hair.

Test whether the apparent association between shape of head and hair colour is significant.

2 In an investigation of how women held a baby against them as compared with how they held an inanimate object, the following counts were obtained:

When handed a baby 49 women placed it first against the left side of their chest and 7 placed it against the right. When handed a large parcel 33 women placed it first against the left side of their chest and 31 placed it against the right.

Test whether the apparent difference in handling a baby and a parcel is significant.

3 The mark distributions of boys and girls in a school examination were as follows:

Mark range	0–20	21–30	31–40	41–50	51–60	61–70	71–80	81–100
Boys	6	8	9	13	10	8	6	4
Girls	5	4	7	10	14	10	5	5

Test whether the difference between the sexes is significant. [The calculation can be carried out using the general formula in 4.2.2, but a simpler form applicable to the $2 \times k$ table is shown in Table 53 (p. 373).]

4 Four vaccines for mumps (A, B, C, D) were compared with a placebo in a clinical trial; the numbers of children uninfected, mildly infected and severely infected in the following 24-month period were as follows:

	Placebo	A	B	C	D
Uninfected	100	146	149	146	147
Mild	71	32	28	37	33
Severe	29	17	16	17	18

Use the χ^2-test
  (a) to compare the 5 groups,
  (b) to compare the 4 vaccines,
  (c) to compare the totals for the 4 vaccines with the placebo.

5 In 5 samples of tissue the frequencies in 100 cells of 6 cell-types were as follows:

Sample	Cell-types					
	P	Q	R	S	T	U
1	3	24	18	13	32	10
2	7	25	16	11	27	14
3	5	28	14	10	31	12
4	7	22	19	11	26	15
5	17	19	13	8	25	18

Test whether the differences between the samples are too large to regard as random.

6 The numbers of plants of 6 species in samples of sward from 6 old pastures were as follows:

Pasture site	Species					
	F	G	H	J	K	L
I	14	10	6	21	12	9
II	3	2	2	10	26	3
III	6	4	4	14	5	5
IV	11	6	9	17	10	8
V	14	9	8	24	15	14
VI	20	10	10	28	16	13

Test whether the sward differed significantly in composition between pastures.

7 The numbers of clusters of $1, 2, \ldots$ fruit from plants of 6 varieties of strawberries were as follows:

Variety	Number of fruit/cluster				
	1	2	3	4	5+
U	13	5	20	11	1
V	10	16	6	14	4
W	6	15	11	14	4
X	4	10	22	14	0
Y	6	12	7	17	8
Z	5	11	18	12	4

Test whether the differences between varieties are significant. The expected frequencies in the $5+$ column are rather small for a reliable test; it is better to combine the 4 and $5+$ columns to obtain a new $4+$ column. The degrees of freedom of the test must of course take account of the reduced number of columns.

8 The distribution of litter size was recorded for sows from 4 breeds, with the following results:

Breed	Number of young in litter					
	$\leqslant 3$	4	5	6	7	$\geqslant 8$
P	4	8	11	12	8	0
Q	3	7	12	10	6	2
R	5	11	13	13	8	2
S	2	5	7	7	4	13

Test whether the differences between breeds are significant. The comment on expected frequencies in the last exercise is again relevant, but to combine columns rather reduces the apparent difference between breeds. Ideally more observations should be taken to help elucidate the differences.

9 A taste panel scored bacon cured by 5 processes; the numbers of sides receiving different scores were as follows:

	Taste score					
Cure	5	6	7	8	9	10
A	10	25	8	6	8	3
B	9	7	12	13	11	8
C	9	8	12	15	9	6
D	10	5	14	15	9	5
E	11	9	12	13	9	6

Test whether the differences between cures are significant.

10 An anthropologist collected information on family size in 4 socio-economic groups: the table shows the numbers of families in the different categories.

	Number of children					
Group	0	1	2	3	4	5+
R	14	20	53	23	23	9
S	39	60	43	76	61	39
T	18	33	53	29	26	15
U	33	49	36	46	39	23

Test whether the differences between groups are significant.

11 Use the χ^2-test to compare the distributions obtained in different groups in exercise 3.7(2).

12 Use the χ^2-test to compare the distributions obtained in different groups in exercise 3.7(4).

13 Use the χ^2-test to compare the distributions obtained in different groups in exercise 3.7(5).

14 Use the χ^2-test to compare the distributions given in exercise 3.7(15).

The χ^2-test used as in exercises 11–14 serves a purpose similar to the Kolmogorov–Smirnov two-sample test discussed in 3.5. The latter test may be preferred if the difference between the two distributions follows a simple pattern, but the χ^2-test may be more useful in detecting irregular differences; in addition, this test can be used to compare more than two distributions in a single analysis, and this can be an important consideration.

Exercises 15–21 are concerned with association between measured variables for which the only method of analysis so far considered is Kendall's coefficient of rank correlation, described in 4.3.1 et sqq. In each exercise the first step should be to plot the relationship, since the coefficient is less useful for a curved scatter. When the coefficient has been calculated it is necessary to consider whether a one- or two-sided test of significance is appropriate.

15 Investigate the relationship between the frequencies of cell types *B* and *C* in exercise 2.6(17).

16 Investigate the relationship between the frequencies of gulls of species *B* and *D* in exercise 2.6(18).

17 Investigate the relationship between duration of response and size in exercise 2.6(22).

18 Investigate the relationship between growth of mould and manganese concentration in exercise 2.6(23).

19 Investigate the relationships between measurements *A*, *B* and *H* in exercise 2.6(24). [This exercise illustrates not only the use of the rank correlation coefficient, but also its limitations as an aid to understanding several simultaneous relationships.]

20 Investigate the relationship between the weight gains in replicates 1 and 2 in exercise 3.7(34).

21 Investigate the relationship between variables *x* and *y* in Table 49 (p. 318).

22* Calculate the distribution of S under the null hypothesis when $n = 5$ (see 4.3.1.2; 60 rank orderings have to be considered).

23* Calculate the distribution of S under the null hypothesis when $n = 6$ (see 4.3.1.2; 360 rank orderings have to be considered, so that the exercise is more suitable for class work than for an individual).

5 CHOOSING BETWEEN ACTIONS

5.1 The problem

In the preceding chapters we have developed the inductive process to the point where we can make statements about single parameters, or differences between parameters, with a certain degree of confidence, in the sense that if we use our procedures repeatedly they will give us correct answers in a known proportion of occasions of use. For certain purposes this form of knowledge is not adequate. Suppose for example that, being concerned in the manufacture of micro-pipettes, we want to know whether to allow a certain batch from the production line to be sent out to wholesalers. We could set up a test procedure to measure the average volume given by the pipettes in our batch, with confidence limits for this average, and we could be sure that the calculation leading to these limits would enable us to enclose the true average in a specified proportion of batches. However this information by itself would be of rather limited use; we still would have to decide whether or not to accept the batch, and this decision would depend on our judgement of the seriousness of the difference between the desired volume and the batch average volume. In other words, we ought to use our prior ideas about how important any particular deviation was, and it would seem sensible to try to combine these prior ideas with the knowledge (and uncertainty) obtained from the test procedure so as to have an objective rule to tell us what to do. Such a rule is called a *decision procedure*; the purpose of this chapter is to discuss some of the principles which are used in designing such procedures.

This part of statistical theory is fairly modern in development but in certain fields of application it has already become very important. These applications have mostly been industrial and

136

the spread of decision theory methods into the experimental sciences has been relatively slow. To some extent this is because the choice of a course of action is not always a very relevant end to an experiment, but there is undoubtedly scope for more applications than have so far been attempted; one useful example is given in 6.5.

5.2 Utility

The first requirement must clearly be a rational basis on which to choose between the possible courses of action; in 5.1 this meant choosing whether or not to accept the batch of micro-pipettes. This choice must be made after consideration of the consequences of the actions; we require first a scale on which we can evaluate these consequences.

A common first thought is that simple financial considerations should be enough to guide us; thus if we reject the pipettes unnecessarily we lose the cost of their production and the profit we could make from them but if we allow them to be sold when they are wrongly calibrated we lose, at the very least, our customer's good will. However, even if we leave aside the problem of how to put a cash value on good will, it is clear that this simple comparison is not sufficient, because monetary value is an inadequate guide. Two pairs of examples will help to illustrate this point.

Suppose first that as my friend X and I leave a cafe after lunch he pays for both of us, leaving me owing him £4. I attempt to pay my debt, but he, in expansive mood, offers to toss 'double or quits'; if I win the debt is settled, but if I lose I owe him £8. If my mood also is fairly expansive I am likely to accept this bet, and let the throw of a coin settle the issue. If, however, my bank manager made me the same offer in respect of my overdraft of £500 I would be most unlikely to accept the bet; release from a debt of £500 would be welcome, but to incur an extra £500 liability would be intolerable! In both cases the bet offered was fair, but whilst I could accept the first because the loss of £4 meant no more than an equal gain, the second was quite unacceptable because the equal losses and gains offered (£500) were not of equal importance to me.

For the second pair of examples suppose first that my friend Y has with much effort scraped together £35 to provide an evening out with his fiancée; no more money is in prospect until the end of the week. If I offer him 'a double or quits' bet he is most unlikely to accept it because, whilst another £35 would add a few luxuries to the evening, the loss of his own £35 would make all his plans impracticable. If, however, I met him outside the ground where the Cup Final was being played, again with £35 in his pocket but now with a tout offering him a ticket at £60 and no less, he might well be willing to accept the same bet; in this case the possible extra £35 would be of great value – it would enable him to get in to see the game – but his present £35 is, for this object, of relatively little value and so its loss might be worth risking. In these two cases the amounts of money involved in the two bets (both fair) are the same, but the different circumstances change my friend's reaction to the prospect.

These two pairs of examples show that both the amount of money and the circumstances in which it is considered affect our value judgements; it is easy to produce further examples which all emphasize the conclusion that simple monetary value is not enough to serve as a basis in choosing between actions. Our preferences are in fact decided by a complex and personal procedure and we need to use our own personal scale to measure them. The use of a personal assessment is at first sight undesirable, because our whole scientific training leads us to require objective determinations of any quantity, but it is only too easy to see that no non-personal scale can ever be used for preferences; the Cup Final means much more to my friend Y than it does to me, and there is no hope of he and I agreeing on the worth of the bet I offered to him – in fact, I would not have offered it him had I not been fairly sure in advance that our assessments of its worth would not be the same. We require therefore a personal measure of preference; this is known as *utility*.

Because of its personal nature no precise definition of utility is possible, but if it is to be a useful concept we can specify certain properties which our judgements of preferences must have. A full discussion of these properties is outside the scope of our elementary approach, but three simple requirements can be stated:

(*a*) Faced with a choice between two possible prospects *A* and *B*, we can determine which of the following three statements is true, that we prefer *A* to *B*, that we prefer *B* to *A*, or that we have no preference between *A* and *B*.

(*b*) Our preferences can be ordered, so that if we prefer *A* to *B* and *B* to *C* then we will also prefer *A* to *C*.

(*c*) There are no prospects which are so wonderful or so terrible that the smallest possibility of them will upset our system of preferences. To see the meaning of this requirement consider the case of my friend *Z*, who is afflicted with an excessive fear of being knocked down while crossing the road; setting out to visit the shops in Main Street, he is well aware that the supermarket across the street is giving a new Rolls-Royce with all purchases over £50, but he cannot bring himself to cross because his phobia overrules his well known addiction to Rolls-Royces. No workable system of utilities can include such overriding emotions.

These conditions do not enable us to define utility, but they do ensure that we can use the concept without leading ourselves to choices which we regard as irrational.

5.2.1 A possible utility–money relationship

It is illuminating to consider a little further the distinctions between utility and monetary value. Suppose I decide, after suitable introspection, that the relationship between these quantities can at this time be described (for me) by a curve of the form shown in Fig. 15; what does this form show about my views on money? To consider this the only point on the axes that need be specified is that with monetary value zero; the curve shows a sharp change of slope at this point because even a small debt (monetary value less than zero) means more to me that an equal credit balance. We discuss the interpretation of three portions of the curve.

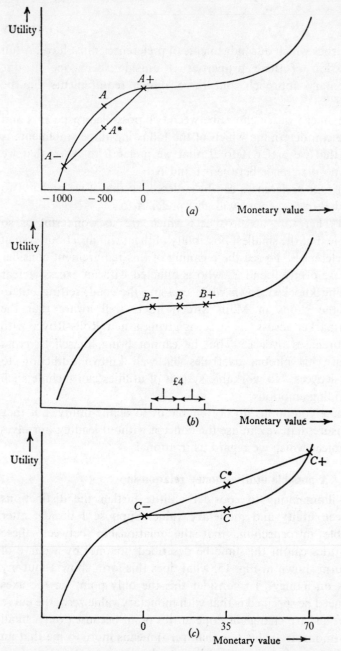

Fig. 15. Possible utility–money relationships to illustrate various decision situations; see 5.2.1. The three diagrams are drawn with different monetary scales so as to increase clarity.

If I consider the overdraft example of 5.2 I find myself at a position such as A on the curve, and give this point monetary value $-£500$ (Fig. 15a). If I were to take the bank manager's bet and to win on the throw of the coin I would find myself at A_+, but if I were to lose I would find myself at A_- with monetary value $-£1000$. Because the chances of winning and losing are equal, the best estimate I can make of where I shall find myself after the throw is exactly midway between A_+ and A_-, that is at A^*. The bet is fair (monetary value and odds exactly balanced) so that A^* has the same monetary value as my starting point A, but the utility of A^* is markedly below that of A. In other words I estimate that if I take the bet I shall lose utility; this is in agreement with the suggestion in 5.2 that I would not take this bet.

The bet over lunch can be considered to centre round B, with B_+ and B_- denoting the positions to be reached after winning or losing £4 (Fig. 15b). Once again the best estimate I can make of my final position is the mid-point of B_+B_-; because the utility–money relationship near B is effectively a straight line this mid-point brings me back to B. I estimate therefore that my utility will be unchanged by taking the bet, so that I have no preference for or against taking the bet.

The situation of my friend Y planning his evening out is essentially the same as mine when negotiating with the bank manager. In both cases it is more serious to lose than to win; this corresponds to the utility–money relationship being convex upwards in the neighbourhood of the starting point. However when Y is trying to get in to the Cup Final his starting position may be taken at C with monetary value £35 (Fig. 15c); if he loses the bet he has monetary value zero and is at C_- but if he wins he moves to C_+. His estimated finishing point is C^*, which shows a considerable increase in utility, in agreement with his inclination to take the bet. The utility–money relationship around C is concave upwards.

5.2.1.1 Backing horses and buying insurance

An interesting extension of the ideas of the previous section is to the case of a man contemplating backing a horse. Taking the same general form of utility–money relationship (Fig. 16) we may suppose his present position to be D. If his horse loses he forfeits his stake and finishes at D_-, but if the horse wins at good odds he will finish at a point such as D_+. If the bet were fair his estimate of his finishing would be D^*, but since the bookmaker has his living to earn the bet cannot be fair and the estimated finishing point must be, say, D^{**}, nearer to the losing result that D^*. However, even though the expected monetary result is disadvantageous the utility of D^{**} is greater than that of D so that the punter considers his bet worthwhile.

Similar considerations apply to the purchase of an insurance policy; this is an unfair bet, but we estimate that we will increase our utility by taking it, because the eventuality against which we insure is very serious. On the other hand the gambler who is already in debt (at E, say) and attempts to retrieve his position, may expect to finish at E^{**} and is clearly ill-advised to bet.

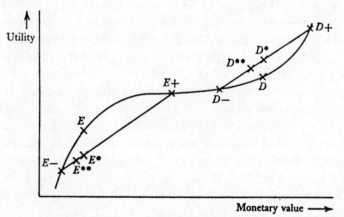

Fig. 16. Further interpretations of the utility–money relationship used in Fig. 15; see 5.2.1.1.

5.2.2 Savage's approach

An interesting approach to the justification for introducing utility is given by Savage (1972). He sets out to define a 'rational' man, summarizing his behaviour in a set of five definitions and seven postulates. The book is somewhat mathematical in presentation but this is so because of the need for precision rather than because the ideas set out are inherently abstract. By way of illustration we will describe two of the postulates (P_1 and P_3); the examples used are given by Savage, but the description here is necessarily less precise. The sure-thing principle, which is closely related to P_1, is illustrated by the following example:

A businessman contemplates buying a certain piece of property. He considers the outcome of the next presidential election relevant to the attractiveness of the purchase. So, to clarify the matter for himself, he asks whether he would buy if he knew that the Republican candidate were going to win, and decides that he would do so. Similarly, he considers whether he would buy if he knew that the Democratic candidate were going to win, and again finds that he would do so. Seeing that he would buy in either event, he decides that he should buy, even though he does not know which event obtains, or will obtain, as we would ordinarily say.

This principle, briefly stated, is that if we prefer a certain action whether or not some event occurs then we decide on the action without considering the event. So stated, it hardly seems a matter for discussion; but it is precisely Savage's object to define his rational man in the simplest terms.

P_3 is rather more complex; essentially it is that 'knowledge of an event cannot establish a new preference among consequences or reverse an old one'. This principle is partly a statement of how a rational man behaves and partly a definition of terms; Savage illustrates it in the following way:

Before going on a picnic with friends, a person decides to buy a bathing suit or a tennis racket, not having at the moment enough money for both. If we call possession of the tennis racket and possession of the bathing suit consequences, then we must say that the consequences of his decision will be independent of where the picnic is actually held. If the person prefers the bathing suit, this decision would presumably be reversed, if he learned that the picnic were not going to be held near water . . . . But under the interpretation of 'act' and 'consequence' I am trying to formulate, this is not the correct analysis of the situation. The possession of the tennis racket and the possession of the bathing suit are to be regarded as acts, not consequences. (It would be equivalent and more in accordance with ordinary discourse to say that the coming into possession, or the buying, of them are acts.) The consequences relevant to the decision are such as these: a refreshing swim with friends, sitting on a shadeless beach twiddling a brand new tennis racket while one's friends swim, etc.

Thus Savage is asserting that what we consider as consequences must come at the end of a sequence of actions and events, that we can decide our preferences amongst these consequences, that these preferences are determined by the consequences and not by the events leading up to them so that, finally, we can sensibly choose between actions, not knowing events, by considering possible consequences. Putting it very crudely, we can reasonably plan ahead even if future events are uncertain.

The other postulates are all attempts to set out simple ideas like P_1 and P_3, but they require increasingly sophisticated presentation. Nevertheless they make very interesting reading to anyone concerned with understanding the principles of choosing actions. Using only these seven postulates Savage shows *inter alia* that a rational man acts as if he had constantly in mind the concept of utility, and chooses his actions so as to increase his utility. His formal approach thus gives a broader basis to the rather intuitive ideas we reached in earlier sections.

5.3 Choosing actions when we know utility

The introduction of utility gives a criterion by which we can judge our preferences between consequences, but in most cases these preferences are not enough to determine our choice of actions. Consider for example Savage's discussion of the man choosing between a tennis racket or a bathing costume (5.2.2). This man finds himself with four consequences to consider: (*a*) costume bought, picnic on the beach; (*b*) racket bought, picnic on the beach; (*c*) costume bought, picnic nowhere near water; (*d*) racket bought, picnic nowhere near water.

Let us suppose that by a process of introspection he decides the utilities of these consequences to be (*a*) 7, (*b*) 2, (*c*) 4 and (*d*) 6. These values mean that his order of increasing preference for the consequences is (*b*), (*c*), (*d*), (*a*); (*b*) is very unattractive because, although he will have a racket for future use, the picnic comes far short of his hopes; (*c*) is a little preferable, partly because he would rather have the costume for future use and partly because it is worse not to be able to bathe than not to be able to play tennis; (*d*) is better because he will get the most out of a picnic away from water, but (*a*) is best because he really does enjoy swimming. The scale on which he measures utility and the precise values that he gives to it are of course entirely personal matters.

In expressing this kind of judgement it is often convenient to reverse the utility scale. For each action in turn, we define the *regret* associated with any one of its consequences as the utility for the most preferred consequence of that action less the utility for the specified consequence. Thus for the present example we have regrets (*a*) 0, (*b*) 4, (*c*) 3 and (*d*) 0; we now want to choose actions which, in some sense, make our regret small (near to zero). The mere knowledge of the regrets is however not always enough to decide our choice of action, because if future events are uncertain we do not know which regrets to consider. If we were to choose our action after the site of the picnic had been announced, the known regrets would be enough to determine the action, but they are not sufficient when the location is uncertain. In the face of uncertainty we need a rule telling us how to use the information provided by the regrets.

5.3.1 The minimax principle

One simple rule for choosing actions is the *minimax principle*. To use this we consider for each action the various possible regrets (depending on what events occur) and note particularly the event which gives maximum regret; thus for the action buy a costume the possible regrets are (*a*) 0 and (*c*) 3, so the maximum possible regret is 3, whilst for the action buy a racket the possible regrets are (*b*) 4 and (*d*) 0 with maximum possible regret 4. The minimax principle is that we choose the action for which the maximum regret is a minimum; in the present case this principle requires us to buy a costume (maximum regret 3) rather than to buy a racket (maximum regret 4).

An important advantage of this principle is that it can be applied as soon as we have evaluated the regrets; no other knowledge – for example, which event is more likely – is needed. It has however the disadvantage that it is essentially a pessimistic approach because it is based on the consideration of the worst that can happen. This is a reasonable procedure when we are considering a game against a skilled opponent – and some military actions can be included under this description – but it is not clear that it is useful for everyday life. Are we in fact prepared to behave as if fate were essentially malevolent, and the worst always happens? In many cases I believe that we are not ready to act on this view.

5.3.2 The Bayes action

A method of avoiding the difficulties of the minimax approach is available if we are prepared to make some assessment of how likely the various possible events are. In the picnic example this would require us to estimate the relative chances of the two locations, in fact to assign *prior probabilities* to them. This at once raises difficulties of definition; what can we mean by the probability that the picnic will be on the beach? The frequency approach to probability (2.1.1.1) is very little help because it is not useful to invoke the concept of a long series of days when a picnic is planned but its location is uncertain; we need a quantitative statement that we can use now, before the present location is decided.

This difficulty can be resolved only by extending the concept of probability to include the assessment of degree of belief, or, as Savage describes it, *personal probability*. This usage fits in with everyday practice, for we regularly assign prior chances to events, sometimes quite precisely – 'I am 90 per cent certain that . . .' – and sometimes more vaguely – 'I think it is more likely that . . .'. However it is not at once obvious that the probabilities quoted in statements of this kind can be used statistically in the same way as frequency probabilities. Savage's view, which I share, is that they can, but it is not appropriate to attempt a full discussion here; readers who want to study this point further can usefully read the very clear argument in chapter 1 of Lindley (1980).

Suppose then that we decide, by considering what we know of the tastes of our friend who is planning the picnic, that suitable prior probabilities are 0.6 for the location being at the beach and 0.4 for it being away from water.† We use this assessment to calculate what average regret we expect to have consequent upon the two possible actions, taking the average as the sum of the individual regrets weighted by their prior probabilities.‡ We refer to this form of average as the *expected regret*. For the action buy a costume the regrets were 0 and 3, so that

$$\text{expected regret} = 0 \times 0.6 + 3 \times 0.4$$
$$= 1.2.$$

For the action buy a racket the corresponding regrets were 4 and 0, so that

$$\text{expected regret} = 4 \times 0.6 + 0 \times 0.4$$
$$= 2.4.$$

The criterion for choice is to select the action for which the expected regret is least; in this case we are (again) led to buy a costume. The principle of choice is thus to select that action

† Putting it very roughly, the interpretation for this statement is that if we were offered odds greater than 6 to 4 that the picnic would be away from the beach we would be inclined to accept the bet, with odds less than 6 to 4 we would be inclined not to accept and with odds of exactly 6 to 4 we would be indifferent to the bet.

‡ This form of average is closely analogous to that used above in calculating the mean of a sample. If for example we have a discrete variable taking values $r_1, r_2, \ldots$ with frequencies $f_1, f_2, \ldots$ then it is easy to show that the sample average $\bar{r}$ (total of all observations/number of observations) can be calculated as $\sum r_i f_i / \sum f_i$, the sums being taken over the full range of the variable. If we now consider, in the manner of 2.1.1.1, the behaviour of this expression as the sample size ($\sum f_i$) increases we find that

$$\bar{r} = \sum r_i \left(\frac{f_i}{\sum f_i} \right)$$
$$= \sum (r_i \times \text{proportional frequency of } i\text{th value})$$

which approximates to

$$\sum (r_i \times \text{frequency probability of the } i\text{th value}).$$

This last quantity is known as the *expectation* of r and can be regarded as a population average; it is in fact the formal definition of the population mean which we have often used. The present expression is of the same form as the expectation but uses personal probabilities; this is clearly correct if we accept that these probabilities can be used in statistical procedures.

which, taking account of how likely the various uncertain events are, gives us least average regret (or greatest average utility).

The action selected by this principle is known as the *Bayes action*; Thomas Bayes gave much thought to the concept of prior probability and he also proved a theorem in the manipulation of probabilities which now bears his name.

5.4 Making use of observations

We have so far considered how to select a sensible course of action given only the statement of the problem and, if we want to use the Bayes action, our personal assessment of the prior probabilities. These two elements are essential starting points, but in many problems – as in this one – we would expect to be able to obtain extra information of some kind which would help us in our decision; in experimental terms, we would expect to take one or more observations relevant to the investigation. In the present problem a simple observation would be obtained by asking the organizer of the picnic where he intended to go. This observation, like all those we consider, would be subject to random variation – the organizer might not yet be sure in his own mind – and we need therefore to consider how to use uncertain information in selecting a course of action. More formally, when we have decided what experiment to undertake, we need to consider what observations are possible, and to select a rule telling us to which action each possible observation should lead: such a rule is called a *strategy*.

In the costume/racket example let us suppose that the organizer will give one of three answers, that the picnic will be at the beach or that it will be away from the water or that he does not know where it will be. Given that these are the only possible observations, we can list the possible strategies, remembering that whatever observations we get we have only two actions open to us. The possible strategies are given in Table 22, and we see at once that they use the observations in very different ways. Thus S_1 and S_8 require us to ignore the observation; S_2 and S_5 use it in ways that seem at first sight to be sensible, whilst the remaining strategies show either perverseness on our part or else a profound distrust of the organizer's truthfulness.

To go beyond this rather vague comparison between the strategies we need to consider the regrets to which they may lead us, and in calculating these we need to assess how likely the possible answers (observations) are for each state of nature – in loose terms we require an estimate of how reliable the observations are. We express this assessment by giving, for each state of nature separately, the probabilities of the various observations; we refer to these as *conditional probabilities* because they depend on the state of nature. In the present case we would have to use personal, subjective, probabilities because neither past experience nor theory are available to lead us to frequency probabilities, but in some decision problems these conditional probabilities would admit of a frequency interpretation. Let us suppose that the values we assign are those given in Table 23; for example we judge that if the picnic is eventually at the beach, there is a 0.6 (60 per cent) chance that the organizer will be able to tell us this when we ask him. The values in the table suggest that we do not regard our observation as very reliable.†

We use these probabilities in calculating the regrets appropriate to each strategy and state of nature, and as a first step we obtain the *action probabilities* which tell us how likely each action is for a given strategy and state of nature. For S_1 the probabilities of buying a costume and a racket are respectively 1.0 and 0 for both states of nature; S_8 is equally easy to deal with, but for the non-trivial strategies the calculation is a little less direct. For S_2, given that the ultimate location is the beach, the action probability of buying a costume is the sum of the conditional probabilities (Table 23) of the observations which lead to this

† We note that for each state of nature the probabilities in the table sum to 1.0. The summation corresponds to considering the overall probability of one or other observation, in the same way that the probability of throwing either a 5 or a 6 with a fair die is $1/3$, the sum of the probabilities ($1/6$ each) of the individual events; more generally, the *summation law of probability* states that the probability that we observe an event belonging to some subset of the possible mutually exclusive events, is the sum of the probabilities of the events in that subset. The overall probability for each state of nature is 1.0 because we have summed over all the possible observations, so that we can be certain that one or other of the subsets considered will be obtained.

Table 22. *Possible strategies in the costume/racket example. Each strategy is a rule telling us which action (buy costume, C, or buy racket, R) to take after we have obtained an observation (the organizer's answer).*

	Observation (organizer's answer)		
Strategy	At the beach	Don't know	Away from water
S_1	C	C	C
S_2	C	C	R
S_3	C	R	C
S_4	R	C	C
S_5	C	R	R
S_6	R	C	R
S_7	R	R	C
S_8	R	R	R

Table 23. *Conditional probabilities for the observations given the state of nature.*

State of nature (location of picnic)	Observation (organizer's answer)		
	At the beach	Don't know	Away from water
Beach	0.6	0.3	0.1
Away from water	0.2	0.4	0.4

action; these answers are 'at the beach' and 'don't know', with conditional probabilities 0.6 and 0.3, so that the action probability is 0.9. That corresponding to buying a racket is, of course, 0.1, whilst if the state of nature had been 'picnic away from water' the action probability for buying a costume would be 0.6 ($= 0.2 + 0.4$) and that for buying a racket would be 0.4. The action probabilities for all the strategies are given in Table 24.

We now use the action probabilities to calculate the expected regret for each strategy and state of nature. This is done, as in 5.3.2, by summing the products for each action of regret multiplied by action probability: thus for S_2 the expected regret is

$$0 \times 0.9 + 4 \times 0.1 = 0.4$$

when the state of nature is that the picnic is at the beach, and

$$3 \times 0.6 + 0 \times 0.4 = 1.8$$

when it is away from the water. The full table of expected regrets

Table 24. *Action probabilities for each strategy and state of nature.*

	State of nature			
	Beach		Away from water	
Strategy	Buy a costume	Buy a racket	Buy a costume	Buy a racket
S_1	1.0	0	1.0	0
S_2	0.9	0.1	0.6	0.4
S_3	0.7	0.3	0.6	0.4
S_4	0.4	0.6	0.8	0.2
S_5	0.6	0.4	0.2	0.8
S_6	0.3	0.7	0.4	0.6
S_7	0.1	0.9	0.4	0.6
S_8	0	1.0	0	1.0

is given in Table 25. The remaining problem is to decide how to use these values in choosing between the strategies.

Table 25. *Expected regrets; the overall values are calculated with prior probabilities 0.6 (beach) and 0.4 (away from water). The minimax strategy is S_5 and the Bayes strategy is S_2.*

	State of nature		
Strategy	Beach	Away from water	Overall
S_1	0	3.0	1.20
S_2	0.4	1.8	0.96
S_3	1.2	1.8	1.44
S_4	2.4	2.4	2.40
S_5	1.6	0.6	1.20
S_6	2.8	1.2	2.16
S_7	3.6	1.2	2.64
S_8	4.0	0	2.40

This problem is very similar to that of choosing between actions when no observations are available (5.3.1 and 5.3.2); once again, two of the possible approaches are to use the concept of minimax or to introduce prior probabilities. To select the *minimax strategy* we consider for each strategy what is the largest expected regret and choose the strategy for which this maximum regret is least. Thus if we use S_1 the worst which can happen to us is that the picnic is away from the beach, leaving us with expected regret 3.0; the maximum regrets for each strategy are underlined in Table 25, and we see that the minimax strategy is S_5, with maximum regret 1.6.

If we feel that the essentially pessimistic view implicit in minimax is not appropriate we can select the *Bayes strategy*, that for which the expected overall regret, computed with the use of prior probabilities, is least. If we take, as in 5.3.2, prior probabilities of 0.6 for the picnic being at the beach and 0.4 for it being away from water, we obtain the overall expected regrets shown in Table 25. The least of these is 0.96, corresponding to S_2, and this is therefore the Bayes strategy.

5.4.1 The value of an experiment

We can apply the expected regrets obtained in 5.3.2 and 5.4 to give us useful information about the value of our observation – that is of the enquiry we made of the organizer. The smallest expected regret before taking the observations, that correspond-ing to the Bayes action in 5.3.2, was 1.20. The smallest after the observation, that corresponding to the Bayes strategy in 5.4, was 0.96. The observation is thus worth 0.24 in reducing our expected regret. We can compare this gain with the cost, in terms of utility, of obtaining the observation. If we assess the time and expense of the telephone call to the organizer at less than 0.24 units of utility, then the 'experiment' is worth undertaking, but if the cost is greater than this value we shall actually lose (expected) utility by making the enquiry and should therefore not carry out the experiment. We note that this calculation can be completed before taking any observations: by using the principles described – in very elementary terms – in this chapter we can decide in advance whether an experiment is worth carrying out.

5.5 Further developments

The application of decision theory methods to examples less naïve and artificial than those used above requires, in most cases, more sophisticated methods than can properly be considered in this elementary book. The only further case we shall consider is a decision theory solution to a one-sample problem (6.5). A very helpful and readable account of the further development of the approach is given by Chernoff and Moses (1959).

5.6 Exercises

These exercises are all concerned with situations in which a choice is to be made between possible courses of action. The situations are specified only in general terms, so as to leave the reader free to define the details in a way which is personally meaningful and acceptable; this personal evaluation and choice is an essential part of the decision theory approach. The steps required in each exercise are as follows:

(a) *specify the possible actions;*

(b) *specify the states of nature that are relevant, and hence the prospects that we need to consider – a prospect being the consequence of a specified action taken in a specified state of nature;*

(c) *determine the regrets associated with the various prospects; this is essentially a personal evaluation;*

(d) *decide whether any observation should be considered;*

(e) *if the answer to (d) is 'no', decide how to choose between actions – shall we use the minimax action or the Bayes action;*

(f) *if we want to use the Bayes action, we must assign prior probabilities to the states of nature;*

(g) *decide between actions;*

(h) *if the answer to (d) is 'yes', decide what observation to take, and what results are possible;*

(j) *set out the strategies possible, associating observations with actions;*

(k) *specify the distribution of the observations for each possible state of nature;*

(l) *evaluate the action probabilities and expected regrets for each combination of a state of nature and a strategy;*

(m) *decide how to choose between strategies;*

(n) *if we want the Bayes strategy, assign prior probabilities to the states of nature;*

(p) *decide between strategies;*

(q) take the observation and use the selected strategy to determine the action to be taken.

The situations postulated in the exercises are of different degrees of complexity; the first is very simple and serves only to give practice in the techniques.

1 You are preparing to dress for a party. Some of the people to be there you do not know very well but would like to know better: some that you know well have their own definite ideas on clothes: and you are not very sure what the mood of the party will be. How do you set about deciding what to wear?

Now we will consider something a little more technical.

2 You find in your laboratory cupboard an unlabelled bottle of a colourless, odourless liquid. How do you set out to identify it? Remember that you are not limited to laboratory tests.

3 You are planning your day's work. There are routine jobs to be done, of varying levels of importance. There are some continuing projects you would like to advance a little; again these are not of equal importance. There is bound to be a 'random element'. How do you plan the day?

Next we will take three examples of a rather greater level of complexity.

4 You are planning and carrying out a research programme: as usual neither time nor money is unlimited. The programme has started well, but now the difficulties are beginning to increase. How do you decide on the next experiment? [This could be discussed in general terms but it would be much more satisfactory if you could relate it to a programme with which you have some contact.]

5 You are planning to buy a new piece of specialized apparatus. Several makes are available, offering a range of prices and a range of facilities. How do you choose the best for your purpose? [Here again, as in exercise 4, it is best to try to consider a situation with which you are moderately familiar.]

6 A patient whom you do not know very well has come to you, as a general practitioner, with a rather unusual assortment of symptoms. Some of the symptoms could be associated with serious illness: some are so bizarre as to make you suspect a psychosomatic condition. How do you proceed?

Lastly let us look at a situation from the history of statistics.

7 Soon after World War I, R. A. Fisher received, within a few days of each other, two suggestions concerning possible appointments, one as senior assistant to Professor Karl Pearson and the other at Rothamsted to start a new statistical department. Fisher was approaching 30 years of age and had published some papers on statistical topics. Pearson was the leading statistician in the country, and was not too many years from retirement. How might Fisher have evaluated these prospects?

His choice is of course a matter of history; and later in his life he had very little sympathy with the concepts of decision theory as these developed; but it is nevertheless interesting to play 'Let's pretend', and to try to reconstruct the arguments he used in reaching his decision.

6 THE NORMAL DISTRIBUTION

6.1 Why the normal is important

In the preceding chapters the concepts and methods which we have developed have been useful for more or less any form of distribution. The ideas of the histogram and the frequency function as representations of the sample and the population apply for all forms of population. The confidence limits for a median can be used with any distribution, and those for a difference of medians require only that the two distributions are of the same form; this condition is also sufficient for the Mann–Whitney and Wilcoxon procedures and needs only direct extension to cover the cases of more than two populations.

In contrast most of the methods to be discussed in the remaining chapters will require that the populations belong to one special family, known as the *normal* distribution; such a restriction is in marked contrast to the apparent generality of the earlier discussion, and we must begin by considering why it is introduced. There are two main reasons; one appears to be essentially of a practical nature, whilst the other seems to be theoretical, but in fact they are closely related.

The first reason is that many of the frequency functions applicable to observed distributions do have a normal form; examples of this form are given in Fig. 17. We note at once that they are all symmetrical; each is in fact completely specified when we know its central value and the distance from the central value to the point where the frequency function ceases to curve inwards and begins to curve outwards (the *point of inflexion*). Observable variables which may approximate closely to a normal form are, for example, the height of children of one sex at a specified age, the yield of wheat from plots of a given size and treatment in an experimental area, the logarithm of the threshold of response to digitalis in cats. Many other examples can be found in the literature, but unfortunately there is in most cases no simple criterion by which we can tell in advance whether some new variable will be normal or not; only observation will help us, and this is necessarily an approximate method.

The theoretical reason for the importance of the normal family is that, for nearly any variable, the sampling distribution of the mean is approximately normal. This means that whatever

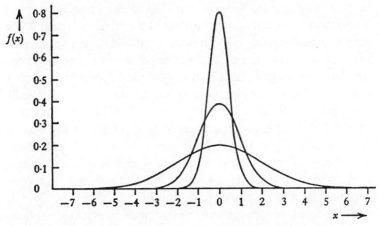

Fig. 17. Three examples of normal distributions; see 6.1.

variable we are using (within very broad limits), if we take the average of a sample of observations and consider the form of distribution appropriate to this average we find that the required form is close to one of the normal family. The goodness of the approximation improves as the samples from which the mean is calculated are made larger; for many initial distributions, the mean of a sample of only 10 observations has a distribution very close to normality. This remarkable and useful result is known as the *central limit theorem*; much ingenuity has been devoted to obtaining elegant and general proofs of it. Its importance for our purpose is in showing that whenever we use an average, which we do very frequently, we are using a quantity with (approximately) a standard form of distribution. Moreover the theorem also applies to most cases of variables whose value can be regarded as the result of a large number of small and independent contributions, as, for example, can experimental error and many variables resulting from a growth process; this explains why so many naturally occurring variables are close to the normal in form.

6.2 Parameters and tables

In the preceding section it was stated that a member of the normal family was completely specified by the central value and the deviation between this value and the point of inflexion; the central value is, of course, the mean and the deviation is known as the *standard deviation*. This definition of standard deviation applies only to the normal family; the term is used, with a mathematically more general definition, for many families of distributions, but its use and interpretation are particularly simple in the case of the normal family and we shall not consider the general definition. The square of the standard deviation is known as the *variance*; in mathematical work it is a more convenient measure of random variation than is the standard deviation.

Since these two parameters (mean and standard deviation) specify the distribution, we expect to be able to relate any property of the distribution to them. One important property which relates to them in a particularly simple way is the probability of an observation lying outside any specified range; the discussions of confidence limits and significance tests in earlier sections indicate the importance of probabilities of this kind. If we wish to know the probability that an observation from a normal distribution with standard deviation σ will deviate by more than, say X, units from the mean of the distribution we have only to express the deviation in standard deviation units – i.e. calculate $x = X/\sigma$ – and obtain from Table A11 the required probability P corresponding to x. The area under the frequency function measured by P is shown in Fig. 18a; if we require the probability for a one-sided deviation – for example, the chance that an observation will exceed the mean by more than x standard deviation units – the symmetry of the distribution implies that

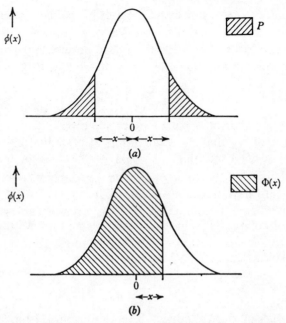

Fig. 18. These diagrams illustrate the relationships between the standardized normal deviate, x, the two-tailed probability P, and the distribution function $\Phi(x)$; see 6.2, 6.2.1, and Table A11.

we halve the tabulated value of P. The quantity x is known as a *standardized normal deviate*.

6.2.1* Frequency and distribution functions

The frequency function of a normal variable z with mean μ and standard deviation σ is

$$f(z) = \frac{1}{\sigma \sqrt{(2\pi)}} e^{-(z-\mu)^2/2\sigma^2}.$$

The variable z occurs only in the term $(z - \mu)^2$, which shows that the distribution is symmetrical about μ; as in all symmetrical distributions the mean and median coincide. The standardized normal deviate corresponding to z is $x = (z - \mu)/\sigma$; it may be shown that the frequency function for this variable is

$$\phi(x) = \frac{1}{\sqrt{(2\pi)}} e^{-\frac{1}{2}x^2}.$$

159

This function does not contain either of the parameters, μ and σ, of the original distribution; this is why one table (Table A11) is sufficient to give probabilities for all normal variables.

The distribution function of x is denoted by $\Phi(x)$, where

$$\Phi(x) = \frac{1}{\sqrt{(2\pi)}} \int_{-\infty}^{x} e^{-\frac{1}{2}t^2} \, dt.$$

From the argument in 3.5.1.1 we see that this expression measures the probability that an observation of a standardized normal deviate will be less than or equal to x, which in turn is the probability that the value of a normal variable will be less than x standard deviation units above its mean; this is illustrated in Fig. 18b. In 6.2 we found that $\frac{1}{2}P$ (P as given in Table A11) gave the probability of a value *more* than x standard deviation units above the mean, so that $\frac{1}{2}P$ and $\Phi(x)$ are clearly connected by the relationship

$$\frac{1}{2}P = 1 - \Phi(x)$$

or

$$P = 2[1 - \Phi(x)].$$

The values of P are obtained, using this formula, from numerical evaluations of the integral for $\Phi(x)$, which are also given in Table A11.

6.3 Point estimates of a single mean and variance

In most applications of a normal distribution we do not know the population values of the mean and standard deviation and need to estimate them from our observations. We will consider first the simple case where we have one sample of n independent observations of a normal variable z; we denote these values by $z_1, z_2, \ldots, z_n$.

The estimation of the mean follows exactly the formula given in 2.2; the point estimate is

$$\bar{z} = \frac{\sum z}{n}.$$

This form suffers from the usual limitations of point estimates and we shall shortly give the expression for the corresponding interval estimate; first however we require the point estimate of the variance.

To calculate this we require not only the sum of the observations $\left(\sum_{i=1}^{n} z_i \right)$ but also the sum of their squares $\left(\sum_{i=1}^{n} z_i^2 \right)$. We note that this latter sum is expressed more fully as

$$\sum_{i=1}^{n} z_i^2 = z_1^2 + z_2^2 + \cdots + z_n^2;$$

in other words, we square each value of z and then obtain the total of these squares. We distinguish carefully between this total and the quantity $\left(\sum_{i=1}^{n} z_i \right)^2$ which is

$$\left(\sum_{i=1}^{n} z_i \right)^2 = (z_1 + z_2 + \cdots + z_n)^2,$$

that is the square of the total of all the observations. This distinction is important because both quantities appear in the formula for the point estimate of variance; we will therefore give a simple arithmetic example before proceeding further. Table 26 shows the thicknesses (mm) of ten slices of bread cut by a slicing machine; for these observations $\sum z$ is 39, so that $(\sum z)^2 = 39^2 = 1521$, whilst $\sum z^2$ is 165.

To estimate the variance we calculate first the *sum of squares*† of the observations, using the formula

$$\text{sum of squares} = \sum_{i=1}^{n} z_i^2 - \frac{1}{n} \left(\sum_{i=1}^{n} z_i \right)^2;$$

the estimate is then given by

$$s^2 = \frac{\text{sum of squares}}{n-1}.$$

The divisor $(n-1)$ is known as the *degrees of freedom* of the estimate; this is a term which we encounter very frequently in analyses using normal variables.‡

---

† This name is at first misleading, because of the possible confusion with $\sum z^2$, the sum of the squares of the individual observations, but it is too well established in the literature to allow the use of any other term; its origin is explained shortly.

‡ The reason for dividing by $(n-1)$ rather than n – as might at first sight be expected – is that the sampling distribution of $s'^2 = (\text{sum of squares})/n$ does not have its mean at the population variance, whilst the sampling distribution of s^2 does. We say that s^2 is an *unbiased* estimate of the population variance; unbiasedness is in most cases a desirable property in an estimate.

In the example in Table 26 we have

$$\text{sum of squares} = 165 - \tfrac{1521}{10}$$
$$= 165 - 152.1$$
$$= 12.9,$$

and
$$s^2 = \frac{12.9}{9} = 1.433.$$

The usual point estimate of the standard deviation is s, the square root of the estimate of variance: in the present case this gives

$$s = \sqrt{1.433} = 1.20 \text{ mm}.$$

We note that s is in the same units as the observed variable.

Table 26. *The thickness of ten slices of bread cut by a slicing machine.*
The calculation of the point and interval estimates of the mean and variance of z is described in 6.3, 6.3.1, 6.4, 6.4.1.

Observations (thickness of slice)	z^2
$z_1 = 4$	16
$z_2 = 5$	25
$z_3 = 4$	16
$z_4 = 3$	9
$z_5 = 2$	4
$z_6 = 3$	9
$z_7 = 5$	25
$z_8 = 4$	16
$z_9 = 3$	9
$z_{10} = 6$	36
$\sum_{i=1}^{10} z_i = 39$	$\sum_{i=1}^{10} z_i^2 = 165$

6.3.1 The meaning of the term 'sum of squares'

The term sum of squares introduced in 6.3 is a contraction of *sum of squares of deviations from the sample mean* which, whilst too long for everyday use, is a good verbal description of the algebraic result

$$\sum_{i=1}^{n} z_i^2 - \frac{1}{n}\left(\sum_{i=1}^{n} z_i\right)^2 = \sum_{i=1}^{n} (z_i - \bar{z})^2.$$

The expression on the right-hand side of this equation is obtained by totalling the squares of the deviations of the individual observations from the sample, and is sometimes given as the definition of the sum of squares on the grounds that it is a conceptually easy measure of the scatter of the observations. Whilst this may be true, it is arithmetically a very inconvenient formula, and for that reason it will not be used here.

Another name which is sometimes used for the sum of squares is the corrected sum of squares: associated with this terminology is the name *correction factor* for the quantity

$$\frac{1}{n}\left(\sum_{i=1}^{n} z_i\right)^2.$$

6.4 Interval estimate of a single mean

The method of obtaining the interval estimate of the mean is based on the important result that if z is a normal variable then $\bar{z}$, the sample average, is also normal.† This means that the distribution of $\bar{z}$ can be completely specified by its mean, which is of course the same as the mean of z, and its standard deviation or, equivalently, its variance. Moreover it may be shown that for any distribution (not just the normal family) the variance of the sample average is related to the variance of the original distribution by the simple formula

$$\text{variance of sample average} = \frac{\text{variance of original distribution}}{\text{sample size}}.$$

We are therefore able to obtain a point estimate of the variance of $\bar{z}$ from the expression

$$\text{estimated variance of } \bar{z} = \frac{s^2}{n},$$

whence we have

$$\text{estimated standard deviation of } \bar{z} = \frac{s}{\sqrt{n}}.$$

We now extend the relationship between standard deviation and probability, described in 6.2, to determine what deviation between $\bar{z}$ and the population mean we can accept as reasonable,

---

† The central limit theorem would lead us to expect this to be true for large samples; the importance of the given result is that it is true for any sample size.

this term being interpreted in the same conventional way as in earlier discussions of interval estimation (for example 2.5.1, 3.2). Using this interpretation, we require the deviation which would be exceeded only on a specified proportion, 5 per cent, of occasions. This deviation could be determined from Table A11 if we knew the population standard deviation, but as we are working with an estimated standard deviation we have to use another distribution and table. This distribution is known as *Student's t-distribution* and relevant values are given in Table A12.† To use the table we need to specify the probability level we choose, which determines the column of the table, and the degrees of freedom of the variance estimate, which determine the row of the table; thus if we wish to obtain the deviation which, in samples of 10 observations, will be exceeded in only 5 per cent of occasions, we use the column headed 5 per cent and the row for degrees of freedom 9, finding the tabulated value 2.262. To calculate the deviation itself we remember that the entries in Tables A11 and A12 are expressed in standard deviation units, so that the tabulated value has to be multiplied by the appropriate standard deviation, in this case that for $\bar{z}$; we thus have the result

$$\text{conventionally reasonable deviation for} \atop \text{the mean of a sample of 10 observations} = 2.262 \times \frac{s}{\sqrt{10}},$$

where s is calculated as in 6.3.

For the example of Table 26, where we had $s = 1.20$ mm, this expression gives

$$\text{reasonable deviation} = 2.262 \times \frac{1.20}{\sqrt{10}}$$
$$= \frac{2.262 \times 1.20}{3.162}$$
$$= 0.86 \text{ mm}.$$

We are thus able to calculate the deviation between $\bar{z}$ and the population mean which we accept, in a conventional sense, as

---

† W. S. Gosset discovered this distribution in 1908 and published under the pseudonym 'Student'; it was the first of the fundamental distributions of modern statistics to be properly understood.

reasonable. To obtain the limits which we would regard as 'reasonably sure' to enclose the population mean, we add this deviation to $\bar{z}$ and subtract it from $\bar{z}$; we thus have

$$\text{upper confidence limit for population mean} = \bar{z} + t\,\frac{s}{\sqrt{n}},$$

$$\text{lower confidence limit for population mean} = \bar{z} - t\,\frac{s}{\sqrt{n}}.$$

In both expressions the value of t is determined by the confidence level we select, and by the degree of freedom $(n-1)$ of the estimate s. It is convenient to write both limits in a single formula as follows

$$\text{confidence limits for population mean} = \bar{z} \pm t\,\frac{s}{\sqrt{n}};$$

the sign $\pm$ is thus an instruction to obtain two values, one by adding and the other by subtracting.

In the example of Table 26. with $\bar{z} = 3.9$ mm, we have

$$\text{confidence limits} = 3.9 \pm 0.86$$
$$= 3.04 \text{ mm}, 4.76 \text{ mm}.$$

The probabilities used in this calculation are all frequency probabilities so that we can give a simple interpretation of the limits, namely that if we use this method of calculation repeatedly we shall be right 95 times out of 100 in claiming that the limits enclose the population mean – provided, as always, that we interpret '95 times out of 100' as true only in the very long run. However, as in all the procedures discussed earlier, the limits will from time to time fail to enclose the population mean and we shall, at least at the time, have no means of knowing when this happens.

6.4(c) Interval estimate of a single mean using Minitab

The commands, data and output for the use of Minitab to obtain point estimates of the mean and the standard deviation together with an interval estimate of the mean for the sample of observations contained in Table 26 are set out in CTable 14. The command tint – an acceptable abbreviation of TINTerval –

operates on one column in the Minitab worksheet and requires the specification of the confidence level to be used in calculating the interval estimate. The values shown in the output section of CTable 14 correspond exactly with those obtained in 6.3 and 6.4.

CTable 14. *Commands, data and output for the use of Minitab to obtain point estimates of the mean and standard deviation together with the 95% confidence interval for the mean for the observations contained in Table 26.*

Commands and data

```
    set cl
    4  5  4  3  2  3  5  4  3  6
    tint 95 cl
    stop
```

Output†

```
  -- tint 95 cl
    Cl   N = 10    MEAN =   3.9000    ST. DEV. =  1.20

    A 95.00 PERCENT C.I. FOR MU IS (    3.0433,    4.7567)
```

† In this and subsequent CTables giving Minitab output only the statistical sections of the output are included; the non-statistical sections – data input and so on – which for illustrative purposes were shown in full in CTable 11 are not included.

6.4.1 Interval estimate of a single variance

The interval estimate for the variance is obtained with the use of the χ^2 distribution: the appropriate values are given in Table A13.† The row we use from this table is determined, as in the t-distribution, by the degrees of freedom of the variance estimate. We need, however, unlike our application of the t-distribution, to use two columns from the table because, whilst the symmetry of the t-distribution implies that one deviation is enough to give limits at both ends, χ^2 is markedly skew. The pair of columns we require is specified by the chance of failure that we are prepared to tolerate in our limits. Thus if, in the example of Table 26 with

† The table used for this distribution in 4.2, 4.2.2 (Table A5) is not appropriate here because we require limits at both ends of the distribution, whilst in the earlier sections only large values of χ^2 were of interest and so all the probability was taken at the upper end of the distribution.

9 degrees of freedom, we require limits which will be right 95 times out of 100, and therefore wrong only 5 times out of 100, we use the pair of columns headed $P = 5$ and obtain tabulated values 2.70 and 19.02.

The limits for the interval estimate of variance are obtained by dividing the sum of squares for the sample by the tabulated values of χ^2. Thus in the example we find limits

$$\frac{12.9}{19.02} = 0.6782 \quad \text{and} \quad \frac{12.9}{2.70} = 4.7778.$$

These limits have the usual probability interpretation. We obtain limits for the standard deviation by taking the square root of the limits for the variance; in the present case this gives

$$\sqrt{0.6782} = 0.82 \text{ mm} \quad \text{and} \quad \sqrt{4.7778} = 2.19 \text{ mm}.$$

6.5 Decision theory approach to one-sample inference

Lindley (1962) has described the application of a decision theory approach to the interpretation of one sample of a normal variable and it is instructive to compare the methods and results of his treatment with those in 6.3 and 6.4. The case he considers is a little simpler than ours because he assumes that the variance, σ^2, is known and does not have to be estimated from the sample; in terms of our presentation the only effects this simplification has are that we use σ in place of s and give to t the number of degrees of freedom, infinity (∞), appropriate to a very large sample.

Lindley expresses his example in terms of a variety trial and supposes that, whilst variety A is in widespread use with known mean yield μ_A, we are interested in the possibility of changing to variety B with unknown mean yield μ_B.† He supposes that we take n observations of the yield of variety B, obtaining sample mean $\bar{x}$. Before considering the details of his approach it is convenient to examine how the methods of 6.3 and 6.4 can be applied to this problem.

---

† This can properly be described as a one-sample problem because the mean of A is assumed known so that we observe only one sample, that of variety B.

The point estimate of the mean yield of B is $\bar{x}$ and the corresponding interval estimate has limits

$$\bar{x} \pm t \frac{\sigma}{\sqrt{n}},$$

where, as explained, t is taken with ∞ degrees of freedom and whatever probability level we choose. We can determine whether the yield of B differs from that of A at this level of confidence by noting whether or not μ_A lies outside these limits; if it does we conclude that the mean yields differ. It is not at all clear, however, how far such a test provides a sensible basis on which to decide whether or not to change to variety B. The significance test is a procedure which will tell us, with a known long-run chance of error, whether we could reasonably regard the two variety means as equal. If the test showed a significant difference we would need to consider in addition whether the effect was large enough to justify the change of practice; for this purpose the interval estimate is not a very helpful device, because it gives a range of values which we ought to regard as reasonable. Some of these values may correspond to a difference of means which is too small to be of any practical importance, but some may imply a large difference that would compel us to make the change. Lindley's approach takes account, under certain simplifying assumptions, of these considerations.

He specifies first the prior knowledge we may have of the mean yield of B, and also the form of the utility. Unless we have the results of earlier experiments available we can probably only express the prior knowledge of B in a rather vague way. Lindley suggests that we may reasonably give an interval estimate at, for example, 95 per cent confidence – this confidence may have to be based on personal probability – saying that we are fairly sure that the yield of B is neither less than some value $\underline{B}$ nor greater than some other value $\bar{B}$. These values can conveniently be linked with the required level of confidence by taking the prior distribution of the mean yield of B to be normal with mean μ_0 and variance τ^2, the values of μ_0 and τ being chosen to give the required interval estimate, that is with

$$\underline{B} = \mu_0 - 2\tau, \quad \bar{B} = \mu_0 + 2\tau.$$

These equations may readily be solved to give

$$\mu_0 = \tfrac{1}{2}(\bar{B} + \underline{B}), \quad \tau = \tfrac{1}{4}(\bar{B} - \underline{B}).$$

The assumption of normality for the prior distribution is made largely for reasons of mathematical convenience; since our prior knowledge is pretty vague, this consideration is as good as any other.

In specifying the utility Lindley makes a series of postulates; he would not claim that these are the only possible postulates but rather that they represent one simple and realistic statement of an experimenter's views.

Postulate 1. If the mean yield of B is the same as that of A we prefer not to change varieties, because this change would involve us in various costs and so would decrease our utility.

Postulate 2. We can find a value $\bar{\mu}$ (greater than μ_A) of the mean yield of B at which we are indifferent as to whether or not we change varieties; with this mean yield the cost (loss of utility) involved in changing is just balanced by the gain (increase in utility) from the higher yield of B. We are free to choose the value of $\bar{\mu}$ to suit our own assessment of costs and gains.

Postulate 3. The change in utility produced by adopting variety B is proportional to the difference between the mean yield of B and the 'don't care' value $\bar{\mu}$. Thus if the mean yield of B exceeded $\bar{\mu}$ by 10 units we should gain twice as much utility by changing as we should if the excess were only 5 units; again if the mean of B fell short of $\bar{\mu}$ by 5 units we should lose just as much by changing as we would gain if it exceeded $\bar{\mu}$ by 5 units. This postulate is introduced to simplify the analysis; it is probably reasonable if we envisage only fairly small difference between the varieties, but is not acceptable if a large difference is possible.

Given these statements of prior knowledge and utility, Lindley shows that we should calculate the quantity X, given by

$$X = \bar{\mu} + \frac{\sigma^2}{n\tau^2}(\bar{\mu} - \mu_0)$$

and should change to variety B only if the value of X is exceeded by the sample mean $\bar{x}$. Suppose for example that we know

$$\mu_A = 10, \quad \sigma = 1,$$

169

that we believe $\qquad \mu_0 = 15, \qquad \tau = 2,$

(that is, we are fairly confident the mean yield of B lies between 11 and 19), that we postulate

$$\bar{\mu} = 11$$

and that we take 10 observations ($n = 10$). The decision procedure is to change to variety B if $\bar{x}$ exceeds

$$11 + \frac{1^2}{10 \times 2^2} (11 - 15) = 11 - \frac{4}{40}$$

$$= 10.9.$$

The value of X for a series of similar cases is given in Table 27; it is instructive to consider the effect of changes in μ_0, τ and n.

We note first that when μ_0 exceeds $\bar{\mu}$ – that is when we start with the belief that B is good enough to be worth changing to – we may be prepared to make the change even if $\bar{x}$ is below $\bar{\mu}$ – that is even if the sample alone suggests that the change is not worth while. On the other hand, if μ_0 is below $\bar{\mu}$ – our prior beliefs about B do not encourage us to change – we need to have $\bar{x}$ above $\bar{\mu}$ before we decide to change.

Secondly, for a specified value of μ_0 an increase in τ brings the critical value (X) nearer to $\bar{\mu}$; in other words, if our prior assessment is not very precise we are guided less by it and more by the simple comparison between the sample mean and the 'don't care' value.

Thirdly, as the sample size (n) increases the critical value (X) comes near to $\bar{\mu}$ in all cases. This means that as we take more observations, so we attach less importance to our prior beliefs.

These three effects – of changes in μ_0, τ and n – thus all agree very well with what common sense requires; the decision theory approach has made possible a more nearly rational view of the choice before us. We note, however, that at no stage in the argument is there any consideration of the reliability of our decision. We have ensured that in the long run this procedure will increase our utility, but we have not considered how often we shall, because of random variation, take a decision that reduces utility; on the decision theory approach this consideration is treated as irrelevant.

Table 27. *Values of* $X = \bar{\mu} + \sigma^2(\bar{\mu} - \mu_0)/n\tau^2$ *for* $\sigma = 1$, $\bar{\mu} = 11$ *and a range of values of* n, μ_0 *and* τ. *The use of* X *is described in 6.5.*

n	μ_0: τ:	10		15	
		1	2	1	2
1		12	11.25	7	10
5		11.2	11.05	10.2	10.8
10		11.1	11.025	10.6	10.9
20		11.05	11.0125	10.8	10.95

A further point which we have not considered here, but which Lindley does discuss, is how great an increase in utility we may expect to obtain. The calculation is fairly involved and we do not give the result here; it is useful however both as a guide to the value of the experiment and at the same time as an indication of how costly an experiment (in terms of utility) is worth while.

6.6 Two-sample comparisons

We require now to extend the one-sample methods described in 6.3 and 6.4 to the two-sample case. In doing this we have to distinguish between the direct two-sample form of experiment and the paired-sample form; this distinction was introduced in 1.3.1 and exemplified in the different conditions of the Mann–Whitney and Wilcoxon procedures (3.2, 3.4). We shall find normal analogues for these two tests.

6.6.1 Difference of means with equal population variances

In the simple two-sample case (more formally we would call this a *single-factor classification with only two levels*) we suppose that we have two normal variables, which we will represent by x and y, and that we do not know the mean or variance of either. In our first discussion we will assume that the two variances are the same; since a normal distribution is completely specified by its means and standard deviation (or variance) this assumption implies that the two populations differ only in mean, that is in location. This is the same type of assumption as was made in developing the Mann–Whitney procedure (3.2.1). We give below

171

(6.6.2) a method which can be used for two normal samples when it is not justified.

Suppose that we have n observations of the variable x and m of the variable y. We will represent these, in keeping with earlier notation, by $x_1, x_2, x_3, \ldots, x_n$ and $y_1, y_2, y_3, \ldots, y_m$. As we are interested in the means of the two populations we calculate first their point estimates

$$\bar{x} = \frac{\sum x}{n} \quad \text{and} \quad \bar{y} = \frac{\sum y}{m}$$

and then the point estimate of their difference, namely $(\bar{x} - \bar{y})$. To obtain the interval estimate for this difference we require, just as in the one-sample case (6.4), the point estimate of the (common) variance; this is calculated by a method which is a simple extension of the one-sample procedure. We obtain first the sums of squares for the two samples,

$$\text{sum of squares for variable } x = \sum x^2 - \frac{(\sum x)^2}{n},$$

$$\text{sum of squares for variable } y = \sum y^2 - \frac{(\sum y)^2}{m};$$

these have respective degrees of freedom $(n-1)$ and $(m-1)$. We now obtain the point estimate of variance by forming the total of the two sums of squares and dividing this by the total of the degrees of freedom. Thus

point estimate of the common variance of the two samples

$$= \frac{\sum x^2 - \frac{(\sum x)^2}{n} + \sum y^2 - \frac{(\sum y)^2}{m}}{n - 1 + m - 1}.$$

This estimate can again be represented by s^2, and we speak of it as having $(n + m - 2)$ degrees of freedom – we shall always find the degrees of freedom of a variance estimate as the denominator in the fraction defining the estimate.

To illustrate these computations, consider the observations in Table 28; these are the yields in pounds per plant over a four-week period of samples of two varieties of tomatoes. The samples each contain few observations, so that the histograms in Fig. 19 can give only very imprecise information about the distributions; they

Table 28. *The yields (lb in 4 weeks) of plants from two varieties of tomatoes.*
The notation and procedure of the calculations are described in 6.6.1.

| Variety A | 3.2 | 4.9 | 3.8 | 4.3 | 4.7 | 3.5 | 3.8 | 3.9 | |
| Variety B | 4.1 | 4.6 | 4.9 | 3.9 | 5.2 | 4.4 | 4.7 | 4.8 | 5.4 |

$$n = 8 \qquad\qquad m = 9$$
$$\sum x = 32.1 \qquad\qquad \sum y = 42.0$$
$$\bar{x} = 4.012 \text{ lb} \qquad\qquad \bar{y} = 4.667 \text{ lb}$$
$$\sum x^2 = 131.17 \qquad\qquad \sum y^2 = 197.88$$
$$\frac{(\sum x)^2}{n} = 128.801 \qquad\qquad \frac{(\sum y)^2}{m} = 196.000$$
$$s^2 = \frac{2.369 + 1.880}{7 + 8} = \frac{4.249}{15} = 0.28327$$
$$s = 0.532 \text{ lb}$$

suggest however that it is not unreasonable to regard the distributions as normal and the variances as equal – a test for this latter assumption is given below (6.6.4). The totals of the observations and of their squares are given in the table and, using these, we obtain the following point estimates:

mean yield of variety A = 4.012 lb,
mean yield of variety B = 4.667 lb,
difference of mean yields = 0.655 lb,
common variance = 0.28327,
common standard deviation = 0.532 lb.

The standard deviation is obtained, as usual, by taking the square root of the variance.

The procedure for obtaining the interval estimate of the difference of means is now very similar to that for the one-sample case (6.4).† We use first the result that the difference between (or, indeed, the sum of) the averages of samples of two normal variables is itself a normal variable, and so can be specified by its mean – which is the population quantity we are interested

† The interval estimate for the common variance is obtained exactly as in the one-sample case 6.4.1, taking χ^2 with $n + m - 2$ degrees of freedom.

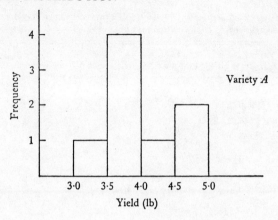

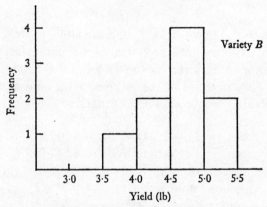

Fig. 19. Histograms of the yields in Table 28; see 6.6.1.

in – and by its standard deviation. This standard deviation is related to those of the two sample averages by the important result†

variance of a difference of two independent quantities
= sum of their separate variances.

---

† This result is a special case of the formula

$$\text{var}(aX + bY) = a^2 \, \text{var}(X) + b^2 \, \text{var}(Y),$$

which is true for any constants a and b provided that X and Y are independent: the present case is obtained by writing $a = 1$, $b = -1$. We note that taking $a = 1$, $b = 1$ gives the same answer, so that we have

$$\text{var}(X - Y) = \text{var}(X + Y) = \text{var}(X) + \text{var}(Y)$$

for independent X, Y.

174

The condition for the validity of this result (independence of the two averages) is satisfied because we have specified that the experiment is of the single factor type, with no other factor to link the two samples in pairs. We already know the estimated variances of $\bar{x}$ and $\bar{y}$, each being the estimated variance of a single observation divided by the appropriate sample size; thus

$$\text{estimated variance } (\bar{x}) = \frac{s^2}{n},$$

$$\text{estimated variance } (\bar{y}) = \frac{s^2}{m}.$$

We have therefore

$$\text{estimated variance } (\bar{x} - \bar{y}) = \frac{s^2}{n} + \frac{s^2}{m}$$

$$= s^2\left(\frac{1}{n} + \frac{1}{m}\right)$$

and   estimated standard deviation $(\bar{x} - \bar{y}) = s\sqrt{\left(\frac{1}{n} + \frac{1}{m}\right)}.$

This estimated standard deviation can be used to give the interval estimate for the difference of means by introducing, as in 6.4, a suitable value from the t-distribution. As before, the value is specified by the level of confidence we require and by the degrees of freedom of the variance estimate, in this case $(n + m - 2)$. We have then

limits of the interval estimate for the difference of means

$$= \bar{x} - \bar{y} \pm ts\sqrt{\left(\frac{1}{n} + \frac{1}{m}\right)}.$$

If, in the example of Table 28, we accept as reasonable a chance of 5 per cent for our limits to be wrong, we find that the value of t is 2.131. The point estimate for the standard deviation of the difference of sample means is

$$s\sqrt{\left(\frac{1}{n} + \frac{1}{m}\right)} = 0.532 \times \sqrt{(\tfrac{1}{8} + \tfrac{1}{9})}$$

$$= 0.532 \times \sqrt{(0.125 + 0.1111)}$$

$$= 0.532 \times \sqrt{0.2361}$$

$$= 0.532 \times 0.486$$

$$= 0.259 \text{ lb}$$

The limits for the interval estimate of the difference of means are then

$$0.655 + 2.131 \times 0.259 = 0.655 \pm 0.552$$
$$= 0.103 \, \text{lb} \quad \text{and} \quad 1.207 \, \text{lb}.$$

As before, the interpretation for these limits is that, if we use this procedure repeatedly, the pairs of limits we obtain will enclose the true difference of means 95 times out of 100 *in the very long run*.

6.6.1.1 Two-sample *t*-test

If we require only a significance test for the null hypothesis that the two population means are equal, and are not interested in the complete interval estimate, we can, as in the Mann–Whitney case, obtain a slightly simpler procedure. We express the observed deviation between the sample averages in standard deviation units and consider whether this value is greater or less than that which we accept as (conventionally) reasonable for a *t*-variable; this is equivalent to considering whether the interval estimate for the difference of means encloses the value zero. Thus we compute

$$\text{observed } t = \frac{\bar{x} - \bar{y}}{s\sqrt{(1/n + 1/m)}}$$

and reject the null hypothesis if this value is numerically greater than the limiting value of t used in 6.6.1. Thus in the example of Table 28 we find

$$\text{observed } t = \frac{0.655}{0.259}$$
$$= 2.53.$$

This exceeds the value 2.131, obtained from the table for 15 degrees of freedom and 5 per cent probability, and so we reject the null hypothesis that the variety means are equal – on the grounds, as usual, that if we claimed that they were equal we should have to agree that a (conventionally) unlikely event had occurred.

This procedure is known as the *two-sample t-test*. We note that the distributional conditions which make it valid – two normal populations with equal variances – are sufficient also for the validity of the rather simpler Mann–Whitney procedure, and that the two tests in fact answer the same question, because the two population medians (compared in the Mann–Whitney procedure) are respectively equal, in the normal case, to the two population means. It is natural therefore to ask whether the t-test has any advantages to justify its somewhat greater complexity. Its main advantage is that it extends to more elaborate problems in a way that the Mann–Whitney test does not, and it is for this reason that it is discussed here, but in the simple case of two normal populations there is actually very little to choose between the two tests and only a small amount of information is wasted if we use the computationally easier method. This same comment applies to later sections, where we give the normal equivalents of the Kruskal–Wallis, Wilcoxon and Friedmann procedures.

6.6.1.2(c) The two-sample comparison using Minitab or SPSS-X

The commands, data and output for the comparison between the two samples in Table 28 are shown in the Minitab version in CTable 15 and in the SPSS-X version in CTable 16. The two

CTable 15. *Commands, data and output for the use of Minitab in the two-sample comparison between the yields of varieties A and B given in Table 28.*

```
Commands and data
        set c1
        3.2  4.9  3.8  4.3  4.7  3.5  3.8  3.9
        set c2
        4.1  4.6  4.9  3.9  5.2  4.4  4.7  4.8  5.4
        pool 95 c1 c2
        stop

Output
  -- pool 95 c1 c2
      C1    N =   8    MEAN =    4.0125    ST.DEV. =   0.582
      C2    N =   9    MEAN =    4.6667    ST.DEV. =   0.485

      DEGREES OF FREEDOM -   15

      A 95.00 PERCENT C.I. FOR MU1-MU2 IS (    -1.2055,    -0.1028)

      TEST OF MU1 - MU2 VS. MU1 N.E. MU2
      T = -2.530
      THE TEST IS SIGNIFICANT AT  0.0231
```

output sections give essentially the same information, both showing the mean and within-sample standard deviation for each sample, together with the observed t-value and its probability level for the two-sample t-test. In addition the Minitab output gives the 95 per cent confidence interval for the difference between the two population means, whilst sections of the SPSS-X output

CTable 16. *Commands, data and output for the use of SPSS-X in the two-sample comparison between the yields of varieties A and B given in Table 28.*

Commands and data

    data list free / variety yield
    variable labels yield 'Yield (lb in 4 weeks)'
    begin data
    1 3.2   1 4.9   1 3.8   1 4.3   1 4.7   1 3.5   1 3.8   1 3.9
    2 4.1   2 4.6   2 4.9   2 3.9   2 5.2   2 4.4   2 4.7   2 4.8   2 5.4
    end data
    t-test groups - variety / variables - yield

Output

-------------------- T - TEST --------------------

GROUP 1 - VARIETY EQ   1.00
GROUP 2 - VARIETY EQ   2.00

VARIABLE	NUMBER OF CASES	MEAN	STANDARD DEVIATION	STANDARD ERROR
YIELD Yield (lb in 4 weeks)				
GROUP 1	8	4.0125	0.582	0.206
GROUP 2	9	4.6667	0.485	0.162

POOLED VARIANCE ESTIMATE

T VALUE	DEGREES OF FREEDOM	2-TAIL PROB.
-2.53	15	0.023

not shown in CTable 16 give the F-test comparison between the two within-sample variances (see 6.6.4) and an approximate version of the t-test using the separate within-sample variances (see Norusis (1983), section 7.5).

6.6.2 Difference of means with unequal population variances

If the assumption of equal variances is not justified the procedures of the preceding sections cannot be followed. In this case we use separate variance estimates for the two samples, each computed according to the method used in 6.3; thus we have

$$\text{variance estimate for first sample} = s_1^2 = \frac{\sum x^2 - \dfrac{(\sum x)^2}{n}}{n-1},$$

$$\text{variance estimate for second sample} = s_2^2 = \frac{\sum y^2 - \dfrac{(\sum y)^2}{m}}{m-1}.$$

Then the estimated variance of the difference between the sample means is

$$\frac{s_1^2}{n} + \frac{s_2^2}{m}$$

and the corresponding standard deviation is, as always, the square root of this variance.

To illustrate these calculations we will consider the observations in Table 29; these are percentage reductions in the growth of a bacterial culture after treatment with one or other of two preparations of penicillin. The histograms (Fig. 20) show that each preparation separately gives a distribution of percentage reduction which can reasonably be regarded as normal, but that the dispersions of the two distributions are very different; in these observations, as is indeed usually the case, percentage values centring around a value near 50 per cent are more variable than those centring near the ends of the percentage range.

For the values given we have the following point estimates:

mean reduction with preparation A = 48.136 per cent,

mean reduction with preparation B = 18.010 per cent,

179

difference of mean reduction $= 30.126$ per cent,

variance for $A = 88.2845$,

variance for $B = 40.6477$.

The estimated standard deviation of the difference between the sample means is then

$$\sqrt{\left(\frac{88.2845}{11} + \frac{40.6477}{10}\right)} = \sqrt{(8.0259 + 4.0648)}$$
$$= \sqrt{12.0907}$$
$$= 3.4772 \text{ per cent.}$$

Table 29. *The percentage reduction in growth of 21 samples of a bacterial culture produced by one or other of two strains of penicillin. The notation and procedure of the calculations are described in 6.6.2.*

Preparation A	41.4	48.3	51.2	30.3	56.8	45.7	51.7	62.4	36.8	57.3	47.6
Preparation B	10.7	16.3	27.2	18.7	21.9	13.2	15.1	8.3	25.6	23.1	

$n = 11$	$m = 10$
$\sum x = 529.5$	$\sum y = 180.1$
$\bar{x} = 48.136$ per cent	$\bar{y} = 18.01$ per cent
$\sum x^2 = 25,371.05$	$\sum y^2 = 3,609.43$
$\dfrac{(\sum x)^2}{n} = 25,488.205$	$\dfrac{(\sum y)^2}{m} = 3,243.601$
$s_1^2 = \dfrac{882.845}{10}$	$s_2^2 = \dfrac{365.829}{9}$
$= 88.2845$	$= 40.6477$

To obtain the interval estimate for the difference of means we use as multiplier for the standard deviation not t but a variable d following the Fisher–Behrens distribution. Values of this quantity are given in Table A14; this table is more complex to use than that of t for we have to specify, in addition to the probability level, the degrees of freedom of the separate variance estimates and also a quantity θ related to these estimates by the expression

$$\tan \theta = \sqrt{\left(\frac{\text{estimated variance of } \bar{x}}{\text{estimated variance of } \bar{y}}\right)}.$$

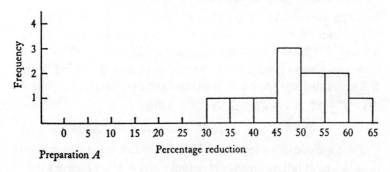

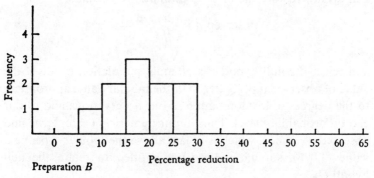

Fig. 20. Histograms of the percentage reductions in bacterial growth given in Table 29; see 6.6.2.

When we have obtained the value of d, we calculate the limits for the difference of means from the formula

limits for the interval estimate of the difference of means
$$= \bar{x} - \bar{y} \pm d \sqrt{\left(\frac{s_1^2}{n} + \frac{s_2^2}{m}\right)}.$$

In the example of Table 29 we obtain
$$\tan \theta = \sqrt{\left(\frac{8.0259}{4.0648}\right)}$$
$$= \sqrt{1.9745}$$
$$= 1.405,$$

so that $\theta = 54° 34'$. This value is not given exactly in Table A14, but we can obtain sufficient accuracy by a rough interpolation

181

between the values for $\theta = 45°$ and those for $\theta = 60°$. With degrees of freedom 10 and 9, and using the 5 per cent probability level, we find $d = 2.244$ for $\theta = 45°$ and $d = 2.228$ for $\theta = 60°$. For the observed value of θ it is thus reasonable to take $d = 2.234$, which will give sufficient accuracy. We then have as values for the limits

$$30.126 \pm 2.234 \times 3.4772 = 30.126 \pm 7.768$$

$$= 22.358 \text{ per cent and } 37.894 \text{ per cent.}$$

If we are content with a significance test for the null hypothesis that the population means are equal – the *Fisher–Behrens test* – and do not require the interval estimate we calculate

$$\text{observed } d = \frac{\bar{x} - \bar{y}}{\sqrt{\left(\dfrac{s_1^2}{n} + \dfrac{s_2^2}{m}\right)}}$$

and reject the null hypothesis of equal population means if the value of observed d is greater than the tabular value appropriate to the degrees of freedom and θ of our observations and to our chosen probability level. Thus in our numerical example we find observed $d = 8.664$, which is in fact above the tabular 1 per cent value (3.096), so that we are not entitled to retain the null hypothesis.

6.6.2.1* Fiducial probability

The theory of the Fisher–Behrens procedure does not allow us to make the usual frequency interpretation of the probability level. The probabilities are in fact of a type not so far introduced, namely *fiducial probabilities*. No definition of this term will be given here; it originated in the work of Sir Ronald Fisher and was for some time a matter of controversy among statisticians. Some workers are indeed still unhappy about results obtained by the fiducial approach, but it is interesting to note that Lindley (1980), using personal probabilities, has found that the method and assumptions which lead, in the case of equal variances, to the two-sample t-test also lead, in the case of unequal variances, to the Fisher–Behrens procedure.

For our purposes it is sufficient to say that this procedure leads to results which we can use – in their appropriate circumstances – in the same way as the t-distribution results, but that the justification for their use is not so simple as in the t-case.

6.6.3 Paired-sample t-test

In the *two-factor classification* situation – the randomized block case – different procedures are needed; in this chapter we consider only the case where the factor of interest is at two levels (cf. the Wilcoxon procedure in 3.4.1). Suppose that we wish to compare two systems of calf feeding and that we have available ten pairs of identical twin calves – this type of experiment has been found to be a particularly worthwhile application of identical twins. We allocate at random one twin from each pair to the two treatment groups and, after keeping all the calves under the same conditions apart from feeding, observe the individual weight gains. We thus have a two-way table in which the gains are classified by treatment and by twin pair (Table 30). For the general case we will denote the observations by

$$x_1, x_2, x_3, \ldots, x_n;$$
$$y_1, y_2, y_3, \ldots, y_n.$$

As before we should satisfy ourselves that the distributions are normal; this is not easy to do in the two-way case and the method used will not be explained until a later section (7.3.1). The simple procedure used in the single factor case – plotting a histogram for each treatment group – is not satisfactory because the histograms would show not only the random variation, which does interest us, but also the between pairs variation, which does not. The variances of the normal distributions have as usual to be the same.†

As in the earlier sections, our objects are to obtain an interval estimate for the difference between the population means for the two groups, and to give a significance test for the null hypothesis that this difference is zero. In doing this we have to use the information that the observations are paired and this is achieved by letting each pair contribute its individual estimate of the

---

† This condition can in fact be relaxed slightly but the increase in generality obtained thereby is small and the possibility is not considered here.

Table 30. *The weight gains (lb) under two systems of feeding of calves of ten pairs of identical twins. The notation and procedure of the calculation are described in 6.6.3.*

Twin pair	1	2	3	4	5	6	7	8	9	10
Weight gains under system $A(x)$	43	39	39	42	46	43	38	44	51	43
Weight gains under system $B(y)$	37	35	34	41	39	37	35	40	48	36
$z = x - y$	6	4	5	1	7	6	3	4	3	7

$n = 10$ $\qquad\qquad \sum z = 46 \qquad\qquad \bar{z} = 4.6\,\text{lb}$

$\sum z^2 = 246 \qquad\qquad \dfrac{(\sum z)^2}{n} = 211.6 \qquad \text{s.s. for } z = 34.4$

$s_z^2 = \dfrac{34.4}{9} = 3.82 \qquad\qquad s_z = 1.955\,\text{lb}$

difference between groups, in other words by forming a single sample of differences $z_1, z_2, z_3, \ldots, z_n$, where

$$z_1 = x_1 - y_1, \quad z_2 = x_2 - y_2, \quad \ldots, \quad z_n = x_n - y_n.$$

The difference of means between the two groups is then the same as the mean of z and so both our objects can be attained by considering this mean. This brings us back to the one-sample situation discussed in 6.3, 6.4, so that we have the following point estimates

$$\text{mean of } z = \frac{\sum z}{n} = \bar{z},$$

$$\text{variance of } z = \frac{\sum z^2 - \dfrac{(\sum z)^2}{n}}{n-1} = s_z^2.$$

These are combined to give the interval estimate of the mean of z as

$$\bar{z} \pm t \frac{s_z}{\sqrt{n}},$$

where t, like s_z^2, has $(n-1)$ degrees of freedom and is taken at the probability level we require.

Considering the values in Table 30 we have the following point estimates:

$$\text{mean of } z = 4.6\,\text{lb},$$

$$\text{standard deviation of } z = 1.955\,\text{lb},$$

limits for the interval
$$\text{estimate of the mean of } z = 4.6 \pm 2.262 \times \frac{1.955}{\sqrt{10}}$$
$$= 4.6 \pm 1.40$$
$$= 3.20 \, \text{lb} \quad \text{and} \quad 6.00 \, \text{lb}.$$

These limits have the usual frequency interpretation, the probability level used in determining t being 5 per cent. They have been derived as limits for the mean of z, but, as explained above, they are to be used as limits for the difference of means between the two treatment groups.

If we require only a significance test for the null hypothesis that this difference is zero, we compute
$$\text{observed } t = \frac{\bar{z}}{s_z/\sqrt{n}}$$

and reject the null hypothesis if this value exceeds that tabulated for t with $(n-1)$ degrees of freedom and our chosen probability level. In the calf example we find
$$\text{observed } t = \frac{4.6}{1.955/\sqrt{10}}$$
$$= 7.44;$$

this is in fact larger than the 0.1 per cent value of t with 9 degrees of freedom (4.781) and so we reject the null hypothesis at the 0.1 per cent level of significance, concluding that the two systems of feeding do differ in the mean weight gains they produce. This procedure is known as the *paired-sample t-test*.

6.6.3(c) The paired-sample comparison using Minitab or SPSS-X

The commands, data and output for the comparison between the paired samples in Table 30 are shown in the Minitab version in CTable 17 and in the SPSS-X version in CTable 18. In Minitab the test cannot be achieved by a single command; it is necessary first to form the sample of differences by the command

let c3 = c1 − c2

and then to use the tint (TINTerval) command as in 6.4(c) followed by ttes (TTESt) against the null hypothesis mean 0. In

185

the SPSS-X version the test requires only the command t-test with the subcommand pairs; the specification of the pairs is by a varlist ($=$ AB) as was used for the two-factor tests under the npar tests command discussed in 3.7(c).

CTable 17. *Commands, data and output for the use of Minitab in the paired-sample comparison between the weight gains under two systems of feeding given in Table* 30.

```
Commands and data

        read c1 c2
        43  37
        39  35
        39  34
        42  41
        46  39
        43  37
        38  35
        44  40
        51  48
        43  36
        let c3 = c1 - c2
        print c3
        tint 95 c3
        ttes 0 c3
        stop

Output

    - - tint 95 c3
        C3    N =   10    MEAN =    4.6000    ST.DEV. -   1.96

        A 95.00 PERCENT C.I. FOR MU IS (     3.2011,     5.9989)

    - - ttes 0 c3
        C3    N =   10    MEAN =    4.6000    ST.DEV. =   1.96

        TEST OF MU =    0.0   VS.  MU N.E.    0.0
        T = 7.440
        THE TEST IS SIGNIFICANT AT    0.0000
```

The Minitab output contains the same information – point estimates of mean and standard deviation, interval estimate of mean and significance test against null hypothesis 0 – about the sample of within-pair differences as was obtained in 6.6.3. The SPSS-X output does not give the interval estimate but does

CTable 18. *Commands, data and output for the use of SPSS-X in the paired-sample comparison between the weight gains under two systems of feeding given in Table 30.*

Commands and data

```
data list free / A B
variable labels A 'Weight gain under system A'
                B 'Weight gain under system B'
begin data
43 37   39 35   39 34   42 41   46 39
43 37   38 35   44 40   51 48   43 36
end data
t-test pairs = A B
```

Output

- - - - - - - - - - - - - - - - - - T - TEST -

VARIABLE	NUMBER OF CASES	MEAN	STANDARD DEVIATION	STANDARD ERROR	*
A	Weight gain under system A				*
		42.8000	3.824	1.209	*
	10				*
		38.2000	4.131	1.306	*
B	Weight gain under system B				*

*(DIFFERENCE) MEAN	STANDARD DEVIATION	STANDARD ERROR	* CORR.	2-TAIL PROB.	*
*			*		*
*			*		*
* 4.6000	1.955	0.618	* 0.882	0 001	*
*			*		*
*			*		*

*	T VALUE	DEGREES OF FREEDOM	2-TAIL PROB.
*			
*			
*	7.44	9	0.000
*			
*			

include the estimate of the (product moment) correlation coefficient between the two treatment groups; this coefficient, which is discussed in 8.2.1, is used here to give some indication of the value in terms of increased precision of the pairing of the individuals – i.e. the calves – on which the observations were made.

6.6.4 Comparing two variances

At several points in the discussion of the various two-sample cases we have been concerned with the possibility that two variances may differ. In the present section we show how to obtain an interval estimate and a significance test relevant to the comparison of two variance estimates; a significance test for comparing several variance estimates is given in 7.2.3.2. For arithmetic illustration we will use the observations in Table 29, where we found

$$s_1^2 = 88.2845 \text{ with 10 degrees of freedom,}$$
$$s_2^2 = 40.6477 \text{ with 9 degrees of freedom.}$$

In comparing variances it is convenient to work not with the difference, as we have done with means, but with the ratio. Thus in the example we have as point estimate of the ratio of the population variances

$$\frac{s_1^2}{s_2^2} = \frac{88.2845}{40.6477}$$
$$= 2.172 \text{ with 10, 9 degrees of freedom.}$$

We note that the degrees of freedom of both variance estimates have to be specified, those for the numerator in the ratio being given first; it is important not to confuse them.

To obtain the corresponding interval estimate we use values of the *variance ratio distribution* (F) obtained from Table A15: to use the table, which refers only to the 5 per cent probability level, we use the column specified by the degrees of freedom of the numerator of our ratio and the row specified by the degrees of freedom of the denominator. For the example of Table 29 these degrees of freedom are 10 and 9, so that reference to Table A15 gives values

$$\underline{F} = 0.265, \quad \bar{F} = 3.98.$$

The limits of the interval estimate of the ratio of the population variances are obtained by dividing these tabulated values into the sample ratio: in our numerical case this gives

$$\text{lower limit} = \frac{\text{sample ratio}}{\bar{F}}$$
$$= \frac{2.172}{3.98}$$
$$= 0.55,$$
$$\text{upper limit} = \frac{\text{sample ratio}}{\underline{F}}$$
$$= \frac{2.172}{0.265}$$
$$= 8.20.$$

For the significance test of the null hypothesis that the two population variances are equal, we have only to note whether the limits just obtained enclose the value 1.0, rejecting the null hypothesis at the 5 per cent level of significance if they do not. Another way of expressing this procedure is to observe whether the sample ratio lies between $\underline{F}$ and $\bar{F}$, rejecting the null hypothesis if it does not.

6.7 Exercises

Exercises 1–4 require the use of Table A11. This enables us to calculate various characteristics of a normal distribution when we know the mean and variance. This is, of course, not a practical situation, but exercises of this type are nevertheless useful in making us familiar with the normal family. Try to sketch the distributions for each exercise and to indicate on the sketch the characteristics specified and also those required: Fig. 18 is relevant.

1 The dry weight of plants of a certain species is a normal variable with mean 8 g and standard deviation 1.5 g – and therefore variance $1.5^2 = 2.25$. [This information can be written more briefly thus: the dry weight is distributed $N(8, 1.5^2)$.] Calculate:

(*a*) the proportion of dry weights within 3 g of the mean,

(*b*) the dry weight below which 70 per cent of the population falls,

(c) the range of values (symmetrical about the mean) which includes 90 per cent of the population,

(d) the dry weight exceeded by 20 per cent of the population.

2 The angle, measured in degrees, between successive segments of the hyphae of a certain fungus is distributed $N(120, 6^2)$. Calculate:

(a) the proportion of angles in the range $(110°, 130°)$,

(b) the angle below which 40 per cent of the population falls,

(c) the range of values (symmetrical about the mean) which excludes 20 per cent of the population,

(d) the angle exceeded by 55 per cent of the population.

3 The concentration of fructose (mg %) in samples of bull semen which have been incubated for 2 hours is distributed $N(80, 20^2)$. Calculate:

(a) the proportion of concentrations in the range (50 mg %, 120 mg %),

(b) the proportion of the population below 73 mg %,

(c) the shortest range of values enclosing half the population – you may prefer to argue from the symmetry of the distribution, or to use trial and error,

(d) the concentration exceeded by half the population.

4 The time (sec) taken by experienced rats to run a maze is distributed $N(140, 30^2)$. Calculate:

(a) the proportions of times in the range (150 sec, 190 sec),

(b) the proportion of times below 170 sec,

(c) the shortest range of values excluding 5 per cent of the times,

(d) the proportion of times above 180 sec.

Each of exercises 5–8 contains observations from a single normal population of unknown mean and variance. The requirement in each case therefore is to obtain point and interval estimates for the mean and variance; the methods of sections 6.3, 6.4 and 6.4.1 are relevant.

5 The dry weights (g) of 25 plants of a certain species were as follows:

9.27	8.38	9.71	6.76	6.43	7.50	5.68	8.04	7.35
9.88	9.62	9.54	9.33	7.66	9.89	6.74	7.23	8.94
6.19	5.48	7.41	8.64	9.60	7.73	10.83		

Draw the histogram of these values and consider whether it appears normal. Assuming normality, obtain point and interval estimates of the mean and variance. [The distribution is normal, and is that specified in exercise 1.]

6 The angles (degrees) between the first two segments of 15 hyphae of a certain fungus were as follows:

| 114 | 118 | 129 | 134 | 122 | 125 | 115 | 122 | 133 |
| 120 | 111 | 123 | 115 | 112 | 112 | | | |

Draw the histogram of these values and consider whether it appears normal. Assuming normality, obtain point and interval estimates of the mean and variance. [The distribution is normal, and is that specified in exercise 2.]

7 The concentrations of fructose (mg %) in 18 samples of bull semen which had been incubated for 2 hours were as follows:

| 115 | 43 | 77 | 99 | 86 | 57 | 83 | 68 | 82 | 35 | 78 |
| 66 | 87 | 73 | 89 | 65 | 71 | 46 | | | | |

Draw the histogram of these values and consider whether it appears normal. Assuming normality, obtain point and interval estimates of the mean and variance. [The distribution is normal, and is that specified in exercise 3.]

8 The times (sec) taken by 30 experienced rats to run a maze were as follows:

115	124	136	133	193	171	149	121	170	94
133	87	146	139	136	147	110	71	201	100
125	95	114	177	123	153	111	178	149	103

Draw the histogram of these values and consider whether it appears normal. Assuming normality, obtain point and interval estimates of the mean and variance. [The distribution is normal, and is that specified in exercise 4.]

Exercises 9 and 10 require the use of the decision theory procedure discussed in 6.5. In each example the reader should consider how useful would be the information conveyed by confidence limits for the mean.

9 The cleanliness of laboratory glassware after washing is scored on a normal scale with standard deviation 0.8. The mean score obtained after using a standard detergent is 6.2. A technician decides that it would be worthwhile to change to a certain new detergent if the mean score after using it exceeded 6.5. His assessment of the new detergent is that he is 95 per cent confident that its mean score lies between 6.5 and 7.3; 15 tests using it gave the following scores:

| 8.5 | 5.5 | 7.2 | 7.3 | 8.7 | 6.5 | 7.0 | 6.4 | 7.9 |
| 7.2 | 5.6 | 7.4 | 5.6 | 7.9 | 7.7 | | | |

Determine whether these results lead the technician to change to the new detergent. Explore the effect of specifying other confidence limits to the range of prior knowledge.

10 The effort in an exercise test (compare question 8 of 3.7) is a normal variable with standard deviation 50. The mean effort for patients receiving a standard drug is 1050: a doctor decides that it would be worthwhile to change to a certain new drug if the mean effort achieved by patients using it exceeded 1090. The doctor is 95 per cent confident that this mean lies between 1060 and 1240: 24 tests on patients who had received the new drug gave the following results:

1141	1039	1190	1246	1135	1072	969	1162
1165	1079	1043	1125	1131	1054	1091	1041
1135	1125	1107	1236	1089	1087	1096	1117

Determine whether these results lead the doctor to adopt the new drug. Explore the effect of specifying other confidence limits to the range of prior knowledge.

Exercises 11 and 12 involve the comparison of the means of two normal populations whose variances are known to be equal. In practice we do not usually know *that the populations are normal or that their variances are equal: later exercises in this section do not allow these assumptions and so are more 'realistic', but these first examples are deliberately simple. The methods of sections 6.6, 6.6.1, 6.6.1.1 are relevant.*

11 The conception rates obtained from artificial insemination of the semen of two bulls were determined on groups of 50 cows;

18 groups were used for bull S and 16 for bull T. The conception rates (per cent) were as follows:

Bull S	74.2	62.1	57.7	71.7	62.0	76.1	70.6	68.3	68.4
	79.8	71.1	70.9	65.5	61.2	60.8	73.9	51.9	63.7
Bull T	49.6	49.2	53.2	56.5	69.1	54.2	80.7	62.7	71.5
	67.5	64.6	75.4	79.6	59.8	68.8	60.2		

Obtain point and interval estimates for the common variance and for the difference between the two means; use the two-sample t-test to determine whether the two mean conception rates differ significantly.

(*The assumptions of normality and equal variances are satisfactory for these two samples because the numbers of trials (cows) for each value are quite large and all equal, and the two means differ only slightly. In general, however, as the discussions in chapter 9 show, proportions do not satisfy these conditions and cannot be analysed by simple normal variable methods.*)

12 The relief obtained from two antitussive drugs is scored on a normal scale; 15 patients used drug P and 16 used drug Q. The scores obtained were as follows:

Drug P	13	11	10	10	11	10	10	10
	12	10	13	10	11	11	12	
Drug Q	12	13	13	14	14	14	14	14
	14	13	15	14	13	11	13	13

Obtain point and interval estimates for the common variance and for the difference between the two means: use the two-sample t-test to determine whether the mean relief given differs significantly between the two drugs.

Exercises 13–15 require two-sample comparisons for which the variances are known to be unequal: the procedure is described in 6.6.2. The reader should draw a separate histogram for each sample and compare visually the two histograms of each exercise: he will perhaps note how difficult it is to detect from the histograms that the population variances are not equal.

13 The live weights (kg) of two groups of young cattle were as follows:

Group A	187.6	180.3	198.6	190.7	196.3	203.8
	190.2	201.0	194.7	221.1	186.7	203.1
Group B	148.1	146.2	152.8	135.3	151.2	146.3
	163.5	146.6	162.4	140.2	159.4	181.8
	165.1	165.0	141.6			

Obtain point and interval estimates for the difference between the two means: use the Fisher–Behrens test to determine whether the means differ significantly.

14 The numbers of grains in individual ears of a certain variety of wheat grown under two sets of conditions were as follows:

Conditions X	36	43	44	39	44	51	39	44	46
	39	42	50	42	46	39	38	49	38
Conditions Y	28	31	25	28	28	27	32	24	33
	33								

Obtain point and interval estimates for the difference between the two means: use the Fisher–Behrens test to determine whether the means differ significantly.

15 The times (sec) required by 18 inexperienced rats to run a maze were as follows:

254	197	209	208	209	211	185	198	189
162	219	253	227	221	259	206	227	136

Obtain point and interval estimates for the difference between the mean of this population and that of experienced rats, sampled in exercise 8. Use the Fisher–Behrens test to determine whether the two means differ significantly.

Exercises 16–18 are paired-sample comparisons from populations satisfying the conditions required for the paired-sample t-test, described in 6.6.3.

16 Fifteen farms co-operated in a field trial in which a normal fattening ration for pigs (ration A) was compared with the same ration supplemented with a small trace of copper (ration B). Each farmer set up two pens of pigs, as similar as possible in all respects, and allocated the two rations at random, one ration to one pen, the other to the other. The mean weight gains per pen (lb/day) were as follows:

Farm	Ration A	B	Farm	Ration A	B	Farm	Ration A	B
1	0.93	1.17	6	1.11	1.15	11	1.11	1.13
2	1.16	1.03	7	0.98	0.96	12	1.18	1.20
3	1.05	1.23	8	0.99	1.02	13	1.02	1.11
4	1.10	1.29	9	0.66	0.95	14	1.05	1.10
5	0.93	1.04	10	1.14	1.25	15	1.17	1.30

Obtain point and interval estimates for the difference between the mean growth rates on the two rations: use the paired-sample t-test to determine whether the two means differ significantly.

17 The distances travelled (cm) by two constituents of a dye in 13 paper chromatographs were as follows:

Sample	Components 1	2	Sample	Components 1	2	Sample	Components 1	2
1	5.8	4.0	6	6.5	5.1	11	5.1	3.8
2	6.6	6.1	7	5.0	5.2	12	5.6	4.3
3	7.3	4.5	8	4.9	5.2	13	6.2	5.7
4	6.3	4.9	9	5.6	5.4			
5	5.9	5.2	10	5.7	5.6			

Obtain point and interval estimates for the difference between the distances travelled by the two constituents. (The significance test is probably not of any practical interest.)

18 The concentration of fructose (mg %) before and after incubation was measured in 12 samples of bull semen; the values obtained were as follows:

Sample	Concentration before	after	Sample	Concentration before	after
1	116	30	7	120	54
2	190	58	8	322	66
3	570	100	9	429	67
4	375	48	10	102	34
5	236	58	11	167	69
6	505	153	12	299	82

By working on the logarithms of these concentrations, obtain point and interval estimates for the proportional fall in fructose concentration during incubation.

Variables are sometimes transformed – by using logarithms or some other functional form – before statistical analysis. This process can serve either or both of two main purposes:

(a) to obtain a variable of increased practical relevance,

(b) to obtain a variable more nearly satisfying the conditions required by some analysis.

These requirements can sometimes be achieved together – as in this example, where the logarithm of fructose concentration is approximately normally distributed – but often they cannot be reconciled, so that choosing a transformation to satisfy one requirement gives a new variable which quite fails to satisfy the other. There is some discussion of the use of transformations in section 9.4 et sqq., but a full consideration is beyond the scope of this book.

Exercises 19–23 require the use of the two-sample variance comparison discussed in 6.6.4.

19 Obtain an interval estimate for the ratio of the variances of the two populations considered in exercise 11.

20 Obtain an interval estimate for the ratio of the variances of the two populations considered in exercise 12.

21 Obtain an interval estimate for the ratio of the variances of the two populations considered in exercise 13.

22 Obtain an interval estimate for the ratio of the variances of the two populations considered in exercise 14.

23 Obtain an interval estimate for the ratio of the variances of the two populations considered in exercises 8 and 15.

Exercises 24–8 involve two-sample comparisons in which no prior knowledge concerning the distribution or the variances is available. The procedure should therefore be:

(a) plot histograms for each sample to see if the distribution appears to be approximately normal;

(b) if so, use the two-sample variance comparison;

(c) if this is non-significant, use the two-sample t procedure: if the variances do differ significantly, use the Fisher–Behrens procedure.

Step (a) is only a very rough check; it should be used in conjunction with any prior knowledge we have about the variable and may be improved by using the Kolmogorov–Smirnov one-sample test described in 7.2.3.1. If this step indicates that the distributions are not normal, we can first consider whether the methods of chapter 3 are relevant; if the forms of the distributions do not permit the use of these procedures, we have no valid methods of analysis available. In such cases the use of a functional transformation, mentioned above, may be considered as a last resort.

24 The fattening periods (days) of two groups of pigs receiving different rations were as follows:

Ration A	125	108	113	122	126	113	116	121
	122	111	117	123	123	119	128	110
Ration B	130	127	116	121	132	131	131	128
	121	127	124	129	126	118	131	

Compare the two populations as fully as possible.

25 The yields (kg per plot) of two varieties of beans were as follows:

Variety X	3.4	4.2	4.8	3.7	4.2	4.3	3.3	3.6	4.2	4.1	4.8
	3.7	4.7	3.9	4.8	4.0	3.8	4.9	4.5	4.0		
Variety Y	4.6	4.5	3.8	4.0	4.7	4.9	4.3	5.0	3.8	4.1	5.1
	5.2	4.9	4.3	4.2	3.9	3.6	4.4	4.9	4.6	4.2	

Compare the two populations as fully as possible.

26 The numbers of hours of sleep obtained by a group of hospital patients after treatment with one or other of two sleep-inducing drugs were as follows:

Drug P	2.7	2.5	16.9	2.3	5.9	6.3	1.6	7.6	5.7
	6.5	3.5	6.4	3.5	5.4	2.6			
Drug Q	4.2	11.6	10.6	4.0	6.0	3.2	5.3	6.5	7.1
	6.1	3.7	4.8						

Compare the two populations as fully as possible.

27 The numbers of brown eggs in batches of 20 eggs from hens of two strains, one selected for brown eggs and one unselected, were as follows:

Unselected strain	2	0	4	6	6	4	3	1
	1	2	3	1				
Selected strain	5	11	10	14	8	11	11	10
	8	11	11	11	12	7		

Compare the two populations as fully as possible.

28 The oxygen uptakes (ml) during incubation of two sets of cell suspensions, one buffered and one unbuffered, were as follows:

Buffered	13.0	13.2	15.0	13.2	14.2	14.8	14.1
suspensions	12.2	12.1	13.4	13.8	13.1	13.5	14.7
	15.1	12.9	15.0	13.9			
Unbuffered	6.7	7.2	9.1	8.3	7.6	9.2	7.8
suspensions	8.7	6.9	6.8	7.7	7.2	8.5	7.5
	7.8	7.5					

Compare the two populations as fully as possible.

7 THE NORMAL VARIABLE IN EXPERIMENTS AND SURVEYS

7.1 The analysis of variance

In the previous chapter we discussed methods that can be used with the normal variable when we are interested in one or two populations; the purpose of the present chapter is to extend these techniques to cases with more than two populations. We shall consider, as usual, both single-factor and two-factor classifications; we shall therefore give, amongst other things, the normal analogues of the Kruskal–Wallis (3.3) and Friedmann (3.4.2) procedures.

In both these cases we shall consider examples of the very general technique known as the *analysis of variance* (*anovar*); it is convenient to explain first the principle of this technique. In spite of its name, the object of the anovar is to provide statistics which are useful in comparing population means; the method can be used only with normal variables, and we shall find indeed that certain further restrictions are also required. The general procedure, applying to all anovars, is to determine how much of the variation in our observations is due to population differences and how much to random variability; by comparing the contributions of these two kinds of variation we can determine the importance of the population differences.

The most convenient measure of variation with a normal variable is, as we already know, a sum of squares and all anovars begin therefore by the computation of a sum of squares to measure the overall variation of all the samples taken together; this is known as the *total sum of squares*. We next calculate how much of the variation can be ascribed to the population differences that interest us and then, by subtracting this variation from the total sum of squares, obtain the *residual sum of squares* which measures the contribution of random effects to the overall variation. The precise steps of the calculation vary according to the arrangement

199

of the experiment material and the populations under study, but the general sequence and logic are the same in every case.

7.2 Single-factor anovar

The single-factor analysis is a convenient first example because the calculations required are only a slight extension of those already discussed in the two-sample case (6.6.1). We will first treat the analysis from this point of view – that is as a direct extension – but we shall find that once the pattern is established it is possible to simplify the calculations; the first part of this discussion should be regarded therefore as an exposition of ideas rather than of arithmetic methods.

We will begin with a numerical example. Table 32 gives the values of the liver weight, expressed as a percentage of body weight, of rats from groups fed on four diets; the design is thus a single-factor classification with four levels.† A visual comparison between the groups is obtained by drawing separate histograms for each group (Fig. 21); as in the example analysed in 6.6.1, these histograms can give no precise indications of the form of the distribution, but they are nevertheless a useful summary.

As we are to compare the means of the four populations it is reasonable to begin by calculating the sample means, obtaining first the totals for each level and dividing them by their appropriate numbers of observations. If we denote the totals by T_1, T_2, T_3, T_4, the numbers of observations by n_1, n_2, n_3, n_4 and the sample means by $\bar{x}_1, \bar{x}_2, \bar{x}_3, \bar{x}_4$ we have

$$\bar{x}_1 = \frac{T_1}{n_1} \qquad \bar{x}_2 = \frac{T_2}{n_2} \qquad \bar{x}_3 = \frac{T_3}{n_3} \qquad \bar{x}_4 = \frac{T_4}{n_4}$$

$$= \frac{26.62}{7} \qquad = \frac{27.44}{8} \qquad = \frac{21.59}{6} \qquad = \frac{31.48}{8}$$

$$= 3.803 \qquad\quad = 3.430 \qquad\quad = 3.598 \qquad\quad = 3.935$$

per cent   per cent   per cent   per cent

We begin the anovar as explained in 7.1, by calculating the

† These observations have already been analysed by the Kruskal–Wallis procedure in exercise 3.7(20).

total sum of squares to measure the variation of all the observations taken together. This means that we require

$$\sum x^2 - \frac{(\sum x)^2}{n} \text{ with } (n-1) \text{ degrees of freedom (d.f.),}$$

where both sums are taken over all the n observations, forgetting for the moment that they come from different populations. The quantities $\sum x$ and n can easily be obtained from the earlier calculations as follows:

$$\sum x = T_1 + T_2 + T_3 + T_4$$

and can conveniently be renamed T;

$$n = n_1 + n_2 + n_3 + n_4.$$

Table 32. *Liver weights, expressed as a percentage of body weight, for rats fed on four different diets; the preliminary calculations of the single-factor anovar are also shown.*[†]

	Level					
	A	B	C	D	Total	
	3.42	3.17	3.34	3.64		
	3.96	3.63	3.72	3.93		
	3.87	3.38	3.81	3.77		
	4.19	3.47	3.66	4.18		
	3.58	3.39	3.55	4.21		
	3.76	3.41	3.51	3.88		
	3.84	3.55		3.96		
		3.44		3.91		
n_i	7	8	6	8	29	n
T_i	26.62	27.44	21.59	31.48	107.13	T
$\bar{x}_i$ (per cent)	3.803	3.430	3.598	3.935		
S_i	101.6106	94.2474	77.8283	124.1280	397.8143	S
$\dfrac{T_i^2}{n_i}$	101.2321	94.1192	77.6880	123.8738	396.9131	$\sum \dfrac{T_i^2}{n_i}$
					395.7530	$\dfrac{T^2}{n}$
s.s. for random variation	0.3785	0.1282	0.1403	0.2542	0.9012	Res: s.s.

† The values in the total column are the sums of those for the four levels (diets), except for $T^2/n = 395.7530$, which is obtained from T ($= 107.13$) and n ($= 29$) in the total column; the steps of the calculations are described in 7.2.

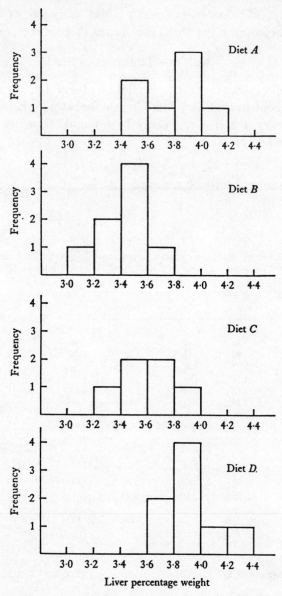

Fig. 21. Histograms of the liver percentage weights in Table 32; see 7.2.

In the example we thus have
$$T = 26.62 + 27.44 + 21.59 + 31.48 = 107.13$$
and
$$n = 7 + 8 + 6 + 8 = 29.$$

To obtain $\sum x^2$ for all the observations it is helpful to sum the squares of the observations one sample at a time, and then to total these sums; we can denote the sums of the squares for the four samples by S_1, S_2, S_3, S_4 and their total by S. We have then

$$S_1 = 101.6106, \quad S_2 = 94.2474,$$
$$S_3 = 77.8283, \quad S_4 = 124.1280,$$
$$S = 397.8143.$$

Finally therefore we calculate:

$$\text{Total sum of squares} = \sum x^2 - \frac{(\sum x)^2}{n}, \text{taken over all the observations}$$

$$= S - \frac{T^2}{n}$$

$$= 397.8143 - \frac{107.13^2}{29}$$

$$= 397.8143 - 395.7530$$

$$= 2.0613$$

with $n - 1 = 29 - 1 = 28$ degrees of freedom.

In the second stage we would normally calculate the amount of variation due to population differences, and obtain the residual sum of squares by subtracting this amount from the total sum of squares. In the present expository treatment it is however convenient to reverse these steps, calculating first the residual sum of squares and then obtaining the population term by subtraction; this is helpful at this stage because it will enable us to see that the residual sum of squares does measure random variation in the way that we already understand. When we give, later, the general formulae we shall return to the usual sequence of steps.

Random variation in this simple case is directly measurable as the variation between observations in the same population, and so, for any one population, can be measured by a sum of squares of the familiar form

$$\sum x^2 - \frac{(\sum x)^2}{n_i}, \text{ with } (n_i - 1) \text{ degrees of freedom},$$

the sums being taken only over the n_i observations in the sample from the one population considered. Thus in our example we can calculate:

random variation in the sample $= \sum x^2 - \dfrac{(\sum x)^2}{n_1}$, taken over the
from the first population
first sample
only,

$$= S_1 - \frac{T_1^2}{n_1}$$

$$= 101.6106 - \frac{26.62^2}{7}$$

$$= 101.6106 - 101.2321$$

$$= 0.3785$$

with $n_1 - 1 = 7 - 1 = 6$ degrees of freedom;

random variation in the sample $= S_2 - \dfrac{T_2^2}{n_2}$
from the second population

$$= 94.2474 - \frac{27.44^2}{8}$$

$$= 94.2474 - 94.1192$$

$$= 0.1282$$

with $n_2 - 1 = 8 - 1 = 7$ degrees of freedom;

random variation in the sample $= S_3 - \dfrac{T_3^2}{n_3}$
from the third population

$$= 77.8283 - \frac{21.59^2}{6}$$

$$= 77.8283 - 77.6880$$

$$= 0.1403$$

with $n_3 - 1 = 6 - 1 = 5$ degrees of freedom;

random variation in the sample $= S_4 - \dfrac{T_4^2}{n_4}$
from the fourth population

$$= 124.1280 - \frac{31.48^2}{8}$$

$$= 124.1280 - 123.8738$$

$$= 0.2542$$

with $n_4 - 1 = 8 - 1 = 7$ degrees of freedom.

The residual sum of squares is obtained by summing the contributions from the four samples and so is

residual sum of squares $= 0.3785 + 0.1282 + 0.1403 + 0.2542$

$$= 0.9012,$$

with degrees of freedom $= 6 + 7 + 5 + 7$

$$= 25.$$

These calculations have been given at length to help in understanding their meaning, but they can usefully be condensed into tabular form as shown in Table 32; the reader should note where all the expressions used above are to be found in this table. (One term in the table, namely

$$\frac{T_1^2}{n_1} + \frac{T_2^2}{n_2} + \frac{T_3^2}{n_3} + \frac{T_4^2}{n_4} = 396.9131,$$

which is found in the 'total' column of the 'T_i^2/n_i' row, has not been mentioned so far, but will be used later.)

The sum of squares measuring level (population) differences is now obtained by subtraction:

Sum of squares measuring $=$ Total sum of squares $-$ residual
level differences sum of squares

$$= 2.0613 - 0.9012$$

$$= 1.1601,$$

with d.f. $=$ d.f. for total $-$ d.f. for residual

$$= 28 - 25$$

$$= 3.$$

To complete the anovar procedure – that is to compare the population and residual variations – we use a standard form of table. This has five columns, with headings that we shall find in every anovar; the different types of anovar differ in their row entries but not in their column layout. The headings for the first three columns are implied by the calculations we have already carried out. They are:

Source of variation — labels to tell us which term is which;

Sum of squares (S.S) — measuring the amount of variation due to each source;

Degrees of freedom (d.f.) — appropriate to the various s.s.

In the two remaining columns we have:

Mean square (M.S.) — variance estimates obtained, as usual, by dividing each s.s. by its appropriate d.f.;

Variance ratio (F) — comparisons between the variance estimates, used in determining the relative importance of the different sources of variation.

The tabular layout for our present example is given in Table 33. We note that no M.S. value is entered in the total row; this is because the total s.s. is — in any anovar — a mixture of variation from many sources, so that its variance estimate is of no interest. Only a single F value, obtained by dividing the M.S. for level differences by the M.S. for residual variation, is needed because we have only a single-factor anovar; in more complex designs several variance comparisons are needed.

To determine whether the population differences are large enough to be important (more formally, whether they are significant) we use the calculated F value — 10.73 in this case — in conjunction with values of known probability level obtained from Table A16, which gives, in three separate tables, the values of F that are significant at 5, 1 and 0.1 per cent. To find the values appropriate to our analysis we use the degrees of freedom of the numerator of the variance ratio (in this case the between level M.S.) to determine the column of the table, and the degrees of freedom of the denominator (in this case the residual M.S.) to determine the row. Thus in the example we use the column headed 3 and the row numbered 25; the critical values for 5, 1 and 0.1 per cent significance are thus 2.99, 4.68 and 7.45. We see that large values of F indicate increasing significance, so that, as our observed value (10.73) is greater than the 0.1 per cent value (7.45), we do in this case have significant evidence of a difference between the level (population) means. More formally, if the population means were equal we would have a chance less than 1 in 1000 of observing F as large as in our case, and so we have (conventionally) good reason to reject the null hypothesis of equal population means.

Table 33. *The anovar of the rat liver weights in Table 32; the calculation is described in 7.2.*†

Source of variation	Sum of squares (s.s.)	Degrees of freedom (d.f.)	Mean square (M.S.)	Variance ratio (F)
Between levels	1.1601	3	0.38670	10.73***
Residual	0.9012	25	0.03605	
Total	2.0613	28		
		Detailed comparisons		
A v. B	0.5190	1	0.5190	14.40***
A v. (B + C)	0.4220	1	0.4220	11.71**

† The detailed comparisons are made as follows:

A. v. B (see 7.2.2.1 and Table 32)

$$\text{s.s.} = \frac{(8 \times 26.62 - 7 \times 27.44)^2}{7 \times 8 \times 15} = \frac{20.88^2}{840} = 0.5190.$$

A v. $(B + C)$ (see 7.2.2.3 and Table 32)

$$\text{s.s.} = \frac{(14 \times 26.62 - 7 \times 49.03)^2}{7 \times 14 \times 21} = \frac{29.47^2}{2058} = 0.4220.$$

The significance levels in the anovar can conveniently be indicated, as in Table 33, by marking the F-values with stars. The convention is that one star (*) indicates significance at the 5 per cent level, two stars (**) at the 1 per cent level and three stars (***) at the 0.1 per cent level. This form of marking makes it easy – especially in a large analysis – to see at a glance those effects which are important.

7.2.1 General formulae

The form of calculation used in the previous section was chosen, as already pointed out, to show the single-factor anovar as a simple extension of earlier calculations with a normal variable. In particular we were able to calculate the residual s.s. directly because, in this case, the random variation can be understood as the variation between observations within the levels. It is now however convenient to rearrange the calculation into a form which extends easily to other types of anovar; this rearrangement leads us to calculate the between levels s.s. directly and then to obtain the residual s.s. by subtraction.

The between levels term was obtained as the difference between the total s.s., $S - T^2/n$, and the residual s.s., this in turn being calculated as:

$$\left(S_1 - \frac{T_1^2}{n_1}\right) + \left(S_2 - \frac{T_2^2}{n_2}\right) + \left(S_3 - \frac{T_3^2}{n_3}\right) + \left(S_4 - \frac{T_4^2}{n_4}\right).$$

This expression can be rearranged into the form

$$S_1 + S_2 + S_3 + S_4 - \left(\frac{T_1^2}{n_1} + \frac{T_2^2}{n_2} + \frac{T_3^2}{n_3} + \frac{T_4^2}{n_4}\right)$$

$$= S - \left(\frac{T_1^2}{n_1} + \frac{T_2^2}{n_2} + \frac{T_3^2}{n_3} + \frac{T_4^2}{n_4}\right).$$

Thus the between groups s.s. is

$$\left\{S - \frac{T^2}{n}\right\} - \left\{S - \left(\frac{T_1^2}{n_1} + \frac{T_2^2}{n_2} + \frac{T_3^2}{n_3} + \frac{T_4^2}{n_4}\right)\right\}$$

$$= S - \frac{T^2}{n} - S + \left(\frac{T_1^2}{n_1} + \frac{T_2^2}{n_2} + \frac{T_3^2}{n_3} + \frac{T_4^2}{n_4}\right)$$

$$= \frac{T_1^2}{n_1} + \frac{T_2^2}{n_2} + \frac{T_3^2}{n_3} + \frac{T_4^2}{n_4} - \frac{T^2}{n}.$$

This is a simple expression and we shall find it convenient to use; it is one of the basic forms found repeatedly in anovar calculations and its pattern should be carefully studied. It consists first of the sum over the levels to be compared of terms of the form 'square of the level total, divided by the number in the level', from which sum is subtracted a term of similar nature for all the levels together, namely 'square of the total over all the levels, divided by the total number in all the levels'. If we use the $\sum$ notation for addition the expression can be condensed to

$$\text{between levels s.s.} = \sum \frac{T_i^2}{n_i} - \frac{T^2}{n}.$$

Table 34. *The preliminary calculations of the single-factor anovar in the general case (p levels); see 7.2.1 and compare Table 32.*

	Levels					Total
	1	2	3	$\cdots$	p	
n_i	n_1	n_2	n_3	$\cdots$	n_p	$n = \sum n_i$
T_i	T_1	T_2	T_3	$\cdots$	T_p	$T = \sum T_i$
$\bar{x}_i$	T_1/n_1	T_2/n_2	T_3/n_3	$\cdots$	T_p/n_p	
S_i	S_1	S_2	S_3	$\cdots$	S_p	$S = \sum S_i$
T_i^2/n_i	T_1^2/n_1	T_2^2/n_2	T_3^2/n_3	$\cdots$	T_p^2/n_p	$\sum \dfrac{T_i^2}{n_i}$
						T^2/n

We can now give formulae for the single-factor anovar that can be applied to all cases and not just to our very simple example. We will suppose that we have samples from p populations, the samples containing $n_1, n_2, n_3, \ldots, n_p$ observations. The preliminary calculations for the anovar are given in Table 34; the reader should compare the general expressions of this table with the particular case of Table 32. We note that the tables begin with similar calculations, but that in Table 34 we do not calculate the contributions of the separate levels to the residual s.s.; this is because we shall obtain the residual s.s. in the anovar by subtraction. The anovar itself is given in Table 35, which may be compared with the example in Table 33. We note that the total s.s. is obtained by the same method as previously, calculating $S - T^2/n$, but that we now obtain the between level s.s. by the formula developed earlier in this section, namely

$$\text{between levels s.s.} = \sum_{i=1}^{p} \frac{T_i^2}{n_i} - \frac{T^2}{n}.$$

The residual s.s. is then calculated as

$$\text{residual s.s.} = \text{total s.s.} - \text{between level s.s.}$$

Table 35. *The single-factor anovar for a total of n observations taken over p levels; see 7.2.1 and compare Table 33.*

Source of variation	S.S.	D.F.	M.S.	F
Between levels	$\sum \dfrac{T_i^2}{n_i} - \dfrac{T^2}{n}$	$p-1$	$\dfrac{\text{S.S.}}{\text{D.F.}}$	$\dfrac{\text{Levels M.S.}}{\text{Res. M.S.}}$
Residual	Total s.s. $-$ Between levels s.s.	$n-p$	$\dfrac{\text{S.S.}}{\text{D.F.}}$	
Total	$S - T^2/n$	$n-1$		

In considering the degrees of freedom it is convenient to rearrange the calculation in the same way as we did that for the s.s. Thus in the worked example we obtain the degrees of freedom for the between level s.s. as

d.f. for total s.s. $-$ sum of d.f. for within level
 contributions to residual s.s.

$$= (n-1) - \{(n_1 - 1) + (n_2 - 1) + (n_3 - 1) + \cdots + (n_p - 1)\}.$$

(In the example we had $p = 4$.) This expression can be rearranged to give

$$(n-1) - (n_1 + n_2 + n_3 + \cdots + n_p - p),$$

the term $(-p)$ in the second bracket arising from the terms (-1) contributed by each of the p samples. Thus we have

$$\text{d.f. of between levels s.s.} = (n-1) - (n-p)$$
$$= n - 1 - n + p$$
$$= p - 1;$$

thus in the example with $p = 4$ we had 3 degrees of freedom between groups. We are thus led to the values in the degrees of freedom column of Table 35. The variance ratio table is thus consulted with $(p-1)$ degrees of freedom for the column heading and $(n-p)$ degrees of freedom for the row.

The null hypothesis tested by this analysis is that the population means of the levels are all equal. The probabilities used in preparing the F-table refer to the possibility of obtaining so large a value of F when the null hypothesis is true. If our observed value of F in the anovar table is smaller than the 5 per cent value in Table A16 we (conventionally) accept the null hypothesis; in other words we are allowed to maintain that the population means are equal. If our observed F is larger than the 5 per cent (1 per cent, 0.1 per cent) value in Table A16 we reject the null hypothesis at the 5 per cent (1 per cent, 0.1 per cent) level of significance because so large an observed value would be unlikely to arise if the null hypothesis were true. In other words we are (conventionally) compelled to conclude that the population means are not all equal. This conclusion is not very helpful if we have several populations, and an important practical problem, which we shall consider later (7.2.2.1–7.2.2.5), is to discover what more precise statement we can make about the differences between the groups; all that the F-test tells us is whether we can reasonably treat the group means as all the same.

7.2.2 The model

We consider in this section the conditions which are necessary if the single-factor anovar is to give us reliable conclusions. We know already that the variables must be normal, but this condition by itself is not sufficient for the validity of the analysis.

We will label the variable from level i as x_i and the normality condition is then expressible in the assumption that x_i is normal with unknown mean μ_i and unknown variance σ_i^2. We have n_i observations of the variable x_i and it is convenient to have distinguishing labels for these; this is achieved by introducing a second suffix j, so that the jth observation of the variable x_i is denoted by x_{ij}. Then j takes values $1, 2, 3, \ldots, n_i$ as we identify the different observations in the group.

A second condition for the reliability of the anovar is that the observations shall all be independent; as far as independence between levels is concerned this requirement is largely equivalent to the basic assumption that we have a single-factor design, but independence within levels is an extra and important condition. It means, for example, that we cannot accept any trends or periodicity in the observations in a level; in terms of experimental design this condition is achieved by the use of a randomization procedure as described in 1.3.

A further condition for the use of the anovar is that the variances in the separate levele, σ_i^2, must be equal, that is

$$\sigma_1^2 = \sigma_2^2 = \sigma_3^2 = \cdots = \sigma_p^2 = \sigma^2, \text{ say.}$$

This condition is already implicit in the method we have used for calculating the residual s.s.; we are justified in totalling the contributions to this term from the separate levels only if the within level variability is the same for all levels.

These conditions can now be combined to give one simple statement of our assumptions. We note first that each observation x_{ij} can be thought of as the sum of two parts, one part, μ_i, being the same for all the observations of level i and the other part, r_{ij}, varying independently and normally from observation to observation with variance σ^2; the mean of r_{ij} is zero, because it measures only deviations from the level mean. Thus we may write

$$x_{ij} = \mu_i + r_{ij},$$

where the r_{ij}'s are normal, independent, with zero mean and variance σ^2. This equation is sufficient specification for the single-factor design, but for convenience in generalizing it to later, more complex, designs we define μ as the mean of x over all the populations under study and write

$$\mu_i = \mu + \tau_i.$$

The quantity τ_i, known as the *level constant*, measures how much the mean of level i deviates from the overall mean of all the populations. Combining these two equations we have the final statement

$$x_{ij} = \mu + \tau_i + r_{ij}$$

for $i = 1, 2, 3, \ldots, p$, and $j = 1, 2, 3, \ldots, n_i$, where the r_{ij}'s are normal, independent, with zero mean and variance σ^2. This representation is called the *model* for the single-factor anovar; all the assumptions about the random behaviour of the observations are concentrated in the statement about the *residuals* r_{ij}. We shall find the specification of the model a very convenient way of defining the conditions of later anovars.

As the properties of the variable are expressed in terms of the parameters of the model, we need to consider what estimates are available for these parameters, and to relate the null hypothesis to them. The point estimates of the level means μ_i are the sample means

$$\bar{x}_1 = \frac{T_1}{n_1}, \quad \bar{x}_2 = \frac{T_2}{n_2}, \quad \ldots, \quad \bar{x}_p = \frac{T_p}{n_p};$$

these formulae can be summarized in the single statement

$$\bar{x}_i = \frac{T_i}{n_i} \quad (i = 1, 2, 3, \ldots, p).$$

The point estimate of σ^2 is the residual M.S., s^2, with $(n - p)$ degrees of freedom. The interval estimate for μ_i, the population mean of level i, is

$$\bar{x}_i \pm t \frac{s}{\sqrt{n_i}}$$

where t is taken with $(n - p)$ degrees of freedom and the probability level of our choice. The interval estimate for $(\mu_i - \mu_j)$, the difference between the means of levels i and j, is

$$(\bar{x}_i - \bar{x}_j) \pm ts \sqrt{\left(\frac{1}{n_i} + \frac{1}{n_j} \right)},$$

where t is specified as in the preceding sentence. The interval estimate of σ^2 is obtained from s^2 as in 6.4.1, taking χ^2 with $(n - p)$ degrees of freedom. We note that these three types of interval estimate follow exactly the formulae used in previous sections; all that we need to do is to insert the correct point estimates and degrees of freedom in the earlier formulae. The null hypothesis tested by F in the anovar is that the level means are equal, and this is simply expressed by noting that it implies that none of the level means deviates from the overall mean, or, more formally, that

$$\tau_1 = \tau_2 = \tau_3 = \cdots = \tau_p = 0.$$

7.2.2.1 The comparison of two levels

The interval estimate given in 7.2.2 for the difference $\mu_i - \mu_j$ can be used, in the same way as was that in 6.6.1, to give a significance test of the null hypothesis $\mu_i = \mu_j$. To do this we calculate

$$\text{observed } t = \frac{\bar{x}_i - \bar{x}_j}{s \sqrt{(1/n_i + 1/n_j)}},$$

with $(n - p)$ degrees of freedom, and reject the null hypothesis if this value is greater than the tabulated value of t corresponding to our chosen significance level.

The calculation which this test requires can also be expressed in another form, which is a simple extension of the anovar calculation; to do this we can calculate a s.s. measuring the difference between levels i and j, using the formula

$$\text{s.s. between levels } i \text{ and } j = \frac{(n_j T_i - n_i T_j)^2}{n_i n_j (n_i + n_j)}.$$

This s.s., contrasting two levels, has $2 - 1 = 1$ degree of freedom and is therefore equal to the corresponding M.S. We now obtain an F with $(1, n - p)$ degrees of freedom by dividing this M.S. by the residual M.S. s^2; the value of this F is referred to the table in the usual way and tests the null hypothesis $\mu_i = \mu_j$. The comparison between levels i and j can thus be set out as an extra row in the anovar table and this is a very convenient procedure; a numerical example of this method is included in Table 33. We shall discuss below how more complex comparisons can be treated similarly (7.2.2.3).

In the case $n_i = n_j$, which is often encountered, the formula for the s.s. simplifies to

$$\frac{(T_i - T_j)^2}{2n_i},$$

which can be remembered as the square of the difference between the totals for the two levels, divided by the number of observations in the two levels together.

7.2.2.2* Two algebraic proofs

In this section we show first the equivalence of the t- and F-tests given in 7.2.2.1 for the null hypothesis $\mu_i = \mu_j$, and secondly that the s.s. used in the F-test is equal to that which the form of the calculations of 7.2.1 would lead us to expect for the contrast between two levels.

Consider first the s.s. between levels i and j

$$= \frac{(n_j T_i - n_i T_j)^2}{n_i n_j (n_i + n_j)}$$

$$= \frac{(T_i / n_i - T_j / n_j)^2}{1/n_i + 1/n_j}$$

dividing numerator and denominator by $n_i^2 n_j^2$

$$= \frac{(\bar{x}_i - \bar{x}_j)^2}{1/n_i + 1/n_j}.$$

As was explained above, the observed F is equal to this s.s. divided by s^2, that is

$$F_{1, n-p} = \frac{(\bar{x}_i - \bar{x}_j)^2}{s^2(1/n_i + 1/n_j)} = t_{n-p}^2.$$

The two tests are thus mathematically equivalent; in fact any F-test with 1 degree of freedom for its numerator is equivalent to a t-test.

To show the equality of the two forms for the s.s., consider first the form which the formulae of 7.2.1. give for comparison of two levels; this is

$$\frac{T_i^2}{n_i} + \frac{T_j^2}{n_j} - \frac{(T_i + T_j)^2}{n_i + n_j}.$$

We note that the subtractive term includes totals only for the levels being compared. This expression

$$= \frac{T_i^2 n_j(n_i + n_j) + T_j^2 n_i(n_i + n_j) - (T_i^2 + 2T_i T_j + T_j^2)n_i n_j}{n_i n_j(n_i + n_j)}$$

$$= \frac{T_i^2 n_j^2 + T_j^2 n_i^2 - 2T_i T_j n_i n_j}{n_i n_j(n_i + n_j)}$$

$$= \frac{(n_j T_i - n_i T_j)^2}{n_i n_j(n_i + n_j)},$$

which is the form given in 7.2.2.1. A calculation of this type is called a *difference method* for the s.s.; such methods, which are very useful for saving arithmetic, are further illustrated in the following section.

7.2.2.3 The general comparison with one degree of freedom

In this section we consider more complex one degree of freedom comparisons between levels. Suppose first that we wish to compare the mean of level i with the overall mean of levels j and k together. The point estimates of these two means are

$$\frac{T_i}{n_i} \quad \text{and} \quad \frac{T_j + T_k}{n_j + n_k}$$

and the interval estimate for their difference is thus

$$\frac{T_i}{n_i} - \frac{T_j + T_k}{n_j + n_k} \pm ts \sqrt{\left(\frac{1}{n_i} + \frac{1}{n_j + n_k}\right)},$$

where, as in the simpler interval estimates given in 7.2.1, t is taken with $n - p$ degrees of freedom and the chosen probability level and s^2 is the residual mean square. This interval estimate leads as usual to a significance test, for which we calculate

$$\text{observed } t = \frac{(T_i/n_i) - [(T_j + T_k)/(n_j + n_k)]}{s\sqrt{(1/n_i + 1/(n_j + n_k))}},$$

rejecting the null hypothesis that the mean of level i is equal to the overall mean of levels j and k together if the value observed for t exceeds that tabulated for the chosen significance level.

This t-test, like that of 7.2.2.1, is equivalent to an F-test, for which we calculate the following s.s. with one degree of freedom

$$\frac{\{(n_j + n_k)T_i - n_i(T_j + T_k)\}^2}{n_i(n_j + n_k)(n_i + n_j + n_k)},$$

as before the M.S. is equal to the s.s. and the observed F, with $(1, n - p)$ degrees of freedom, is calculated by dividing the M.S. by s^2 and is tested in the usual way. An example of this calculation is given in Table 33.

We must now note two features of this s.s., both of which are useful in dealing with more general cases. First the numerator is simply related to the point estimate for that difference of means which interests us; this estimate is

$$\frac{T_i}{n_i} - \frac{T_j + T_k}{n_j + n_k} = \frac{(n_j + n_k)T_i - n_i(T_j + T_k)}{n_i(n_j + n_k)}.$$

The numerator of the s.s. is thus the square of the numerator which we obtain when we express the point estimate relative to a common denominator. This property is general. Thus to compare the mean of levels i and j together with the mean of levels k, l, and m together, we would use point estimate

$$\frac{T_i + T_j}{n_i + n_j} - \frac{T_k + T_l + T_m}{n_k + n_l + n_m},$$

which, relative to a common denominator, is

$$\frac{(n_k + n_l + n_m)(T_i + T_j) - (n_i + n_j)(T_k + T_l + T_m)}{(n_i + n_j)(n_k + n_l + n_m)}.$$

The square of the numerator of this expression is the numerator of the s.s. for the contrast.

The second feature of the s.s. concerns the denominator; it is more easily seen if we simplify the s.s. a little as follows

$$\text{s.s. contrasting level } i \text{ with mean} \atop \text{of levels } j \text{ and } k \text{ together} = \frac{\{(n_j + n_k)T_i - n_i T_j - n_i T_k\}^2}{(n_j + n_k)^2 n_i + n_i^2 n_j + n_i^2 n_k}.$$

In this simplification, we have separated the contributions to both numerator and denominator of the three levels involved in the comparison. We now see that the denominator consists of the sum over these levels of terms of the form

$$\begin{matrix} \text{(coefficient of level total} \\ \text{in the numerator)}^2 \end{matrix} \times \begin{matrix} \text{(number of observations} \\ \text{in level total).} \end{matrix}$$

This form is general in one degree of freedom comparisons. Thus in the comparison of levels i and j together with k, l and m together the numerator of the s.s. expands to

$$\{(n_k + n_l + n_m)T_i + (n_k + n_l + n_m)T_j$$
$$- (n_i + n_j)T_k - (n_i + n_j)T_l - (n_i + n_j)T_m\}^2$$

so that the denominator for this s.s. is

$$(n_k + n_l + n_m)^2 n_i + (n_k + n_l + n_m)^2 n_j + (n_i + n_j)^2 n_k$$
$$+ (n_i + n_j)^2 n_l + (n_i + n_j)^2 n_m.$$

In their most general form these results can be expressed as follows. If the numerator of the point estimate for a one degree of freedom comparison between levels $1, 2, \ldots, p$ is

$$a_1 T_1 + a_2 T_2 + \cdots + a_p T_p,$$

where the coefficients $a_1, a_2, \ldots, a_p$ may be positive or negative or zero (if not all the levels are involved), then the s.s. has as its numerator the square of this expression and as its denominator

$$a_1^2 n_1 + a_2^2 n_2 + \cdots + a_p^2 n_p.$$

The s.s. is therefore

$$\frac{(a_1 T_1 + a_2 T_2 + \cdots + a_p T_p)^2}{a_1^2 n_1 + a_2^2 n_2 + \cdots + a_p^2 n_p}.$$

7.2.2.4 Comparisons with more than one degree of freedom

If we wish to make a comparison with more than one degree of freedom, we cannot approach through point and interval estimates, but must use a s.s. of the general form given in 7.2.1. If for example, we wish to compare the mean of level i, the mean of levels j and k together and the mean of levels l and m together, we use the following s.s.

$$\frac{T_i^2}{n_i} + \frac{(T_j + T_k)^2}{n_j + n_k} + \frac{(T_l + T_m)^2}{n_l + n_m} - \frac{(T_i + T_j + T_k + T_l + T_m)^2}{n_i + n_j + n_k + n_l + n_m}.$$

The s.s., comparing 3 means, has $3 - 1 = 2$ degrees of freedom; when the corresponding M.S. is divided by the residual M.S. we obtain an F-test with $(2, n - p)$ degrees of freedom. The null hypothesis of this test is

mean of level i = mean of levels j and k together

= mean of levels l and m together;

any form of departure from these equalities should increase the value observed for F, so that if the null hypothesis is rejected we can say only that there is some difference among these three means.

7.2.2.5 Selected comparisons may not be valid

Some caution is needed in using the methods for comparing levels described in 7.2.2.1–7.2.2.4, because if we pick out for testing only those differences which look large and interesting we may by this selection distort the logic of the significance test procedure and so make the results unreliable. This happens because in any test the probability level calculated is that of a difference as large as the sample difference occurring by chance when the null hypothesis is true. If we select only large differences we cannot properly use a procedure designed to interpret differences occurring by chance.

Two rules which give some protection against this danger are:

(a) In most cases we make detailed comparisons between levels only when the overall comparison between all the levels – by the F-test in the first form of the anovar – is significant.

(*b*) We may depart from rule (*a*) to make comparisons which are implicit in the design of the experiment, and are not simply chosen because their observed values are large enough to appear interesting. For example, we might set up an experiment to compare 5 forms of an insecticide with a control; the overall comparison between the levels of this factor would have $6 - 1 = 5$ degrees of freedom, and two comparisons that we should wish to make whatever the result of the overall comparison would be

(i) between the 5 forms of the insecticide, with $5 - 1 = 4$ degrees of freedom, and

(ii) between the control mean and the overall mean of the different forms of the insecticide, with $2 - 1 = 1$ degrees of freedom.

It should not be assumed that if the overall comparison is not significant then none of the detailed comparisons can be either. Suppose that in the insecticide example just quoted the level totals of the variable measured had been 24, 26, 25, 24 and 27 for the 5 forms of the treatment and 18 for the control, each of these totals being taken over 5 observations, and that the residual mean square with 24 degrees of freedom had been 1.56. The s.s. between the 6 levels of the factor is

$$\frac{24^2 + 26^2 + 25^2 + 24^2 + 27^2 + 18^2}{5} - \frac{144^2}{30} = 10.00,$$

that between the 5 forms of the insecticide is

$$\frac{24^2 + 26^2 + 25^2 + 24^2 + 27^2}{5} - \frac{126^2}{25} = 1.36$$

and that between insecticide and control is

$$\frac{(24 + 26 + 25 + 24 + 27 - 5 \times 18)^2}{5 + 5 + 5 + 5 + 5 + 25 \times 5} = 8.64;$$

the reader should note that this last term is calculated by the method described in 7.2.2.3. The anovar is shown in Table 36; the overall comparison between levels gives a variance ratio of 1.28 on $(5, 24)$ degrees of freedom, which is far from significant but the comparison of the control with the average over the five forms of the insecticide gives a ratio of 5.54 on $(1, 24)$ degrees of freedom, which is significant at the 5 per cent level.

Another approach to the problem of making valid detailed comparisons, which is useful when we are interested in comparing many levels in pairs, is to introduce a new significance test in place of that based on the t-distribution. Several of these *multiple range tests* have been developed; they will not be considered here, but useful discussions will be found in Federer (1955) and Pearce (1965).

Table 36. *Anovar for the insecticide example (see 7.2.2.5).*†

Source of variation	S.S.	d.f.	M.S.	F
Between levels	10.00	5	2.00	1.28
Residual	37.44	24	1.56	
Total	47.44	29		
Forms of insecticide	1.36	4	0.34	—
Control v. insecticide	8.64	1	8.64	5.54*

† Although the overall comparison between levels is far short of significance, the one degree of freedom contrast 'control v. insecticide' is significant at the 5 per cent level. [When an F value is less than one, as for the 'forms of insecticide' term, there is no need to enter its value.]

7.2.3 Checking on the assumptions

The assumptions set out in 7.2.2 about the distribution of the observed variable are essential for the validity of the anovar and of the point and interval estimates associated with it. It is true that small departures from the conditions do not materially affect the reliability of the results – in the jargon, the anovar is a *robust* procedure – but it is not easy to apply this fact because we require to specify what is a 'small' departure from the conditions. We need therefore to have some method of checking on the truth of the assumptions we want to make; in particular we would like to be sure that the distributions for individual levels are normal, and that their variances are equal. We will give in this section two approximate methods for checking these assumptions; two more precise tests are given in 7.2.3.1 and 7.2.3.2.

One approximate method has already been mentioned, namely to draw a histogram for each sample. This should be standard practice, and will help in detecting gross departures from either normality or constancy of variance; usually, however, our samples are too small for lesser departures to be detected with any certainty. Moreover, visual inspection of histograms is necessarily subjective and requires considerable experience if it is to be at all reliable.

A second method is to examine the possibility that the variability within the levels is related to the level means; such a relationship is one of the commoner ways in which our assumptions fail, and may be detected quite easily in the single-factor design. We need only to calculate the standard deviation for each level – the sample means are already available – and to plot standard deviation against sample mean. The standard deviations for the separate levels are calculated as the square roots of the within group variances

$$s_1^2 = \frac{1}{n_1 - 1}\left(S_1 - \frac{T_1^2}{n_1}\right), \quad \ldots, \quad s_p^2 = \frac{1}{n_p - 1}\left(S_p - \frac{T_p^2}{n_p}\right);$$

these are thus readily obtained by adding rows for 'within level s.s.', 'within level M.S.' and 'within level standard deviation' at the bottom of the calculations of Table 34. For the s.s. row we subtract for each level the entry in the T_i^2/n_i row from that in the S_i row, for the M.S. (s_i^2) row we divide each s.s. by its within level degrees of freedom ($=$ number in the level -1) and for the standard deviation (s_i) row we take the square roots of the entries in the M.S. row. These calculations for the example of Table 32 are shown in Table 37.

Table 37. *The calculation of the within level standard deviations for the observations of Table* 32. *The calculation is described in 7.2.3; see also Fig.* 22.

	Level			
	A	B	C	D
s.s.	0.3785	0.1282	0.1403	0.2542
d.f.	6	7	5	7
s_i^2	0.06308	0.01832	0.02806	0.03632
s_i	0.2512	0.1353	0.1675	0.1906
$\bar{x}_i$	3.803	3.430	3.598	3.935

We now plot these standard deviations against their sample means; the points obtained for the numerical example are shown in Fig. 22, but this figure can be no more than an illustration of method since four points can never make clear any form of relationship which can be summarized by a smooth curve through the origin; the most commonly found 'smooth curve' is in fact a straight line. We do not usually find a precise relationship but the general form can often be discerned. If we find a relationship,

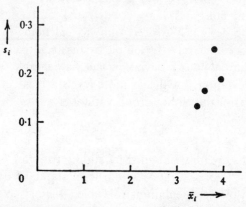

Fig. 22. The relationship between within-level standard deviation (s_i) and within-level mean ($\bar{x}_i$) for the observations of Table 32; see 7.2.3 and Table 37.

indicating not only that the residual variance is not constant between levels but further that it is related to the level means, the anovar procedure cannot be used. In certain cases the procedure can be made valid by working with some *transformation* of the variable – for example, analysing $\log x$ rather than x – but the choice of transformation and the interpretation of the results obtained with it are subjects which we can do no more than consider briefly in section 9.4 *et sqq*.

7.2.3.1 The Kolmogorov–Smirnov one-sample test

We describe in this section a test which, whilst it is sensitive both to departure from normality and to inequality of variance, is more useful in detecting the former effect; this is the *Kolmogorov–Smirnov one-sample test*. It is computationally very similar to the two-sample test described in 3.5. We shall carry out the test on the estimated residuals; this will introduce a slight approximation, because these estimates are not independent, but this is not likely to affect the reliability of our conclusions unless the individual samples are very small.

To obtain the estimates of the residuals we use the model introduced in 7.2.2,

$$x_{ij} = \mu_i + r_{ij}$$

which may be rewritten

$$r_{ij} = x_{ij} - \mu_i.$$

The values of the x_{ij} are the observations and the estimates of the μ_i are the level means $\bar{x}_i$, so that the estimates of the residuals are the deviations of the observations from their level means, that is $(x_{ij} - \bar{x}_i)$. As in the two-sample test (3.5) we arrange the residuals in ascending order; in doing this we must take account of sign, so that the negative residuals (in reverse order of numerical magnitude) come first in our ascending order. The residuals for the example of Table 32 are calculated in Table 38 and are given in ascending order in Table 39. We now group the residuals and obtain cumulative proportional frequencies as we did in the two-sample test; these calculations for the example are given in Table 39.

According to the model the residuals are normally distributed with zero mean and with variance estimated by the residual M.S. s^2, with $(n - p)$ degrees of freedom; we now calculate, for a normal variable with zero mean and variance s^2, the cumulative proportional frequencies corresponding to those for the observed residuals. To do this we standardize the boundaries between the groups of residuals, dividing each boundary value by s, the square root of the residual M.S. The normal cumulative proportional frequencies corresponding to the standardized boundaries are obtained from Table A11. These calculations for the numerical example are shown in Table 39.

The test statistic is, as in the two-sample case, the largest value of the differences between the two sets of cumulative proportional frequencies. The probability level for this observed difference is

223

determined from Table A17; large differences indicate significance, so that if our observed largest difference exceeds the tabulated value we conclude that the residuals cannot reasonably be regarded as normal. The value we take for the sample size in using Table A17 is $(n - p)$, the residual degrees of freedom; this is a convenient – but approximate – way of allowing for the fact that the residuals are not independent. In the example in Table 39, the largest difference is 0.1032 whereas the value for 5 per cent significance in a sample of 25 is 0.27; in this case we therefore have no reason to doubt that the within level distributions are normal.

Table 38. *The calculation of the residuals for the observations of Table 32; the calculation is described in 7.2.3.1. The 'sum of residuals' should be zero for each level (to within rounding errors) and is used as a check.*

	Level			
	A	B	C	D
Observations	3.42	3.17	3.34	3.64
	3.96	3.63	3.72	3.93
	3.87	3.38	3.81	3.77
	4.19	3.47	3.66	4.18
	3.58	3.39	3.55	4.21
	3.76	3.41	3.51	3.88
	3.84	3.55		3.96
		3.44		3.91
$\bar{x}_i$	3.803	3.430	3.598	3.935
Residuals	−0.383	−0.260	−0.258	−0.295
	0.157	0.200	0.122	−0.005
	0.067	−0.050	0.212	−0.165
	0.387	0.040	0.062	0.245
	−0.223	−0.040	−0.048	0.275
	−0.043	−0.020	−0.088	−0.055
	0.037	0.120		0.025
		0.010		−0.025
Sum of residuals	−0.001	0	0.002	0

7.2.3.2 Bartlett's test

The second test we describe is, like the Kolmogorov–Smirnov test, sensitive both to departure from normality and to inequality of variance, but it is used mainly when approximate normality

Table 39. *The Kolmogorov–Smirnov one-sample test applied to the residuals calculated in Table 38 (see 7.2.3.1).*†

Interval	Observations (residuals)	Frequency	Cumulative frequency	Cumulative proportional frequency	x	Φ(x)	|Φ(x) − cumulative proportional frequency|
−0.419 — −0.382	−0.383	1	1	0.0345	−2.0	0.0228	0.0117
−0.381 — −0.344			1	0.0345	−1.8	0.0359	0.0014
−0.343 — −0.306	−0.295		1	0.0345	−1.6	0.0548	0.0203
−0.305 — −0.268		1	2	0.0690	−1.4	0.0808	0.0118
−0.267 — −0.230	−0.260, −0.258	2	4	0.1379	−1.2	0.1151	0.0228
−0.229 — −0.192	−0.223	1	5	0.1724	−1.0	0.1587	0.0137
−0.191 — −0.154	−0.165	1	6	0.2069	−0.8	0.2119	0.0050
−0.153 — −0.116			6	0.2069	−0.6	0.2743	0.0674
−0.115 — −0.078	−0.088	1	7	0.2414	−0.4	0.3446	0.1032
−0.077 — −0.040	−0.055, −0.050, −0.048, −0.043, −0.040	5	12	0.4138	−0.2	0.4207	0.0069
−0.039 — −0.001	−0.025, −0.020, −0.005	3	15	0.5172	0	0.5000	0.0172
0 —0.038	0.010, 0.025, 0.037	3	18	0.6207	0.2	0.5793	0.0414
0.039–0.076	0.040, 0.062, 0.067	3	21	0.7241	0.4	0.6554	0.0687
0.077–0.114			21	0.7241	0.6	0.7257	0.0016
0.115–0.152	0.120, 0.122	2	23	0.7931	0.8	0.7881	0.0050
0.153–0.190	0.157	1	24	0.8276	1.0	0.8413	0.0137
0.191–0.228	0.200, 0.212	2	26	0.8965	1.2	0.8849	0.0116
0.229–0.266	0.245	1	27	0.9310	1.4	0.9192	0.0118
0.267–0.304	0.275	1	28	0.9655	1.6	0.9452	0.0203
0.305–0.342			28	0.9655	1.8	0.9641	0.0014
0.343–0.380			28	0.9655	2.0	0.9772	0.0117
0.381–0.418	0.387	1	29	1.0000	2.2	0.9861	0.0139

† The boundaries of the intervals are chosen as simple multiples (in this case 0.2, 0.4, 0.6 ...) of the residual standard deviation (0.18986 = $\sqrt{0.03605}$; see Table 33). The standardized deviate x refers to the upper boundary of the interval; Φ(x) is obtained from Table 11.

is established and we wish to test for constancy of variance; it is *Bartlett's test for homogeneity of variance*. To use it we need to calculate estimates of the residual variance for the separate levels; these are the s_i^2's introduced in 7.2.3 and calculated for the numerical example in Table 37. We take the logarithms of the separate s_i^2's and of the pooled residual M.S. s^2 and compute a statistic M, given by the formula

$$M = 2.3026\left\{(n-p)\log s^2 - \sum_{i=1}^{p}(n_i-1)\log s_i^2\right\};$$

this means that we multiply each logarithm of a variance estimate by its degrees of freedom and subtract the sum of the products for the separate estimates from the product for the pooled estimate, finally multiplying the difference by 2.3026.† The value of M for our numerical example is obtained in Table 40.

Table 40. *Bartlett's test of homogeneity of variance applied to the variances calculated in Table 37 (see 7.2.3.2).*‡

Level	s_i^2	Degrees of freedom	$\log s_i^2$
A	0.06308	6	−1.2001
B	0.01832	7	−1.7371
C	0.02806	5	−1.5519
D	0.03632	7	−1.4398
Overall	0.03605	25	−1.4431

$$M = 2.3026(-36.0775 + 37.1984) = 2.5810,$$

$$C = 1 + \frac{1}{3\times 3}\left(\frac{1}{6}+\frac{1}{7}+\frac{1}{5}+\frac{1}{7}-\frac{1}{25}\right) = 1.0680,$$

$$\chi_3^2 = \frac{M}{C} = 2.417.$$

‡ The logarithms cannot be used in the form adopted for most calculations, where the mantissa is always positive: thus log 0.06308 would usually be written $\bar{2}.7999$, which means $-2 + 0.7999 = -1.2001$, the form given here.

† Some readers may recognize that the use of this factor is equivalent to taking Naperian (natural) logarithms rather than common logarithms; in fact we could write

$$M = (n-p)\ln s^2 - \sum_{i=1}^{p}(n_i-1)\ln s_i^2.$$

We compute next

$$C = 1 + \frac{1}{3(p-1)} \left\{ \sum_{i=1}^{p} \frac{1}{n_i - 1} - \frac{1}{n - p} \right\},$$

where, as before, p is the number of samples. The term in the bracket is obtained by summing the reciprocals of the degrees of freedom of the separate variance estimates and subtracting the reciprocal of the degrees of freedom of the pooled estimate; the calculation of C for the numerical example is given in Table 40.

The test statistic is M/C and is distributed approximately as χ^2 with $(p-1)$ degrees of freedom; values for various probability levels can thus be found in Table A5. In our example we have

$$\frac{M}{C} = \frac{2.5810}{1.0680}$$
$$= 2.417,$$

and referring to Table A5 with 3 degrees of freedom we find that this value is less than that for 5 per cent probability (7.81), so that we are (conventionally) entitled to accept the null hypothesis that the residual variance is the same in all the groups.

In the general case, if we wish to compare independent variance estimates $s_1^2, s_2^2, \ldots, s_p^2$ with degrees of freedom $f_1, f_2, \ldots, f_p$ we calculate first the pooled variance estimate

$$s^2 = \frac{f_1 s_1^2 + f_2 s_2^2 + \cdots + f_p s_p^2}{f_1 + f_2 + \cdots + f_p} = \frac{\sum f_i s_i^2}{\sum f_i},$$

with degrees of freedom $f = f_1 + f_2 + \cdots + f_p = \sum f_i$.

The quantities M and C are then obtained as

$$M = 2.3026 \left\{ f \log s^2 - \sum f_i \log s_i^2 \right\}$$

and

$$C = 1 + \frac{1}{3(p-1)} \left\{ \sum \frac{1}{f_i} - \frac{1}{f} \right\};$$

finally the approximate test statistic is

$$\chi_{p-1}^2 = M/C.$$

7.2.4(c) Single-factor anovar using Genstat

Of the languages we have already illustrated, Minitab offers rather limited facilities for anovar but SPSS-X has a good selection of very flexible commands; however, we will use anovar as the occasion to introduce a further computational language, Genstat. This was developed at the Rothamsted Experimental Station; the most recent version – Genstat 5 – is described fully in the Genstat Manual (1988) and an introductory account is given by Lane, Galwey and Alvey (1988). The language contains a wide range of statistical and related computations but is especially interesting in its methods for the analysis of designed experiments; in this section we will use it in two single-factor examples from the exercises to this chapter, numbers 5 and 8. Before considering the Genstat methods the reader should explore how these analyses would be undertaken using the desk calculator methods already described.

In the experiment reported in Exercise 5, there were six levels of the single factor referred to as mineral supplement; the numbers of sheep in the six groups were not all the same but this presents no problems in a single-factor anovar. The exercise suggests three particular comparisons between the six treatments, two – (a) and (c) – with two degrees of freedom each and the other – (b) – with one; the desk calculator computations for these comparisons are all direct applications of the methods discussed in 7.2.2.3 and 7.2.2.4. The explanatory text following Exercise 5 describes one procedure for determining whether the comparisons in any particular set are mutually independent and it is easy to check that this set does satisfy the conditions of that procedure.

The commands and data for a Genstat analysis of this exercise are given in CTable 19; in Genstat terminology, instructions are referred to as *statements* and commands specifying data structures are known as *declarations*. In this example the first declaration

units [nvalues = 46]

specifies the total number of individual observations (*units*) in the data; it is not essential to make a unit declaration but subsequent declarations and statements are often made easier and shorter by doing so. The part of the declaration in square brackets constitutes an *option*; very many Genstat declarations and statements contain one or more options, always enclosed in square brackets. The next three declarations introduce quantities

CTable 19. *Commands and data for the Genstat analysis appropriate to Exercise 5 in 7.5.*

```
units [nvalues - 46]
factor [levels - 3] ms1
factor [levels - 5] ms2
factor [labels - lt(A, B, C, D, E, F)]    supplements
read       supplements, weight
1 61.8   2 65.4   3 65.7   4 20.8   5 51.0   6 43.3
1 46.0   2 53.0   3 48.1   4 54.5   5 58.5   6 54.2
1 21.7   2 50.9   3 64.2   4 41.0   5 58.3   6 51.5
1 25.2   2 28.6   3 67.9   4 51.1   5 57.7   6 77.3
1 48.8   2 81.5   3 42.3   4 42.5   5 71.3   6 70.1
1 47.7   2 44.7   3 62.0   4 35.5   5 73.6   6 69.5
1 39.0   2 64.5   3 44.3   4 78.1   5 67.6   6 63.1
1 37.2   2 38.0            4 57.3            6 83.2 :
calculate    ms1 - newlevels (supplements;  l(1,2,2,2,3,3))
     &       ms2 - newlevels (supplements;  l(1,2,3,4,5,5))
treatments   ms1/ms2/supplements
anova [fprob - yes]   weight
```

that will be used in specifying the comparisons that are to be carried out in the analysis; they are discussed below.

The read statement is

read   supplements, weight;

it does not need to contain any information about the number of terms to read because a units declaration has already been made. The statement gives the instruction that for each unit two quantities, referred to as supplements and weight, are to be read; the first quantity is already known to Genstat because its name has been used as the *identifier* in one of the declarations but the second has not been mentioned and is therefore taken to be a *standard variate*, a sequence of values containing the number of elements declared in the unit statement. The read statement could be qualified by several possible options and *parameters*† but none is stated explicitly and in consequence Genstat gives each its default value; this is the standard procedure with unstated options and parameters. In the present case, two of these defaults need

---

† A parameter is a quantity whose value – from a specified set – determines one aspect of how an instruction is carried out.

to be appreciated because they determine certain features of the layout of the data. One specifies that the two quantities to be read must be presented in parallel as opposed to serially, so that supplement and weight for the first sheep are input first – in that order – followed by the values for the second sheep, then for the third sheep and so on; in serial input obtained by inserting the option

[serial = yes]

all the values of supplement would be input before the weights. The second default means that the supplement identifiers are to be given as integers starting at 1 and not as names or letters or other types of numbers; if this were not the case the parameter specification

frepresentation = *string*

would have to be given with the string setting out how the supplement identifiers in the data were to be interpreted.

The data table follows immediately after the read statement and is ended by the default terminator, a colon; the data must be included at this stage because Genstat obeys commands as it reads them and the statements that follow require the data to operate upon. The reader should note how the requirements of the option and parameter defaults discussed in the last paragraph are illustrated in the layout of the data table.

Turning now to the analysis required for this experiment we remember first that the design is a single-factor so that the only source of random variation that we have to consider is that between sheep receiving the same treatment; in the terminology of Genstat we say that the only source of random variation is between units. In this situation one of the two main types of Genstat statement concerning an analysis is not needed. These two types are the *block* statement and the *treatment* statement; when the only source of random variation is between units we do not need to make a block statement.

The object of the treatment statement is to specify the treatments and the comparisons required between them. In the present case the treatments are the levels of a single factor but we are requested to consider certain particular contrasts between them. In order to do this we use the factor declarations

> factor [levels = 3] ms1
> factor [levels = 5] ms2
> factor [labels = !t(A, B, C, D, E, F)] supplements

and the calculate statements

> calculate ms1 = newlevels (supplements; !(1, 2, 2, 2, 3, 3))
> & ms2 = newlevels (supplements; !(1, 2, 3, 4, 5, 5))

to set up the structures Genstat will need. We will consider first the interpretation of these commands.

Each of the factor declarations consists of

> structure name option identifier

Of these, only the options require comment. In the first two declarations the option specifies the number of levels that the factor has so that the factor to be known as ms1 will have three levels and that known as ms2 will have five. In the third declaration not only is the number of levels indicated for the factor to be known as supplements but the labels to be used for these levels in the output are also given; the structure !t(. . .) used to declare the labels is made up as follows:

- the *round brackets* (. . .) indicate that the contents are to be used together and not one by one on separate occasions;
- the indicator t shows that the contents of the brackets are *text* and not numbers and so cannot be used as arguments in calculation;
- the *special symbol* ! indicates an unnamed structure.

In the two calculate statements no options are given but in each a parameter sets the expression defining the calculation. Both calculations use the *function* newlevels, which has the structure

> newlevels (existing factor name; variate with the same number of values as the existing factor has levels)

The effect of the function is to group the levels of the existing

231

factor to form the levels of a new factor; thus ms1 is defined to be formed by dividing the supplements into three groups containing

A   BCD   EF

whilst ms2 has five levels containing

A B C D EF

We note that the specification of the variate in newlevels uses the structure !(...), which is similar to that used above in the declaration of labels but does not of course include the text indicator. In addition, rather than repeat the statement calculate we use the special symbol &; however, we should remember that the use of & implies repetition not only of the command but also of any options stated with it so that in the factor declarations above its use would have been wrong.

The reason for selecting these particular groupings of the supplement levels becomes clear when we consider the contrasts between supplements that Exercise 5 requests us to make

    (a) between B, C and D,
    (b) between E and F,
    (c) between A, the average of B, C and D and the average of E and F.

Contrast (c) requires us to divide the supplements into the three groups defined by ms1, whilst contrast (a) corresponds to the extra variation, over and above that between those three groups, measured when we compare the five groups defined by ms2; finally, contrast (b) corresponds to the extra variation, over and above that between those five groups, measured when we compare all six supplements. These correspondences can be quantified if we consider expressions for sums of squares in the manner of 7.2.2.3 and 7.2.2.4. Denoting the supplement totals by $T_A, \ldots, T_F$ and the overall total by T the s.s.'s for the three contrasts are:

$$(a)\quad \frac{T_B^2}{8} + \frac{T_C^2}{7} + \frac{T_D^2}{8} - \frac{(T_B + T_C + T_D)^2}{23},$$

$$(b)\quad \frac{T_E^2}{7} + \frac{T_F^2}{8} - \frac{(T_E + T_F)^2}{15},$$

(c) $\dfrac{T_A^2}{8} + \dfrac{(T_B + T_C + T_D)^2}{23} + \dfrac{(T_E + T_F)^2}{15} - \dfrac{T^2}{46}.$

Similarly the s.s.'s corresponding to ms1, ms2 and all six supplements are:

(ms1) $\dfrac{T_A^2}{8} + \dfrac{(T_B + T_C + T_D)^2}{23} + \dfrac{(T_E + T_F)^2}{15} - \dfrac{T^2}{46},$

(ms2) $\dfrac{T_A^2}{8} + \dfrac{T_B^2}{8} + \dfrac{T_C^2}{7} + \dfrac{T_D^2}{8} + \dfrac{(T_E + T_F)^2}{15} - \dfrac{T^2}{46},$

(all six) $\dfrac{T_A^2}{8} + \dfrac{T_B^2}{8} + \dfrac{T_C^2}{7} + \dfrac{T_D^2}{8} + \dfrac{T_E^2}{7} + \dfrac{T_F^2}{8} - \dfrac{T^2}{46}.$

We see at once that in terms of s.s.'s

$(c) = (\text{ms1}),$

$(a) = (\text{ms2}) - (\text{ms1}),$

$(b) = (\text{all six}) - (\text{ms2}),$

which are the correspondences described above. To instruct Genstat to carry out these calculations we make the treatment statement as follows:

> treatment   ms1/ms2/supplements

The / notation means that the treatment s.s.'s are to be calculated by subtracting the s.s.'s for the factors ms1, ms2 and supplements in the manner just explained; in Genstat terminology the factors are described as being *nested*, with ms2 within ms1 and supplements within ms2.

The remaining statement to be considered is

> anova [fprob = yes] weight

which instructs Genstat to use weight as the variate in the analysis of variance appropriate to the treatment comparisons and design already specified; the option commands that the probability levels associated with the values of any F-statistics calculated should be included in the output.

The output of this analysis is given in CTable 20; it is divided into three sections

> analysis of variance,
>
> tables of means,
>
> standard errors of differences of means.

233

The first section shows the analysis of variance table with a layout differing from that with which we are familiar only in the appearance of an extra column, the sixth in the Genstat table, which shows the probability commanded by the option to the anova statement. The headings of the other columns are as we

CTable 20. *Output for the Genstat analysis of the observations in exercise 5 of 7.5.*

************* Analysis of variance *************

Variate: weight

Source of variation	d.f.	s.s.	m.s.	v.r.	F pr.
ms1	2	2747.5	1373.8	7.18	0.002
ms1.ms2	2	300.3	150.2	0.79	0.463
ms1.ms2.suppleme	1	7.9	7.9	0.04	0.840
Residual	40	7648.5	191.2		
Total	45	10704.3			

************* Tables of means *************

Variate: weight

Grand mean   53.9

ms1	1	2	3
	40.9	52.3	63.3
rep.	8	23	15

ms1	ms2	1	2	3	4	5
1		40.9				
	rep.	8				
2			53.3	56.4	47.6	
	rep.		8	7	8	
3						63.3
	rep.					15

	suppleme	A	B	C	D	E	F
ms1	ms2						
1	1	40.9					
	rep.	8					
2	2		53.3				
	rep.		8				
	3			56.4			
	rep.			7			
	4				47.6		
	rep.				8		
3	5					62.6	64.0
	rep.					7	8

CTable 20 (*continued*)

********** Standard errors of differences of means ***********

Table	ms1	ms1 ms2	ms1 ms2 suppleme	
rep.	unequal	unequal	unequal	
s.e.d.	6.91X	7.39X	7.39	min.rep
	5.68	6.33	7.16	max-min
	4.08X	5.05X	6.91	max.rep

(No comparisons in categories where s.e.d. marked with an X)

The full output also contains some general headings and a numbered listing of the commands.

have seen earlier except for the use of v.r. (variance ratio) for F. The sources of variation into which the between-treatments term is divided should be read as

> ms1
> ms2 within ms1
> supplements† within ms2

which the argument above showed to be equivalent to

> contrast (*c*)
> contrast (*a*)
> contrast (*b*).

The mean squares for each of these contrasts are tested against the residual mean square to give the *F*-values and probabilities shown in the table.

In the tables of means section the means for each level of each of the three Genstat factors – ms1, ms2 within ms1, and supplements within ms2 – are arranged by factors. Thus, for factor ms1 the three means given are for supplements A, B + C + D and E + F, which are the three levels of this factor.

---

† In reproducing any name (other than text) containing more than eight characters Genstat uses only the first eight.

235

The number of replicates – i.e. the number of observations – in each mean is shown immediately below the mean itself.

In the third section of the output the three columns correspond to the three factors. The entries unequal against each factor in the replicates row indicate that the numbers of observations in the separate levels of each factor are not the same; this implies that the possible differences between the level means within a factor do not all have the same standard deviation. Genstat therefore gives a selection of standard errors of differences (s.e.d.'s) to indicate the range of theoretically possible values, namely:

   (i) s.e.d. for two means each based on the minimum number of replicates in any level,

  (ii) s.e.d. for two means when one is based on the minimum number of replicates and the other on the maximum number,

 (iii) s.e.d. for two means each based on the maximum number of replicates.

If one of these theoretical possibilities does not actually occur for the levels of the factor under consideration the s.e.d. for that possibility is marked with an X. Thus, in ms1 with level means based on 8, 23 and 15 observations there is no possible comparison between two levels each based on eight observations nor one between two levels each based on 23 observations and so both the type (i) and the type (iii) s.e.d.'s are marked with X. The reader should verify that the standard errors given do follow from the expression

$$s\sqrt{\left(\frac{1}{n_i}+\frac{1}{n_j}\right)}$$

used in 7.2.2.

In the second single-factor example, Exercise 8 of 7.5, the eight treatments formed a factorial arrangement and there were four replicates of each treatment; we will use the between-treatment comparisons (b), (c) and (d) proposed in Exercise 7. The commands and data for a Genstat analysis for Exercise 8 are given in CTable 21.

CTable 21. *Commands and data for the Genstat analysis appropriate to exercise 8 in 7.5.*

```
units [nvalues = 32]
factor [labels = !t('0', '10', '25', '40')]   doses
factor [levels = 2]   days
factor [levels = 4]   ewes
read         days, doses, ewes, cycle
1 1 1 17    1 2 1 15    1 3 1 12    1 4 1  8
1 1 2 18    1 2 2 15    1 3 2 12    1 4 2  9
1 1 3 17    1 2 3 14    1 3 3 11    1 4 3 11
1 1 4 17    1 2 4 16    1 3 4 11    1 4 4  6
2 1 1 18    2 2 1 16    2 3 1 16    2 4 1 12
2 1 2 20    2 2 2 14    2 3 2 14    2 4 2 13
2 1 3 17    2 2 3 16    2 3 3 11    2 4 3 12
2 1 4 14    2 2 4 16    2 3 4 14    2 4 4 12 :
matrix [rows = 3; columns = 4; values =  0, -1,  0,  1, \
                                          0,  1, -2,  1, \
                                         -3,  1,  1,  1] dmat
treatments   reg(doses; 3; dmat)*days
anova [fprob = yes]   cycle
```

Apart from the statements concerning treatments and the comparisons between them, the commands are similar to those in the previous example, but we note that in the first factor statement the figures to be used as labels for the dose levels have to be enclosed in single quotes in order that Genstat may accept them as text. The proposed treatment comparisons require the use of a matrix statement and the factorial arrangement of the treatments leads to a different form of treatment statement from that used in the analysis for Exercise 5. The matrix statement is:

$$\text{matrix [rows} = 3; \text{columns} = 4; \text{values} = \begin{array}{rrrr} 0, & -1, & 0, & 1, \\ 0, & 1, & -2, & 1, \\ -3, & 1, & 1, & 1 \end{array} \text{]} \ dmat$$

This sets up an array of numbers with three rows and four columns which is to be known as dmat. We note that the three options (rows, columns and values) are separated by semi-colons; this separator is always to be used when there is more than one option qualifying a command. We note also the use of backslash (\) to indicate that the command is continued on the next line; without this, 'new line' automatically terminates a command (unless the

237

set command is in operation). We see that the three rows of the array, used as coefficients of dose level totals, define the contrasts between these levels proposed in Exercise 7; thus in the notation of the discussion to Exercise 8 the first row of dmat leads to

$$-C_{10} + \qquad C_{40}$$

which is used in contrast (d), the second row leads to

$$C_{10} - 2C_{25} + C_{40}$$

which is used in contrast (c), and the third row leads to

$$-3C_0 + C_{10} + \quad C_{25} + C_{40}$$

which is needed for contrast (b).

The treatment statement is

    treatment   reg(doses; 3; dmat) * days

This statement shows first that the factor doses is to be analysed by the use of the *submodel function* regression†, that this function is of *order* 3 – i.e. three terms are to be fitted – and that the one degree of freedom contrasts defining these terms are specified in the array known as dmat. The statement also shows, by the use of the operator *, that the factors doses and days combine factorially in defining the treatments used in the experiment.

The output of this analysis is shown in CTable 22; it contains the same three sections as the previous output (CTable 20). The analysis of variance section gives for this analysis information corresponding to that contained in the previous table. We note that the doses and doses × days terms are given both as overall comparisons (each with three degrees of freedom) and also broken down into the contrasts (each with one degree of freedom) specified by the matrix statement.

The tables of means section is similar to that in CTable 20, differing in that in the present case the treatments have a factorial arrangement – so that means are given both for the levels of the main effects and in a two-way table for the interaction – and in that the numbers of replicates are not given because in each subtable they are the same for every level. This equality of

† This use of the term *regression* is not quite the same as that introduced in 2.4.2.

CTable 22. *Output for the Genstat analysis of the observations in Exercise 8 of 7.5.*

************ Analysis of variance ************

Variate: cycle

Source of variation	d.f.	s.s.	m.s.	v.r.	F pr.
doses	3	216.750	72.250	33.34	<.001
Reg1	1	95.062	95.062	43.87	<.001
Reg2	1	0.187	0.187	0.09	0.771
Reg3	1	121.500	121.500	56.08	<.001
days	1	21.125	21.125	9.75	0.005
doses.days	3	17.625	5.875	2.71	0.067
Reg1.days	1	10.562	10.562	4.88	0.037
Reg2.days	1	0.021	0.021	0.01	0.923
Reg3.days	1	7.042	7.042	3.25	0.084
Residual	24	52.000	2.167		
Total	31	307.500			

************ Tables of means ************

Variate: cycle

Grand mean   13.87

doses	0	10	25	40
	17.25	15.25	12.63	10.38

days	1	2
	13.06	14.69

doses	days	1	2
0		17.25	17.25
10		15.00	15.50
25		11.50	13.75
40		8.50	12.25

*********** Standard errors of differences of means ***********

Table	doses	days	doses days
rep.	8	16	4
s.e.d.	0.736	0.520	1.041

numbers of replicates also makes rather simpler than in CTable 20 the section for standard errors of differences of means, but the present case necessarily takes account of the factorial arrangement.

These two examples, although dealing only with single-factor analyses, illustrate many of the procedures used in Genstat and provide a sufficient background for the consideration of the further designs discussed in the remainder of this chapter.

7.3 Two-factor anovar

In the two-factor classification we cannot follow all the calculations from first principles, as we could in the single-factor anovar, but the formulae developed from first principles for that case can readily be generalized to the two-factor case; the reader may therefore wish to look again at 7.2.1 before considering the present section.

As an illustration of the two-factor anovar we will use the results given in Table 41; these are values of the weight gain (g) of chicks fed one of 16 diets, the diets differing in amino-acid content. Each tabulated value is the total weight gain of a cage of 3 chicks over the period 10 to 20 days of age; the cages were arranged in 6 groups of 16 in two batteries, the 16 cages in a group (replicate) being taken close together so as to be as similar as possible in environmental conditions. The diets were allocated to cages at random within replicates. We have thus a two-factor arrangement with 16 levels in one classification and 6 in the other; each cell of the two-factor classification contains a single observation, so that we have 96 observations in all.

In the general case we suppose we have factors at p and q levels, giving pq observations in all. Denoting the total of all the observations by T, we have for the total s.s., measuring the variation between all the pq observations together, the expression

$$\text{total s.s.} = \sum x^2 - \frac{T^2}{pq},$$

with $(pq - 1)$ degrees of freedom. The calculation of this, and the other, s.s. for the chicken data, is shown in Table 42; the general formulae are brought together in Table 43.

The s.s. for diets (rows of the table of results) and replicates (columns) are similar in form to the between levels s.s. in the single-factor anovar; this is as we would expect because both

Table 41. Weight gains (g) of cages of 3 chicks in the period 10 days to 20 days of age.†

Diet	Replicates						Totals	Means	Estimated constants
	1	2	3	4	5	6			
A	125	95	121	92	80	87	$R_1 = 600$	100.0	−93.3
B	201	169	152	174	141	128	$R_2 = 965$	160.8	−32.5
C	251	216	209	231	226	230	$R_3 = 1363$	227.2	33.9
D	332	323	310	317	320	291	$R_4 = 1893$	315.5	122.2
E	224	170	176	193	163	153	$R_5 = 1079$	179.8	−13.5
F	294	290	268	279	274	267	$R_6 = 1672$	278.7	85.4
G	206	187	172	180	170	147	$R_7 = 1062$	177.0	−16.3
H	298	237	281	291	267	184	$R_8 = 1558$	259.7	66.4
J	116	101	103	146	94	80	$R_9 = 640$	106.7	−86.6
K	135	137	129	138	121	131	$R_{10} = 791$	131.8	−61.5
L	171	160	156	207	171	144	$R_{11} = 1009$	168.2	−25.1
M	262	277	233	249	213	221	$R_{12} = 1455$	242.5	49.2
N	165	155	135	165	145	124	$R_{13} = 889$	148.2	−45.1
P	222	196	184	200	164	167	$R_{14} = 1133$	188.8	−4.5
Q	180	156	187	187	162	157	$R_{15} = 1029$	171.5	−21.8
R	247	264	211	247	222	229	$R_{16} = 1420$	236.7	43.4
Totals	$C_1 = 3429$	$C_2 = 3133$	$C_3 = 3027$	$C_4 = 3296$	$C_5 = 2933$	$C_6 = 2740$	$T = 18{,}558$		0.3
Means	214.3	195.8	189.2	206.0	183.3	171.2		193.3	193.3
Estimated constants	21.0	2.5	−4.1	12.7	−10.0	−22.1	0		

† The totals R_i, C_j and T are used in the anovar calculation (7.3 and Table 42); the estimates of the constants ρ_i, γ_j and μ are explained in 7.3.1. The totals of the estimates for the ρ_i and the γ_j should both be zero (apart from rounding errors) and are used as checks.

241

measure the variation between levels of a single factor. Denoting the row totals by $R_1, R_2, R_3, \ldots, R_p$ and the column totals by $C_1, C_2, C_3, \ldots, C_q$ we have first the addition check

$$\sum_{i=1}^{p} R_i = \sum_{j=1}^{q} C_j = T;$$

this check should always be carried out, as indeed should any that is available. The diet and replicate s.s. are then calculated as follows:

$$\text{Diet (row) s.s.} = \frac{\sum_{i=1}^{p} R_i^2}{q} - \frac{T^2}{pq} \text{ with } p-1 \text{ d.f.,}$$

$$\text{Replicate (column) s.s.} = \frac{\sum_{j=1}^{q} C_j^2}{p} - \frac{T^2}{pq} \text{ with } q-1 \text{ d.f.}$$

We note, as previously, that each total is divided after squaring by the number of observations on which it is based.

These terms are entered in the anovar tables (Tables 42 and 43) and account for all the variation which we can ascribe to population differences; that which remains is the random, residual, variation, which we calculate by subtracting the diet (row) and replicate (column) s.s.'s from the total s.s. The degrees of freedom for the residual are likewise obtained by subtraction, giving in the chicken example

$$\begin{aligned} \text{degrees of freedom of residual} &= 95 - 15 - 5 \\ &= 75. \end{aligned}$$

We notice that this is the product of the degrees of freedom of the diet (row) and replicate (column) terms, and it is easy to verify that this result is general; thus

$$\begin{aligned} \text{degrees of freedom of residual} &= (pq-1) - (p-1) - (q-1) \\ &= pq - 1 - p + 1 - q + 1 \\ &= pq - p - q + 1 \\ &= (p-1)(q-1). \end{aligned}$$

For the moment this result is no more than a curiosity, but we shall find it meaningful in later sections.

Table 42. *The two-factor anovar for the chick weight gain observations in Table* 41. *The calculation is described in* 7.3 *and general formulae are given in Table* 43.

$$\sum x^2 = 3,964,876 \qquad \frac{\sum R^2}{6} = 3,927,852.33$$

$$\frac{\sum C^2}{16} = 3,606,885.25 \qquad \frac{T^2}{96} = 3,587,493.38$$

Source of variation	s.s.	d.f.	m.s.	F
Between diets (rows)	340,358.95	15	22,690.60	96.52***
Between replicates (columns)	19,391.87	5	3,878.37	16.50***
Residual	17,631.80	75	235.09	
Total	377,382.62	95		

The M.S.'s are obtained, as always, by dividing the separate S.S.'s by their degrees of freedom. To compare the M.S.'s we have two variance ratios, obtained by dividing the diet (row) and replicate (column) M.S.'s by the residual M.S.; significance of these two ratios indicates, for the former, that we cannot reasonably regard the row population means as equal, and, for the latter, that we cannot treat the column population means as equal. The two tests are quite independent of each other, and enable us to draw conclusions about one set of means without suspecting any hidden influence of the other.

7.3.1 The model

The model for the two-factor anovar is a simple extension of that for the single-factor. We denote by x_{ij} the observation in the cell of the table specified by row i and column j; the interpretation of this symbol in the present case should be distinguished from that in the single-factor model, where i specified the level of the (single) factor and j specified which of the n_i observations of this level we were considering. In the two-factor model we suppose that there is only one observation in each cell of the table. We then take as model

$$x_{ij} = \mu + \rho_i + \gamma_j + r_{ij},$$

where ρ_i, the *row constants*, measure the deviation of the row population means from the overall population mean, μ, and γ_j,

243

Table 43. *General formulae for the two-factor anovar; see 7.3 and the worked example in Tables 41 and 42.*

Rows	Observations Columns					Totals
	1	2	3	$\cdots$	q	
1						R_1
2		Individual observations, x_{ij}				R_2
3		(this notation is explained in 7.3.1)				R_3
$\vdots$						$\vdots$
p						R_p
Totals	C_1	C_2	C_3	$\cdots$	C_q	T

Preliminary calculations

$$\sum x^2 \qquad \frac{\sum R^2}{q}$$

$$\frac{\sum C^2}{p} \qquad \frac{T^2}{pq}$$

Two-factor anovar

Source of variation	S.S.	d.f.	M.S.	F
Between rows	$\dfrac{\sum R^2}{q} - \dfrac{T^2}{pq}$	$p-1$	s.s./d.f.	$\dfrac{\text{Row M.S.}}{\text{Res. M.S.}}$
Between columns	$\dfrac{\sum C^2}{p} - \dfrac{T^2}{pq}$	$q-1$	s.s./d.f.	$\dfrac{\text{Col. M.S.}}{\text{Res. M.S.}}$
Residual	Total s.s. − Rows s.s. − Col. s.s.	$(p-1)(q-1)$	s.s./d.f.	
Total	$\sum x^2 - \dfrac{T^2}{pq}$	$pq-1$		

the *column constants*, measure the deviation of the column population means from the overall means. [The letters ρ (rho) and γ (gamma) are the Greek r and c.] As in the single-factor case, the distributional assumptions required for the validity of the anovar are expressed with reference to the residuals r_{ij}, which we require to be independent, normal, with zero means and variance σ^2, this variance being the same in all parts of the table. A further implicit but important assumption is that the row and column effects can be combined by addition; for example, if one diet gives a weight gain 2 g above the average (i.e. $\rho = 2$) and one replicate (set of cages) gives a gain 3 g below average (i.e. $\gamma = -3$) then we assume that the cage specified by this diet and replicate will give a gain of

$$\rho + \gamma = 2 - 3 = -1$$

relative to the overall mean, that is 1 g below the overall average, except in so far as this deviation is subject to random variation. We say that we assume an *additive model*; this would not be permissible if, for example, we suspected that the diets or replicates produced changes constant in proportion rather than in amount.

The point estimates of these constants are as follows:

estimate of overall mean $\mu = \dfrac{T}{pq}$, the overall mean of the sample;

estimate of row constant $\rho_i = \dfrac{R_i}{q} - \dfrac{T}{pq}$, the deviation of the mean of row i from the overall mean;

estimate of column constant $\gamma_j = \dfrac{C_j}{p} - \dfrac{T}{pq}$, the deviation of the mean of column j from the overall mean;

estimate of the residual variance $\sigma^2 = s^2$, the residual M.S.

The values of these constants for the numerical example are shown in Table 41.

Interval estimates are usually given for the row and column means ($\mu + \rho_i$, $\mu + \gamma_j$) rather than for the corresponding constants (ρ_i, γ_j). Thus we have

interval estimate for the mean of row $i = \dfrac{R_i}{q} \pm t \dfrac{s}{\sqrt{q}}$,

interval estimate for the mean of column $j = \dfrac{C_j}{p} \pm t \dfrac{s}{\sqrt{p}}$,

where t is taken with the probability level required and the degrees of freedom of the residual standard deviation s, namely $(p-1)(q-1)$. Estimates for differences between row means or between column means are given by

interval estimate for the difference between the means of rows i_1 and $i_2 = \dfrac{R_{i_1} - R_{i_2}}{q} \pm ts \sqrt{\dfrac{2}{q}}$,

interval estimate for the difference between the means of columns j_1 and $j_2 = \dfrac{C_{j_1} - C_{j_2}}{p} \pm ts \sqrt{\dfrac{2}{p}}$.

The reader should verify that these formulae all follow precisely the forms used for corresponding purposes in earlier sections (6.4, 6.6.1, 7.2.2). The interval estimate for σ^2 is obtained from s^2 by the use of the χ^2 distribution (Table A13) with $(p-1)(q-1)$ degrees of freedom.

The null hypotheses tested by the two variance ratios in the anovar are

$$\rho_1 = \rho_2 = \cdots = \rho_p = 0$$

and

$$\gamma_1 = \gamma_2 = \cdots = \gamma_q = 0.$$

Because of the additivity of the model we are able, as explained previously, to draw separate and independent conclusions about these two hypotheses.

Detailed comparisons, both as interval estimates and significance tests, can be made between the levels of either factor alone in exactly the same way as for the single-factor case (7.2.2.1–7.2.2.5).

To check on the distributional assumptions we need first, as in the single-factor case, to estimate the residuals. We do this by rewriting the model in the form

$$r_{ij} = x_{ij} - \mu - \rho_i - \gamma_j$$

and substituting for μ, ρ_i and γ_j their estimated values, obtained by the methods described earlier in this section. The calculation of the residual estimates is thus straightforward but requires considerable care, because it is very easy to make a mistake of signs. The estimates for the chicken example are given in Table 44; two values are set out fully in the caption to this table, and the reader should obtain others for himself.

Table 44. *The residuals for the two-factor anovar of the observations in Table* 41; *the method of calculating the residuals is described in* 7.3.1.†

Diet	Replicate 1	2	3	4	5	6	Total
A	4.0	−7.5	25.1	−20.7	−10.0	9.1	0
B	19.2	5.7	−4.7	0.5	−9.8	−10.7	0.2
C	2.8	−13.7	−14.1	−8.9	8.8	24.9	−0.2
D	−4.5	5.0	−1.4	−11.2	14.5	−2.4	0
E	23.2	−12.3	0.3	0.5	−6.8	−4.7	0.2
F	−5.7	8.8	−6.6	−12.4	5.3	10.4	−0.2
G	8.0	7.5	−0.9	−9.7	3.0	−7.9	0
H	17.3	−25.2	25.4	18.6	17.3	−53.6	−0.2
J	−11.7	−8.2	0.4	26.6	−2.7	−4.6	−0.2
K	−17.8	2.7	1.3	−6.5	−0.8	21.3	0.2
L	−18.2	−10.7	−8.1	26.1	12.8	−2.1	−0.2
M	−1.5	32.0	−5.4	−6.2	−19.5	0.6	0
N	−4.2	4.3	−9.1	4.1	6.8	−2.1	−0.2
P	12.2	4.7	−0.7	−1.5	−14.8	0.3	0.2
Q	−12.5	−18.0	19.6	2.8	0.5	7.6	0
R	−10.7	24.8	−21.6	−2.4	−4.7	14.4	−0.2
Total	−0.1	−0.1	−0.5	−0.3	−0.1	0.5	−0.6

† The values for diet A, replicates 1 and 2 are obtained as follows from the observations and estimated constants in Table 41:

replicate 1 of diet A; residual $= 125 - (-93.3) - 21.0 - 193.3 = 4.0$;

replicate 2 of diet A; residual $= 95 - (-93.3) - 2.5 - 193.3 = -7.5$.

The row and column of totals should all be zero (except for rounding errors) and are used as checks.

We can apply the Kolmogorov–Smirnov and Bartlett tests to these residual estimates, much as in 7.2.3.1 and 7.2.3.2. In the former test we take the sample size as $(p-1)(q-1)$, that is we use the residual degrees of freedom rather than the number of observations; this allows, to some extent, for the fact that the residuals are not independent. In Bartlett's test we group the residuals in some convenient way; if we do this by rows or by columns we find that the calculation of the separate variance estimates is simplified because the row and column totals of the residuals are zero. This means that the s.s. used in the variance estimates are simply the sums of the squares of the appropriate residuals, because the correction factor is zero. In this test also the non-independence of the residuals causes difficulties. An approximate method of allowing for this is to share out the residual degrees of freedom between the variance estimates; thus if we take an estimate for each row we allow to each estimate $(p-1)(q-1)/p$ degrees of freedom. Approximate procedures of this kind must be used with caution, but they are often sufficient: since the anovar is a fairly robust procedure we are only concerned with substantial departures from the conditions, and even approximate tests are likely to be enough to detect these.

7.3.2 Residual and interaction

The purpose of this section is to explore more fully the interpretation of the residual in the two-factor anovar; we have so far considered this simply as a measure of the random variation, but it is possible to give a rather more precise meaning to it. This may conveniently be approached by two numerical examples.

Suppose first that we wish to analyse the two-factor array shown in Table 45; this is a convenient starting point for the

Table 45. *Two-factor anovar with one factor* (rows) *at two levels. The differences between rows within columns are used in 7.3.2. in interpreting the meaning of the residual s.s.*

						Totals
	8	6	9	4	11	38
	5	7	5	6	4	27
Totals	13	13	14	10	15	65
Differences	3	−1	4	−2	7	11

$$\sum x^2 = 469 \qquad \frac{\sum R^2}{5} = 434.6$$

$$\frac{\sum C^2}{2} = 429.5 \qquad \frac{T^2}{10} = 422.5$$

Source of variation	s.s.	d.f.	m.s.	F
Between rows	12.1	1	12.10	1.77
Between columns	7.0	4	1.75	—
Residual	27.4	4	6.85	
Total	46.5	9		

present discussion because one of the factors (rows) has only two levels. The anovar, calculated exactly as usual, is shown in the table. We consider now whether we can obtain a measure of how much the row difference varies between columns. This question is quite different from any that we have asked before; at first glance the table implies ideas not unlike those of rank correlation (4.3.1), but it is really not at all the same, because there we had two variables whilst here we have only one variable, but two factors influencing it.

To approach the problem we must first estimate the row difference for each column; this is easily done by subtracting the values in one row from those in the other. The results are given in the table. We now have to measure how much these differences vary between the columns of the table, that is to measure how much the difference between the levels of one factor varies between the levels of the other. In the light of the measures of variation that we have used earlier, we expect to calculate a s.s., and the question is simply what form the s.s. between the row differences should take. It is *not* sufficient just to sum the squares of the differences and subtract an appropriate term, because this would not take account of the fact that each difference is not a single observation but is calculated from two observations; we need appropriate divisors. The values of these divisors follow the general rule given in 7.2.2.3, and this is not surprising because each within-column comparison between the two rows may be regarded as having one degree of freedom. Applying this rule, we note that the two terms in each difference are single observations and have coefficients 1 and -1; hence the divisor for each individual squared difference is

$$1^2 \times 1 + (-1)^2 \times 1 = 2.$$

In determining the divisor for the square of the total difference, we note first that this total is also the difference between the row totals, each of which is based on 5 observations; thus the required divisor is

$$1^2 \times 5 + (-1)^2 \times 5 = 10.$$

The s.s. measuring how much the row difference varies between columns is then

$$\frac{3^2 + (-1)^2 + 4^2 + (-2)^2 + 7^2}{2} - \frac{11^2}{10} = \frac{9 + 1 + 16 + 4 + 49}{2} - \frac{121}{10}$$

$$= 39.5 - 12.1 = 27.4.$$

We note at once that this quantity is exactly the value of the residual s.s., and naturally ask whether this is coincidence, consequent upon a carefully chosen example, or is a general result. It is easy to show (7.3.2.1) that it applies to any table with two rows, so that in that case we may conclude that the residual s.s. measures how much the difference between rows varies between columns. We see moreover that the degrees of freedom we would expect to require between the 5 differences, $5 - 1 = 4$, are the same as those for the residual s.s.; this result also applies to all two-row tables. We now consider how the approach must be extended to enable us to investigate tables with more than two rows.

For this purpose we will use the example given in Table 46; the anovar is shown in the table. We cannot now measure the difference between rows directly by subtraction, but we can obtain a s.s. between rows within each column of the table. These s.s.'s follow the usual pattern; for example that for the first column of the table is

$$12^2 + 9^2 + 11^2 - \frac{32^2}{3} = 346 - 341.33 = 4.67.$$

The values of these s.s.'s are given in the table; we now consider how to use them in measuring how much the row differences vary between columns. We note first that if we sum these s.s.'s we shall have a measure which includes all the variation between rows, and that this can be regarded as made up of two parts, one the row variation averaged over columns and the other the variation of the row differences between columns. The former of these two parts is measured by the s.s. between rows in the anovar; this s.s. is calculated from column totals, and so does not include any of the variation of the row differences between columns, which can be calculated only by considering individual columns. Thus we can measure this second part of the total variation between rows by subtracting the row s.s. in the anovar from the sum of the within-column between row s.s.'s; in the example this gives

s.s. measuring the variation of the row $= 40.68 - 30.53$
  differences between columns
$$= 10.15.$$

Again we find that this is the same (except for rounding errors) as the residual s.s. in the anovar, and again it is easy to show (7.3.2.1) that this is a general result. As in the case of two rows, the degrees of freedom for the two approaches agree; thus the degrees of freedom for each within-column s.s. are $3 - 1 = 2$, so that those for the s.s. measuring the variation of these terms between columns are $2 + 2 + 2 + 2 + 2 - 2 = 8$, the same as the value for the residual.

Table 46. *Two-factor anovar, showing the* s.s.*'s measuring the variation between rows for individual columns; the relationship between these* s.s.*'s and the residual* s.s. *is explained in 7.3.2.*

						Totals
	12	14	9	15	13	63
	9	10	6	12	9	46
	11	12	7	13	15	58
Totals	32	36	22	40	37	167
$\sum x^2$	346	440	166	538	475	
$\dfrac{\text{(Col. total)}^2}{3}$	341.33	432.00	161.33	533.33	456.33	
s.s. within column	4.67	8.00	4.67	4.67	18.67	

$$\sum x^2 = 1965 \qquad \frac{\sum R^2}{5} = 1889.80$$

$$\frac{\sum C^2}{3} = 1924.33 \qquad \frac{T^2}{15} = 1859.27$$

Source of variation	s.s.	d.f.	m.s.	F
Between rows	30.53	2	15.265	12.05**
Between columns	65.06	4	16.265	12.84**
Residual	10.14	8	1.267	
Total	105.73	14		

The arguments we have used for the two-row and general cases are in fact the same, because the squared differences with their appropriate divisors, used in the two-row case, are just the forms that the difference method of 7.2.2.1 leads to for the s.s.'s between rows. Thus in all cases we find that the residual s.s. measures how much the difference between rows for individual columns varies between columns. Clearly, the same approach could be used to determine how much the difference between columns for individual rows varies between rows; once again we should arrive at the residual s.s., so that we have only a single measure of how the two factors defining rows and columns vary together. We refer to this joint variation as the *interaction* between the two factors; the result we have established in this section is that in a two-factor anovar the residual s.s. is the same as the *interaction* s.s.

This result is useful because it helps us to understand more clearly the form in which random variation manifests itself in a two-factor table with one observation in each cell. It is not, as in the single factor case, possible to interpret it simply as the variation between replicate observations; this type of variation is present – as it always is – but it is combined with, and cannot be separated from, the interaction variation of row differences between columns and column differences between rows. The presence of both kinds of variation in the residual s.s. is of great importance, and many of the developments of experimental design depend upon it. We cannot here consider these extensions at all fully, but we will give two examples showing the importance of the concept of interaction (7.3.3).

7.3.2(c) Two-factor anovar using Genstat

Many of the basic features of anovar in Genstat were illustrated in the two examples discussed in 7.2.4(c), but because these were both single-factor experiments there was no need to describe the design at all; in each case there was only one source of random variation – between sheep – and Genstat was able to deduce this from the units and treatment statements and the pattern of the data. As soon as we consider the two-factor case we find both that we need a statement for the particular purpose of setting out how the separate sources of random variation are incorporated in the design and that the use of this statement leads to the appearance of new features in the output. These changes would be evident even in the analysis of the simplest two-factor design, such as that discussed in 7.3, but we will in fact consider the slightly more complex example in Exercise 19 of 7.5.

The discussion to this exercise shows that the experiment has a two-factor design – farms × varieties – with a further level of replication between trees within plots; the sources of random variation are thus

> between farms,
> between varieties (i.e. plots) within farms, and
> between trees within plots.

These three levels are nested in the manner of the within-treatment terms in the first example in 7.2.4(c) and we find this indicated in the statement used to describe the random variation – the blocks statement – in the commands shown in CTable 23; in the Genstat terminology these three levels of random variation are referred to as three *strata*. We notice that the other statements are all very similar to those used in the earlier examples and that

CTable 23. *Commands and data for a Genstat analysis appropriate to Exercise* 19 *in* 7.5.

```
units [nvalues - 168]
factor [levels - 7] farms
factor [levels - 6] trees
factor [labels - !t(X, R, S, T)] varieties
read        farms, varieties, trees, skin
1 1 1  4.0  1 2 1  7.9  1 3 1  8.8  1 4 1  11.2
1 1 2  4.0  1 2 2  8.8  1 3 2  7.7  1 4 2  11.3
1 1 3  4.7  1 2 3  5.6  1 3 3  10.2  1 4 3  10.8
      ..... and so on through the data to .....
7 1 5  8.7  7 2 5  6.8  7 3 5  9.4  7 4 5  11.6
7 1 6  7.3  7 2 6  4.6  7 3 6  9.2  7 4 6  10.3 :
blocks        farms/varieties/trees
treatments  varieties
anova [fprob - yes] skin
```

in particular the treatment statement does not need to be in any way modified to take account of the increase in the complexity of the design. We note also that it is acceptable – and indeed desirable – to use the factor varieties, which might be thought to be a factor to be used only in describing the treatments, as a level in the statement of the random variation nesting.

The output of this analysis is given in CTable 24; the same section headings occur as in the analyses discussed earlier but the anovar table in the first section is in the present case divided into three to correspond to the three strata in the blocks

CTable 24. *Output for the Genstat analysis of the observations in Exercise* 19 *of* 7.5.

************ Analysis of variance ************

Variate: skin

Source of variation	d.f.	s.s.	m.s.	v.r.	F pr.
farms stratum	6	83.697	13.950		
farms.varietie stratum					
varietie	3	701.616	233.872	62.49	<.001
Residual	18	67.366	3.743		
farms.varietie. trees stratum	140	483.000	3.450		
Total	167	1335.680			

************ Tables of means ************

Variate: skin

Grand mean 7.27

varietie	X	R	S	T
	4.95	5.68	8.42	10.01

*********** Standard errors of differences of means ***********

Table	varietie
rep.	42
s.e.d.	0.422

statement. The names given to these three strata – farms, farms.varietie and farms.varietie.trees – may usefully be read as

between farms,
between varieties within farms, and
between trees within varieties,

although we should remember that varieties (within farms) and plots (within farms) are synonymous; this notation and terminology should be compared with that in the first example in 7.2.4(c). The reader will find it helpful when seeking to list the sources of variation in an anovar to work out the decomposition of the degrees of freedom (d.f.) between the strata and to consider which terms of the anovar correspond to each stratum; thus in the present case we see that:

- the *between farms* stratum contains $7 - 1 = 6$ d.f. – because there are seven farms – and this comparison appears as a single term in the anovar;
- the *between varieties* (plots) *within farms* stratum contains $4 - 1 = 3$ d.f. from each farm – because there are four plots on each farm – and so has in total $7 \times (4 - 1) = 21$ d.f. made up in the anovar by:

varieties   with $4 - 1 = 3$ d.f.
and farms × varieties   with $(7 - 1) \times (4 - 1) = 18$ d.f.

In Rothamsted practice this latter term is always used as the appropriate measure of random variation against which to test varieties, without any of the step-by-step considerations set out in the discussion to Exercise 19, and so the term is labelled residual in the anovar table;

- the *between trees within varieties* stratum contains $6 - 1 = 5$ d.f. from each of the $7 \times 4 = 28$ variety/plot units and so has a total of $7 \times 4 \times (6 - 1) = 140$ d.f. which form a single term in the anovar measuring the variation within plots.

The reader should consider how to calculate all these terms using a desk calculator.

The information given in the second and third sections of the output follows exactly the pattern of the two examples discussed in 7.2.4(c). The analysis given here does not include any comparisons between varieties individually or in groups. The commands needed would be similar to those in the earlier examples – depending of course on the comparisons chosen – and would not need to be modified just because the present design is two-factor rather than single-factor.

7.3.2.1* Two algebraic proofs

In this section we prove that the interaction s.s., measuring the variation between columns of the difference between rows within columns, is equal to the residual s.s.; we consider first the case of two rows. Let $x_{11}, x_{12}, \ldots, x_{1q}$ denote the observations in the first row of the table, and $x_{21}, x_{22}, \ldots, x_{2q}$ those in the second; let the row totals be $R_1 = \sum_{j=1}^{q} x_{1j}$, $R_2 = \sum_{j=1}^{q} x_{2j}$. The residual s.s., calculated as in 7.3, is

Total s.s. − Row s.s. − Column s.s.

$$= x_{11}^2 + x_{12}^2 + \cdots + x_{1q}^2 + x_{21}^2 + x_{22}^2 + \cdots + x_{2q}^2 - \frac{(R_1 + R_2)^2}{2q}$$

$$- \left[\frac{R_1^2}{q} + \frac{R_2^2}{q} - \frac{(R_1 + R_2)^2}{2q} \right]$$

$$- \left[\frac{(x_{11} + x_{21})^2}{2} + \frac{(x_{12} + x_{22})^2}{2} + \cdots + \frac{(x_{1q} + x_{2q})^2}{2} - \frac{(R_1 + R_2)^2}{2q} \right],$$

since the total of the $2q$ observations is $R_1 + R_2$. The s.s. may be rearranged as

$$x_{11}^2 + x_{21}^2 - \frac{(x_{11} + x_{21})^2}{2} + x_{12}^2 + x_{22}^2 - \frac{(x_{12} + x_{22})^2}{2}$$

$$+ \cdots + x_{1q}^2 + x_{2q}^2 - \frac{(x_{1q} + x_{2q})^2}{2} - \left[\frac{R_1^2}{q} + \frac{R_2^2}{q} - \frac{(R_1 + R_2)^2}{2q} \right]$$

$$= \frac{(x_{11} - x_{21})^2}{2} + \frac{(x_{12} - x_{22})^2}{2} + \cdots + \frac{(x_{1q} - x_{2q})^2}{2} - \frac{(R_1 - R_2)^2}{2q}.$$

The differences $x_{11} - x_{21}, x_{12} - x_{22}, \ldots, x_{1q} - x_{2q}$ are precisely the differences used in estimating the row effect in the individual columns, and the s.s. thus follows exactly the form used in 7.3.2 to calculate the s.s. measuring the variation of the row difference between columns, that is the interaction s.s. Each difference has one degree of freedom and so the interaction, that is the comparison between the q differences, has $(q-1)$ degrees of freedom, the same number as the residual in the two-row case.

In the general case of a table with p rows and q columns, denote by x_{ij} the observation in row i and column j, and let the row, column and overall totals be represented by R_i, C_j and T, as in 7.3. The residual s.s. is then calculated as

Total s.s. $-$ Row s.s. $-$ Column s.s.

$$= \left(\sum_{ij} x_{ij}^2 - \frac{T^2}{pq} \right) - \left(\sum_i \frac{R_i^2}{q} - \frac{T^2}{pq} \right) - \left(\sum_j \frac{C_j^2}{p} - \frac{T^2}{pq} \right)$$

$$= \left(\sum_{ij} x_{ij}^2 - \sum_j \frac{C_j^2}{p} \right) - \left(\sum_i \frac{R_i^2}{q} - \frac{T^2}{pq} \right)$$

$$= \sum_j \left(\sum_i x_{ij}^2 - \frac{C_j^2}{p} \right) - \left(\sum_i \frac{R_i^2}{q} - \frac{T^2}{pq} \right).$$

The expression $\sum_i x_{ij}^2 - C_j^2/p$ measures the variation between rows in column j, and the rearranged form of the residual s.s. is thus equivalent to

Total over columns of the s.s.'s measuring the variation between rows within columns $-$ s.s. measuring the difference between rows averaged over columns.

This is the form used to approach the interaction in 7.3.2. and so in the general case we find that the interaction s.s. is the same as that previously called the residual s.s. Each within-column s.s. has $(p-1)$ degrees of freedom, as has that between the row totals, so that the interaction s.s. has degrees of freedom

$$(p-1) + (p-1) + \cdots + (p-1) - (p-1),$$

with one term in the sum for each of the q columns and one term subtracted for the row totals; the interaction degrees of freedom are thus $(p-1)(q-1)$, the same as those for the residual.

7.3.3 Combining two-factor analyses

To illustrate some of the consequences of the dual nature of the residual s.s. in the anovar of a two-factor table with one observation in each cell, we consider the analysis required when several of these experiments have been carried out and are to be combined. This requirement arises, when, for example, a field trial involving treatments and blocks has been performed at several centres, or a laboratory experiment with a group of technicians and several instruments – each technician using each instrument – has been repeated using samples of material from different patients. The more complex anovar procedures, of which these two cases provide simple examples, form a large part of the methods of biometry and their use requires some care; thus these two examples appear at first sight to be analytically the same, but we shall find that they are not.

We consider first the example of the field trial repeated at several centres. Suppose we have the p treatments laid out, with suitable randomization, in q blocks at each of r centres; at each centre the anovar takes the form described in 7.3, with degrees of freedom as follows

Treatments	$p-1$
Blocks	$q-1$
Residual = treatments × blocks interaction	$(p-1)(q-1)$
Total	$pq-1$

In the combined analysis we have pqr observations and so the total s.s. will have $(pqr-1)$ degrees of freedom; we have now to divide these according to the various sources of variation.

The simplest source is the difference between centres, which will have $(r-1)$ degrees of freedom. When this term is removed we would expect all the remaining variation to be within centres, and this is confirmed by the degrees of freedom, for if we add to the $(r-1)$ degrees between centres the total of all the degrees within centres – $(pq-1)$ from each centre – we find

$$(r-1)+r(pq-1)=r-1+pqr-r=pqr-1,$$

which is the total degrees of freedom. After removing the between-centres term we have therefore only to consider how to combine the analyses for the separate centres.

Each of these analyses contributes $(p-1)$ degrees of freedom between treatments, giving an apparent total of $r(p-1)$ for treatments. However, there are only p treatments and so the treatment term in the combined analysis can have only $(p-1)$ degrees of freedom. We thus have

$$r(p-1)-(p-1)=(r-1)(p-1)$$

degrees of freedom, apparently associated with treatment, still to account for. The s.s. associated with these degrees of freedom will be obtained by adding the s.s.'s between treatments within centres and subtracting the s.s. between treatment totals, that is by a calculation exactly parallel to that of the degrees of freedom. These calculations are in fact precisely those for an interaction, as described in 7.3.2; thus the $r(p-1)$ degrees of freedom obtained by totalling the treatment terms from the individual centre anovars are redivided into

| Treatments | $p-1$ |
| Treatments × centres interaction | $(p-1)(r-1)$ |

If we total the degrees of freedom for blocks from the individual anovars we get $r(q-1)$ degrees of freedom, but it is clear that these cannot be divided into terms for blocks and blocks × centres because we can give no meaning to overall block totals; the blocks are just sub-areas of the experiment and in no sense does block 1 at centre A relate to block 1 at any other centre.

The same comment applies to the $r(p-1)(q-1)$ degrees of freedom obtained by totalling the treatments × blocks interactions from the centre anovars, and so this term also is left undivided. The combined anovar therefore contains the following terms and degrees of freedom

Centres	$r-1$
Treatments	$p-1$
Treatments × centres interaction	$(p-1)(r-1)$
Blocks	$r(q-1)$
Treatment × blocks interactions	$r(p-1)(q-1)$
Total	$pqr-1$

The s.s. for these terms can all be calculated by methods described earlier. Those for centres and treatments are calculated from the centre and treatment totals in the same way as the row and column s.s.'s were calculated from row and column totals in the two-factor anovar, each squared total being divided by the number of observations on which it was based. The treatment × centre interaction s.s. is obtained, as explained above, by subtracting the treatment s.s. from the sum of the treatment s.s.'s for the individual centres. The blocks and treatments × blocks interactions s.s. are obtained by totalling the corresponding terms in the anovar for the separate centres, and the total s.s. is calculated – as always – as a single sample s.s. for all the observations together. A worked example of this series of calculations is given in Tables 47a, b and c; the reader should follow these in detail.

The tests of significance are not as straightforward in this analysis as in the two-factor case; indeed, they present difficulties in all but the very simplest forms of anovar, and much work is still in progress in this field. We shall here give only a very limited discussion of the subject; in particular we shall assume that we are interested in testing only the treatments term, and that we want our conclusion to apply to a wide range of centres, of which we are prepared to regard these in the present experiment as a random selection. Clearly these two assumptions limit our questions very considerably, leaving out as they do all consideration of whether the centres differ, whether the use of blocks was worthwhile, and so on; to include those further questions would however, lengthen the discussions far beyond our present scope.

For the test of the overall treatment effect we divide the treatment M.S. by the treatments × centres interaction M.S., obtaining a variance ratio with $(p - 1)$, $(p - 1)(r - 1)$ degrees of freedom. This tests whether there are any differences at all between treatments; detailed comparisons, either as significance tests or interval estimates, can be obtained as in 7.2.2.1–7.2.2.5, but we must be careful in all these procedures to use the treatments × centres interaction M.S. as the appropriate estimate of random variation. This is because we want our conclusions to apply throughout the population of centres, so that we must allow for the random variation between centres in our estimate, and cannot use simply the treatments × blocks interaction M.S., as we did in the case of a single two-factor experiment.

When we consider the second example, a comparison of p technicians each using the same q instruments to make measurements on samples from a set of r patients, we have for

Table 47a. Data for a worked example of the analysis of an experiment carried out at r (= 4) centres with a two-factor layout at each centre for p (= 5) treatments in q (= 3) blocks. The analysis is discussed in 7.3.3 and carried out in Tables 47b and 47c.

Centre	1				2				3				r = 4				Total
Block	1	2	q = 3	Total	1	2	q = 3	Total	1	2	q = 3	Total	1	2	q = 3	Total	Total
Treatments 1	2.0	11.4	10.2	23.6	2.5	11.4	10.7	24.6	4.5	9.0	12.0	25.5	11.1	12.0	2.1	25.2	98.9
2	3.6	2.6	11.7	17.9	3.1	3.1	11.7	17.9	4.1	10.6	3.6	18.3	10.7	11.4	3.5	25.6	79.9
3	4.2	3.7	2.1	10.0	4.2	3.2	2.6	10.0	3.7	11.4	7.2	22.3	2.2	6.1	4.6	12.9	55.2
4	5.5	8.8	7.3	21.6	5.0	8.8	7.3	21.2	4.5	8.7	11.9	25.1	6.9	7.2	5.1	19.2	87.0
p = 5	5.7	10.4	8.7	24.8	6.2	10.9	8.2	25.3	5.2	2.1	4.1	11.4	7.7	2.0	4.5	14.2	75.7
Total	21.0	36.9	40.0	97.9	21.0	37.4	40.5	98.9	22.0	41.8	38.8	102.6	38.6	38.7	19.8	97.1	396.5

261

Table 47b. *Two-factor anovars for the separate centres calculated from the data in Table 47a.*†

Centre 1

$\sum x^2 = 807.67$ $T^2/15 = 638.96$
$\sum R^2/3 = 686.32$ $\sum C^2/5 = 680.52$

	s.s.	d.f.	m.s.	F
Rows	47.36	4	11.84	1.19
Columns	41.56	2	20.78	2.08
Residual	79.79	8	9.97	
Total	168.71	14		

Centre 2

$\sum x^2 = 821.67$ $T^2/15 = 652.08$
$\sum R^2/3 = 703.62$ $\sum C^2/5 = 696.00$

	s.s.	d.f.	m.s.	F
Rows	51.54	4	12.88	1.39
Columns	43.92	2	21.96	2.37
Residual	74.13	8	9.27	
Total	169.59	14		

Centre 3

$\sum x^2 = 868.68$ $T^2/15 = 701.78$
$\sum R^2/3 = 747.47$ $\sum C^2/5 = 747.34$

	s.s.	d.f.	m.s.	F
Rows	45.69	4	11.42	1.21
Columns	45.56	2	22.78	2.41
Residual	75.65	8	9.46	
Total	166.90	14		

Centre 4

$\sum x^2 = 800.53$ $T^2/15 = 628.56$
$\sum R^2/3 = 675.70$ $\sum C^2/5 = 675.94$

	s.s.	d.f.	m.s.	F
Rows	47.14	4	11.78	1.22
Columns	47.38	2	23.69	2.45
Residual	77.45	8	9.68	
Total	171.97	14		

† The notation in each analysis is based on that in Table 43, the R's being the row (treatment) totals, the C's the column (block) totals and T the overall (centre) total. Thus the row s.s. measures the variation between treatments, the column s.s. that between blocks and the residual s.s. that due to the treatments × blocks interaction.

Table 47c. *The calculation of the combined anovar for the data analysed centre by centre in Tables 47a and 47b; the steps of the calculation are described in 7.3.3.*†

$$\text{s.s. for total in combined analysis} = \sum (\text{within-centre} \sum x^2) - \frac{(\text{overall total})^2}{60}$$

$$= 3298.55 - 2620.20$$

$$= 678.35$$

$$\text{s.s. between centres} = \frac{\sum (\text{centre total})^2}{15} - \frac{(\text{overall total})^2}{60}$$

$$= 2621.38 - 2620.20$$

$$= 1.18$$

s.s. between treatments based on treatment totals over centres
$$= \frac{98.9^2 + 79.7^2 + 55.2^2 + 87.0^2 + 75.7^2}{12} - \frac{396.5^2}{60}$$

$$= 2706.65 - 2620.20$$

$$= 86.45$$

Sum of treatment s.s.'s in the centre analyses
$$= 191.73$$

s.s. for treatments × centres interaction
$$= 191.73 - 86.45$$
$$= 105.28$$

Sum of block s.s.'s in the centre analyses
$$= 178.42$$

Sum of the residual s.s.'s (i.e. the treatments × blocks interaction s.s.'s) in the centre analyses
$$= 307.02$$

	S.S.	d.f.	M.S.	F
Centres	1.18	3	0.393	
Treatments	86.45	4	21.612	2.46
Treatments × centres interaction	105.28	12	8.773	
Blocks	178.42	8	22.302	
Treatments × blocks interaction	307.02	32	9.594	
Total	678.35	59		

† The F test is obtained by dividing the treatments M.S. by the treatments × centres interaction M.S.; it thus has 4, 12 degrees of freedom and since it is less than the 5 per cent value (3.26) we have no reason to reject the null hypothesis that the treatment population means are equal.

each patient the anovar

Technicians	$p - 1$
Instruments	$q - 1$
Technician × instrument interaction	$(p - 1)(q - 1),$

and a total s.s. over all patients with $(pqr - 1)$ degrees of freedom.

Once again $(r-1)$ of these degrees of freedom are due to the differences between the mean measurements for the r patients so that all we have to do is to redistribute the $r(pq-1)$ degrees of freedom obtained by totalling the analyses for the individual patients.

The $r(p-1)$ degrees of freedom for technicians, obtained by totalling the technician terms from the separate patient analyses, divide, in the same way as those for treatment in the first example, into

Technicians $\qquad\qquad\qquad\qquad\quad p-1$
Technicians × patients interaction $\quad (p-1)(r-1)$.

The $r(q-1)$ degrees of freedom for instruments, obtained by totalling the instrument terms from the separate patient analyses, are however unlike the blocks term in the first example and can be divided into

Instruments $\qquad\qquad\qquad\qquad\quad q-1$
Instruments × patients interaction $\quad (q-1)(r-1)$.

This division is justified because the same q instruments were used by all the technicians on the samples from all the patients, whereas the blocks at the different centres were not related. Similarly the $r(p-1)(q-1)$ degrees of freedom obtained by totalling the technicians × instruments interaction s.s.'s for the separate patients can be divided by removing a term with $(p-1)(q-1)$ degrees of freedom corresponding to the average (over patients) technicians × instruments interaction, leaving a term with $(p-1)(q-1)(r-1)$ degrees of freedom corresponding to the variation of this interaction between patients. This last term can be regarded as the interaction of the technicians × instruments interaction with patients and is described as the *three-factor interaction* for technicians × instruments × patients. It may be shown, like the *two-factor interaction* for rows × columns in 7.3.2, to be symmetrical in all the factors entering into it. The complete breakdown of the degrees of freedom in the analysis over all patients is thus

Patients	$r-1$
Technicians	$p-1$
Technicians × patients interaction	$(p-1)(r-1)$
Instruments	$q-1$
Instruments × patients interaction	$(q-1)(r-1)$
Technicians × instruments interaction	$(p-1)(q-1)$
Technicians × instruments × patients interaction	$(p-1)(q-1)(r-1)$
Total	$pqr-1$

The calculation of the s.s. for these terms follows precisely the patterns described above and may be followed in the numerical example worked in Tables 48a, b, c and d; the reader is advised to repeat these computations in full. This form of analysis is known as a *three-factor anovar*; it is distinguished from the first example principally by the appearance of the three-factor interaction.

Table 48a. *Data for a worked example of the analysis of an experiment undertaken by p ($=3$) technicians each using q ($=5$) instruments on material from r ($=4$) patients. The analysis is discussed in 7.3.3 and carried out in Tables 48b, 48c and 48d.*

Patients	Technicians	Instruments					Total
		1	2	3	4	$q=5$	
1	1	7.6	11.9	2.6	3.5	5.9	31.5
	2	12.0	12.0	3.1	7.8	10.1	45.0
	$p=3$	6.3	5.1	3.0	6.9	8.4	29.7
	Total	25.9	29.0	8.7	18.2	24.4	106.2
2	1	11.7	9.4	7.6	3.2	2.0	33.9
	2	3.1	7.2	2.8	5.6	8.4	27.1
	$p=3$	10.6	11.9	3.4	12.0	10.2	48.1
	Total	25.4	28.5	13.8	20.8	20.6	109.1
3	1	5.2	11.7	2.1	4.7	10.6	34.3
	2	7.9	12.0	6.4	5.1	10.8	42.2
	$p=3$	4.1	6.2	3.8	3.2	9.2	26.5
	Total	17.2	29.9	12.3	13.0	30.6	103.0
$r=4$	1	8.2	2.7	10.6	7.9	11.6	41.0
	2	10.7	4.9	12.0	11.1	11.9	50.6
	$p=3$	3.1	3.2	8.5	6.4	3.8	25.0
	Total	22.0	10.8	31.1	25.4	27.3	116.6

265

Table 48b. *Two-factor anovars for observations for the separate patients, calculated from the data in Table 48a.*†

Patient 1

	$\sum x^2 = 906.52$		$T^2/15 = 751.90$	
	$\sum R^2/5 = 779.87$		$\sum C^2/3 = 838.03$	
	S.S.	d.f.	M.S.	F
Rows	27.97	2	13.98	2.76
Columns	86.13	4	21.53	4.25*
Residual	40.52	8	5.06	
Total	154.62	14		

Patient 2

	$\sum x^2 = 982.03$		$T^2/15 = 793.52$	
	$\sum R^2/5 = 839.45$		$\sum C^2/3 = 834.95$	
	S.S.	d.f.	M.S.	F
Rows	45.93	2	22.96	1.82
Columns	41.43	4	10.36	—
Residual	101.15	8	12.64	
Total	188.51	14		

Patient 3

	$\sum x^2 = 857.38$		$T^2/15 = 707.27$	
	$\sum R^2/5 = 731.92$		$\sum C^2/3 = 815.50$	
	S.S.	d.f.	M.S.	F
Rows	24.65	2	12.32	5.73*
Columns	108.23	4	27.06	12.59**
Residual	17.23	8	2.15	
Total	150.11	14		

Patient 4

	$\sum x^2 = 1078.68$		$T^2/15 = 906.37$	
	$\sum R^2/5 = 973.27$		$\sum C^2/3 = 986.10$	
	S.S.	d.f.	M.S.	F
Rows	66.90	2	33.45	10.42**
Columns	79.73	4	19.93	6.21*
Residual	25.68	8	3.21	
Total	172.31	14		

† The notation in each analysis is based on that in Table 43, the R's being the row (technician) totals, the C's the column (instrument) totals and T the overall (patient) total. Thus the row s.s. measures the variation between technicians, the

Table 48b *(continued)*

column s.s. that between instruments and the residual s.s. that due to the technicians × instruments interaction.

[In examining these anovar tables the reader should be struck by the rather large differences between the four residual M.S.'s. In fact they do not quite differ significantly – Bartlett's test gives $\chi_3^2 = 6.99$ – but so large a variation should always be noted, and possible explanations considered.]

Table 48c. *The calculation of the combined (three-factor) anovar for the data analysed patient by patient in Tables 48a and 48b; the principles of the calculation are described in 7.3.3.*†

$$\text{s.s. for total in combined analyses} = \sum (\text{within-patient } \sum x^2) - \frac{(\text{overall total})^2}{60}$$

$$= 3824.61 - 3152.30$$

$$= 672.31$$

$$\text{s.s. between patients} = \frac{\sum (\text{patient total})^2}{15} - \frac{(\text{overall total})^2}{60}$$

$$= 3159.06 - 3152.30$$

$$= 6.76.$$

Totals over patients for the separate technician-instrument levels:

Tech-nicians	Instruments					
	1	2	3	4	$q = 5$	Total
1	32.7	35.7	22.9	19.3	30.1	140.7
2	33.7	36.1	24.3	29.6	41.2	164.9
$p = 3$	24.1	26.4	18.7	28.5	31.6	129.3
Total	90.5	98.2	65.9	77.4	102.9	434.9

The entries in this table are sums of four observations, those in the row (technician) totals sums of twenty and those in the column (instrument) totals sums of twelve.

$$\frac{\sum (\text{technician-instrument total})^2}{4} = 3296.99, \qquad \frac{(\text{overall total})^2}{60} = 3152.30,$$

$$\frac{\sum (\text{technician total})^2}{20} = 3185.35, \qquad \frac{\sum (\text{instrument total})^2}{12} = 3229.62.$$

These calculations lead to the following terms of the anovar, all for terms averaged over patients.

Technicians	33.05	2
Instruments	77.32	4
Technicians × instruments interaction	34.32	8
Technician-instrument total	144.69	14

The technician-instrument total does not appear in the combined analysis and is calculated only to lead to the technicians × instruments interaction.

Sum of technicians s.s.'s in the patient analyses	$= 165.45$
s.s. for technicians × patients interaction	$= 165.45 - 33.05$
	$= 132.40$

267

Table 48c (*continued*)

Sum of instruments s.s.'s in the patient analyses = 315.52
s.s. for instruments × patients interaction = 315.52 − 77.32
= 238.20
Sum of technicians × instruments interaction s.s.'s in = 184.58
the patient analyses
s.s. for technicians × instruments × patients interaction = 184.58 − 34.32
= 150.26

	s.s.	d.f.	m.s.	F
Patients	6.76	3	2.253	
Technicians	33.05	2	16.525	—
Technicians × patients interaction	132.40	6	22.067	
Instruments	77.32	4	19.330	—
Instruments × patients interaction	238.20	12	19.850	
Technicians × instruments interaction	34.32	8	4.290	—
Technicians × instruments × patients interaction	150.26	24	6.261	
Total	672.31	59		

† The F test for technicians, with 2, 6 degrees of freedom is obtained by dividing the technicians m.s. by the technicians × patients interaction m.s.; that for instruments, with 4, 12 degrees of freedom, compares the instruments and instruments × patients interaction m.s.'s; and that for technicians × instruments, with 8, 24 degrees of freedom, compares the technicians × instruments interaction and technicians × instruments × patients interaction m.s.'s. None of the tests shows any evidence of significant differences.

A full discussion of the possible significance tests for this form of anovar is again beyond our present scope, and we consider only the case where we wish to draw conclusions about the particular technicians and instruments that we have tested, but are prepared to regard the patients as a random sample from a population of patients, to all of which our conclusions about technicians and instruments are to apply. In this case we can make three significance tests:

(*a*) A variance ratio with $(p-1)(q-1)$, $(p-1)(q-1)(r-1)$ degrees of freedom obtained by dividing the technicians × instruments interaction m.s. by the technicians × instruments × patients interaction m.s.; this tests the null hypothesis that there is no technician × instruments interaction, that is that any difference between the technicians can be regarded as the same with all the instruments and that any difference between the instruments does not depend on which technician is using them.

(b) A variance ratio with $(q-1)$, $(q-1)(r-1)$ degrees of freedom obtained by dividing the instruments M.S. by the instruments × patients interaction M.S.; this tests the null hypothesis that the instruments do not differ in the average observations they give.

(c) A variance ratio with $(p-1)$, $(p-1)(r-1)$ degrees of freedom obtained by dividing the technicians M.S. by the technicians × patients interaction M.S.; this tests the null hypothesis that the technicians do not differ in the average observations they obtain.

These tests are separate and independent, and the various possible combinations of significance and non-significance can lead to a variety of interpretations of the results. For example, if (c) gives significance whilst (a) and (b) do not we conclude that the technicians differ in their readings, this difference being independent of which instrument they use. Again if (b) and (c) give significance but (a) does not we conclude that there are differences between technicians unrelated to which instruments they use and also differences between instruments irrespective of which technician is operating them. If on the other hand, (a), (b) and (c) all give significance we infer that there are irregular differences between technicians and instruments which need much care in interpretation; one possible explanation is that the additive model relating technicians and instruments is inadequate (7.3.1).

Detailed comparisons between technicians or between instruments can be made by the methods of 7.2.2.1–7.2.2.5, using the technicians × patients M.S. as the measure of random variation when comparing technicians and the instruments × patients M.S. when comparing instruments. Detailed comparisons within the technicians × instruments interaction corresponding to any comparisons between technicians or between instruments can be made by a simple application of the methods given above for calculating an interaction. Thus if we wish to know how much some particular comparison between technicians varies between instruments, we calculate s.s.'s for this comparison for each instrument and for the total over all instruments, and obtain the interaction s.s. by totalling these s.s.'s for the individual instruments and subtracting the s.s. for the total over instruments. An example of such a calculation is given in Table 48d.

Table 48d. An example of how a chosen comparison (between technicians 2 and 3) is calculated as a main effect and then in interaction with another factor (instruments).†

| Technicians | Instruments | | | | | Total |
	1	2	3	4	5	
2	33.7	36.1	24.3	29.6	41.2	164.9
3	24.1	26.4	18.7	28.5	31.6	129.3
Difference	9.6	9.7	5.6	1.1	9.6	35.6

The entries in this table, which is extracted from Table 48c, are sums of four observations, and the row (technician) totals are sums of twenty observations.

$$\text{s.s. technicians 2 v. 3} = \frac{35.6^2}{40}$$

$$= 31.68 \text{ with 1 degree of freedom}$$

Sum over instruments of s.s.'s technicians 2 v. 3 for individual instruments
$$= \frac{9.6^2}{8} + \frac{9.7^2}{8} + \frac{5.6^2}{8} + \frac{1.1^2}{8} + \frac{9.6^2}{8}$$
$$= 38.87 \text{ with 5 degrees of freedom}$$

s.s. for technicians (2 v. 3) × instruments interaction
$$= 38.87 - 31.68$$
$$= 7.19 \text{ with 4 degrees of freedom}$$

The anovar entries corresponding to these terms are:

	s.s.	d.f.	M.S.	F
Technicians (2 v. 3)	31.68	1	31.68	1.44
Technicians (2 v. 3) × instruments interaction	7.19	4	1.797	—

† The F values are calculated as in Table 48c, the value for technicians (2 v. 3) by dividing the M.S. by that for technicians × patients interaction and the value for technicians (2 v. 3) × instruments interaction by dividing the M.S. by that for technicians × instruments × patients interaction.

The models for these two analyses are necessarily more complex than those considered in 7.2.2, 7.3.1 and we shall not discuss them here. Such discussion is necessary however if we want to interpret experiments of these designs in more detail, and if we want to understand the assumptions needed for the validity of the analyses. Consideration of the model is in fact always of great assistance in the understanding of a complex design, because it helps us to specify and appreciate the assumptions and questions that concern us. The ideas introduced in the last few sections – of interaction, of factors which interest us and factors we can regard as random – occur repeatedly in the interpretation of experimental results, but the details of particular cases may require lengthy discussion and we can only develop one or two topics in the exercises.

7.3.3(c)   Two further examples of anovar in Genstat

The two analyses discussed in 7.3.3 were both based on the two-factor anovar. In the first the separate two-factor analyses were blocks × treatment designs, with the same treatments in each; thus, in the usage introduced in the discussion to Exercise 19 and applied in 7.3.2(c), these analyses involved one random and one fixed factor. The individual experiments were then combined with the introduction of the between-centres comparison, this being a further random factor with the blocks factor nested within it. In the second example the individual two-factor analyses involved two fixed factors in the technicians × instruments arrangement; these analyses were combined into a three-factor arrangement by the introduction of the random factor between patients. We shall find that the Genstat analyses of these two experiments do not require any new concepts, even though the two designs appear to be more complex than those considered in previous examples; we shall, however, use them as occasions to illustrate a few more of the facilities available in Genstat.

The commands and data for the analysis of the first example are shown in CTable 25. The block statement closely parallels that in CTable 23 – remembering that in the present case treatments (within blocks) and plots (within blocks) are synonymous in the same way as were varieties (within farms) and plots (within farms) in the earlier design. The treatment statement reflects the fact that there is only one fixed factor in the experiment. The anovar statement, however, displays a new feature in that a print option statement is included; the form of this statement is chosen in order to augment the output used in previous examples by the inclusion of tables of effects and residuals. The default form of the option is:

A, calling for the printing of the anovar table;

I, calling for an 'information summary'; this concept cannot be explained fully here but an illustration is given in the second example in this section;

C, commanding a list of the 'covariates'; this is of no relevance to any of the analyses in this book;

M, requiring the printing of the tables of means; and

Mi, calling for information about any missing values in the data.

CTable 25. *Commands and data for a Genstat analysis appropriate to the first example discussed in 7.3.3.*

```
units [nvalues - 60]
factor [levels - 4] centres
factor [levels - 3] blocks
factor [levels - 5] treatments
read        centres, blocks, treatments,yield
1 1 1  2.0  1 1 2  3.6  1 1 3  4.2  1 1 4  5.5  1 1 5  5.7
1 2 1 11.4  1 2 2  2.6  1 2 3  3.7  1 2 4  8.8  1 2 5 10.4
      ..... and so on through the data to .....
4 2 1 12.0  4 2 2 11.4  4 2 3  6.1  4 2 4  7.2  4 2 5  2.0
4 3 1  2.1  4 3 2  3.5  4 3 3  4.6  4 3 4  5.1  4 3 5  4.5 :
blocks      centres/blocks/treatments
treatments  treatments
anova [fprob - yes; print - A, I, C, E, R, M, Mi] yield
```

If we wish to call for the printing of tables of effects and residuals – by adding E and R to the option string – we have also to list the default values because these are no longer assumed when any form of option statement is given.†

The output for this analysis is shown in CTable 26; because of the values given to the print option there is a new second section

tables of effects and residuals

The three familiar sections require no comment but the new

---

† We could have omitted **I, C** and **Mi** because these have no relevance in the present example; but of course we would have needed to be sure of this before we programmed the analysis.

section does need some explanation; before proceeding, the reader may wish to review the paragraphs relating to residuals in 7.2.2 and 7.3.1.

CTable 26. *Output for the Genstat analysis for the first example in 7.3.3.*

************ Analysis of variance ************

Variate: yield

Source of variation	d.f.	s.s.	m.s.	v.r.	F pr.
centres stratum	3	1.182	0.394		
centres.blocks stratum	8	178.412	22.301		
centres.blocks. treatmen stratum					
treatmen	4	86.448	21.612	2.31	0.073
Residual	44	412.303	9.371		
Total	59	678.346			

************ Tables of effects and residuals ************

Variate: yield

**** centres stratum ****

centres residuals   s.e. 0.140   rep. 15

centres	1	2	3	4
	-0.08	-0.01	0.23	-0.13

**** centres.blocks stratum ****

centres.blocks residuals s.e. 1.724   rep. 5

centres	blocks	1	2	3
1		-2.33	0.85	1.47
2		-2.39	0.89	1.51
3		-2.44	1.52	0.92
4		1.25	1.27	-2.51

**** centres.blocks.treatmen stratum ****

treatmen effects   e.s.e. 0.884   rep. 12

treatmen	1	2	3	4	5
	1.63	0.03	-2.01	0.64	-0.30

273

CTable 26 (*continued*)

centres.blocks.treatmen residuals   s.e.   2.621   rep.   1

centres	blocks	treatmen	1	2	3	4	5
1	1		-3.83	-0.63	2.01	0.66	1.80
	2		2.39	-4.81	-1.67	0.78	3.32
	3		0.57	3.67	-3.89	-1.34	1.00
2	1		-3.33	-1.13	2.01	0.16	2.30
	2		2.29	-4.41	-2.27	0.68	3.72
	3		0.97	3.57	-3.49	-1.44	0.40
3	1		-1.53	-0.33	1.31	-0.54	1.10
	2		-0.99	2.21	5.05	-0.30	-5.96
	3		2.61	-4.19	1.45	3.50	-3.36
4	1		1.75	2.95	-3.51	-1.46	0.28
	2		2.63	3.63	0.37	-1.18	-5.44
	3		-3.49	-0.49	2.65	0.50	0.84

************ Tables of means ************

Variate: yield

Grand mean   6.61

treatmen	1	2	3	4	5
	8.24	6.64	4.60	7.25	6.31

*********** Standard errors of differences of means ***********

Table	treatmen
rep.	12
s.e.d.	1.250

The new section is divided into three according to the strata defined in the block statement. In the first, centres, section the residuals measure the deviations of the centre means from the overall mean; the centre means are

$$6.527 \quad 6.593 \quad 6.840 \quad 6.473,$$

the overall mean is 6.608 and the calculation

$$\text{centre mean} - \text{overall mean}$$

leads to the values

$$-0.081 \quad -0.015 \quad 0.232 \quad -0.135,$$

which are given to two places of decimals in the output. The value of rep. indicates that each centre mean is based on 15 observations. To illustrate the calculation of s.e., consider the residual for centre 1, which can be represented symbolically as

$$\bar{x}_1 - \bar{x}.$$

In considering the variance of this quantity we need to remember that, although the individual centre means are independent, the two terms in this expression are not because

$$\bar{x} = \tfrac{1}{4}(\bar{x}_1 + \bar{x}_2 + \bar{x}_3 + \bar{x}_4).$$

To cope with this difficulty we substitute this expression for $\bar{x}$ in the residual and simplify to obtain

$$\text{residual} = \tfrac{3}{4}\bar{x}_1 - \tfrac{1}{4}(\bar{x}_2 + \bar{x}_3 + \bar{x}_4).$$

For each of the centre means, the estimate of variance is

$$\frac{\text{centres M.S.}}{15} = \frac{0.394}{15} = 0.0263,$$

so that the estimated variance of the residual is

$$[(\tfrac{3}{4})^2 + 3 \times (\tfrac{1}{4})^2] \times 0.0263 = 0.0197$$

and the estimated standard deviation, given as s.e., is $\sqrt{(0.0197)} = 0.140$. This value was calculated by considering the residual for centre 1, but it is easy to see that the same value would be obtained for any centre.

In the next, centres.blocks, section the residuals measure the deviations of the within-centre block means from the centre means; thus for centre 1 the block means are

$$4.200 \quad 7.380 \quad 8.000,$$

the centre mean is 6.527 and the calculation

$$\text{block mean} - \text{centre mean}$$

leads to the values

$$-2.327 \quad 0.853 \quad 1.473$$

corresponding to the centre 1 row in the output table. The value of rep. shows that each block mean is based on five observations and the calculation leading to s.c. is very similar to that given in detail for the centres stratum.

In the final, centres.blocks.treatments, section the first entry deals with the treatment effects; the calculation to obtain these appears to be very similar to that leading to the residuals in the centres stratum but the measure of precision that is given is calculated differently because it is used for a different purpose. Dealing first with the effects, we consider the treatment means

$$8.242 \qquad 6.642 \qquad 4.600 \qquad 7.250 \qquad 6.308$$

and subtract from these the overall mean of 6.608 to obtain the values

$$1.634 \qquad 0.034 \qquad -2.008 \qquad 0.642 \qquad -0.300$$

which are given to two places of decimals in the output; the value of rep. indicates that each treatment mean is based on 12 observations. The chosen measure of precision is not however the estimated standard deviation of each effect – which would be denoted in Genstat by s.e. – because this cannot be used if we wish to compare two effects; and that is probably the main use we would expect to make of the chosen measure. The reason that we could not use the s.e. for that purpose is that the effects are not independent, so that the variance of the difference between two effects is not the sum of their separate variances; the reason that the effects are not independent is that the overall mean is used in the calculation of all of them. To meet this difficulty Genstat calculates an *effective variance* for the effects, obtained in such a way that the (true) variance of the difference between two effects is the sum of the (effective) variances of the two effects – i.e. is twice the effective variance of each. The square root of the effective variance is the chosen measure of precision, labelled as the effective standard error (e.s.e.).

To understand how the e.s.e. is calculated we need first therefore to obtain the variance of the difference between the effects of two treatments, say treatments i and j. The effect for treatment i is estimated as

$$\bar{x}_i - \bar{x}$$

and that for treatment j as

$$\bar{x}_j - \bar{x}$$

so that the difference between the two effects is estimated by

$$(\bar{x}_i - \bar{x}) - (\bar{x}_j - \bar{x}) = \bar{x}_i - \bar{x}_j.$$

The variance of this estimated difference is thus simply the variance of the difference between two treatment means and this, because the means are independent, is just the sum of their separate variances or, more simply, twice their common variance. The variance of a treatment mean is estimated by

$$\frac{\text{residual M.S.}}{12} = \frac{9.371}{12} = 0.7809,$$

so that the estimated variance of the difference between two estimated treatment effects is 2×0.7809; this latter variance, as argued above, is twice the effective variance which we are trying to estimate and so finally we conclude that

$$2 \times \text{effective variance} = 2 \times 0.7809,$$
$$\text{effective variance} = 0.7809,$$
$$\text{e.s.e.} = \sqrt{(\text{effective variance})} = 0.8837,$$

confirming the value of 0.884 given in the output.

The residuals for the centres.blocks.treatments stratum are calculated from the two-way table centres.blocks × treatments in exactly the same way as was illustrated in Table 44. The value of rep. is 1 because each residual measures the deviation of a single observation from the value given by the fitted model and the calculation for s.e. is quite like that given in the centres section above but is more elaborate because we are now dealing with a two-way table.

The commands and data for the second example are in CTable 27. Following the pattern in Tables 48b and c, analyses are set up for individual patients as well as overall; this is not essential – in the desk calculator version the individual patient analyses were carried out as much for convenience as for the information they gave – but gives the opportunity to illustrate another Genstat facility. A further facility is used in the individual patient analyses where only single-factor terms are identified so that the technicians × instruments interaction is not labelled as such but is placed in the residual; in the overall analysis this option is not in use and the interaction term is given separately and identified.

CTable 27. *Commands and data for a Genstat analysis appropriate to the second example in 7.3.3.*

```
units [nvalues = 60]
factor [levels = 4] patients
factor [levels = 3] technicians
factor [levels = 5] instruments
read        patients, technicians, instruments, observations
1 1 1  7.6   1 1 2 11.9   1 1 3  2.6   1 1 4  3.5   1 1 5  5.9
1 2 1 12.0   1 2 2 12.0   1 2 3  3.1   1 2 4  7.8   1 2 5 10.1
            ..... and so on through the data to .....
4 2 1 10.7   4 2 2  4.9   4 2 3 12.0   4 2 4 11.1   4 2 5 11.9
4 3 1  3.1   4 3 2  3.2   4 3 3  8.5   4 3 4  6.4   4 3 5  3.8 :
blocks       patients
treatments   technicians*instruments
for          PA = 1...4
   restrict           observation; patients .eq. PA
   anova [fprob = yes; factorial = 1] observation
endfor
restrict     observation
anova [fprob = yes] observation
```

The statements setting up these aspects of the analyses are

```
for          PA = 1 . . . 4
   restrict           observation; patients.eq.PA
   anova [fprob = yes; factorial = 1] observation
endfor
restrict           observation
anova [fprob = yes] observation
```

The for statement defines a quantity to be known as PA, which can take the values 1, 2, 3 and 4, and the first restrict statement requires that the variable observation shall be considered only for the patient identified by the current value of the label PA. The first anovar statement calls for analysis of the variable observation restricted in this way and also introduces the option factorial = 1. The general form of this option is

factorial = n, where n can take integer values from 1 to 9;

its effect is to limit the calculation of separate terms (main effects, interactions, tables of means, etc.) to those of order $\leq n$; the default setting is $n = 3$. The for loop causes separate analyses to be carried out for each of the values specified by restrict, which in this case means for each of the four patients; the loop sequence of calculations is terminated by the endfor statement. The second restrict statement, in which no qualifier is given (i.e. no expression beginning .eq.), has the effect of removing all restrictions on the values of observation to be included in the analysis so that the following anovar statement applies to all the values.

The output for the analysis for patient 1 is in CTable 28a and that for the overall analysis in CTable 28b. The layout for patient 1 follows the by now familiar pattern, except for the insertion of a new short second section entitled

information summary

This section appears† whenever the information concerning some factor or contrast is incomplete. This occurs in a variety of design circumstances, for example in cases of non-orthogonality (see 9.4.4) or because one or more of the items declared is *aliased*, i.e. is either estimated by the same expression as another item or cannot be estimated at all; in the present analysis, with data from only one patient included, the patients stratum contains no terms. In the analysis of variance, because of the use of the option factorial $= 1$, the s.s. and degrees of freedom corresponding to the technicians × instruments interaction are placed in residual; this would otherwise be empty, in which case v.r.'s could not be calculated for the two main effects nor could s.e.d.'s be given in the final section.

In the analysis of variance in CTable 28b, with factorial at its default setting, the interaction is calculated and is tested, as are the main effects, against the residual for the patients.∗units∗ stratum. These tests differ from those in 7.3.3 where the three interactions involving patients were kept separate and each was used in the test of the fixed factor terms in it. In the author's

---

† As we saw earlier in this section the summary will not appear unless the print option to anova, either by setting or default, includes the value **I**.

judgement, that procedure is to be preferred, because under it each fixed factor effect is tested against a term which, involving only it and the random factor, is more obviously applicable to it than is a measure of variation pooled over all the fixed factor effects.

The remaining sections of CTable 28*b* follow the patterns illustrated in earlier tables of output.

CTable 28a. *Output for the Genstat analysis for patient 1 only in the second example of 7.3.3.*

************ Analysis of variance ************

Variate: observat

Source of variation	d.f.	s.s.	m.s.	v.r.	F pr.
patients.*Units* stratum					
technici	2	27.972	13.986	2.76	0.122
instrume	4	86.137	21.534	4.25	0.039
Residual	8	40.515	5.064		
Total	14	154.624			

************ Information summary ************

Aliased model terms
patients stratum

************ Tables of means ************

Variate: observat

Grand mean   7.08

technici	1	2	3
	6.30	9.00	5.94

instrume	1	2	3	4	5
	8.63	9.67	2.90	6.07	8.13

********** Standard errors of differences of means **********

Table	technici	instrume
rep.	5	3
s.e.d.	1.423	1.837

CTable 28b. *Output for the Genstat analysis over all patients in the second example in 7.3.3.*

```
************ Analysis of variance *************
```

Variate: observat

Source of variation	d.f.	s.s.	m.s.	v.r.	F pr.
patients stratum	3	6.75	2.25		
patients.*Units* stratum					
technici	2	33.05	16.52	1.33	0.275
instrume	4	77.32	19.33	1.56	0.203
technici.instrume	8	34.32	4.29	0.35	
Residual	42	520.87	12.40		
Total	59	672.31			

```
************ Tables of means *************
```

Variate: observat

Grand mean   7.25

technici	1	2	3		
	7.03	8.24	6.46		

instrume	1	2	3	4	5
	7.54	8.18	5.49	6.45	8.57

technici instrume	1	2	3	4	5
1	8.17	8.92	5.72	4.82	7.52
2	8.42	9.02	6.07	7.40	10.30
3	6.02	6.60	4.67	7.12	7.90

```
*********** Standard errors of differences of means ***********
```

Table	technici	instrume	technici instrume
rep.	20	12	4
s.e.d.	1.114	1.438	2.490

7.4 Surveys

The purpose of this section is to give formulae which are related to those developed in chapter 6 and the earlier parts of this chapter but which may be applied in the special circumstances of a survey (1.1.1, 1.2, 1.2.1). All the discussion so far in these chapters has been expressed in terms of the analysis of the results of experiments, and one essential feature of such analysis is that we seek to draw conclusions with wide applicability. In the chick feeding trial with lysine (7.3) we wanted the findings about the diets to be relevant to 'all' young chicks; in the trials discussed in 7.3.3, one repeated over several centres and the other over several patients, we intended our conclusions to apply to 'all' centres and 'all' patients. The term 'all' in these cases can in fact mean only those populations of chicks or centres or patients of which we have taken representative samples, but the implication is that these populations are large.

In contrast to this, when we are conducting a survey we want our conclusion to apply to one definite *target population*, which we are able to specify precisely and which will often be quite small – at least by comparison with the population of 'all' chicks, centres or patients that an experimenter has in mind. Sampling from a small population involves considerations of point and interval estimation and of significance testing exactly similar to those treated above, but we have to make some changes in our formulae to allow for the known size of the population. A considerable literature on sampling has been built up in recent years, and we will here limit our consideration to (*a*) a simple random sample, with no subdivisions in the target population, (*b*) a stratified sample with only one level of stratification (that is, no strata within strata), and (*c*) a two-stage sample, giving only a few simple formulae in each case.

7.4.1 Simple random sample

Suppose that we have taken a random sample of n from a population of N individuals; we shall see that N enters our formulae, and so it is necessary for us to know the size of the target population. We refer to the proportion n/N as the *sampling fraction*, and denote it by f. We denote the values of our observations on the n members of the sample by $x_1, x_2, \ldots, x_n$ – we may of course have measured several different quantities, but we shall here consider only single variable formulae.

The mean of the target population is estimated in the usual

way by the sample average $\bar{x} = \sum x/n$. In estimating the variance of $\bar{x}$ we do not however use the formula s^2/n, where

$$s^2 = \frac{\sum (x - \bar{x}^2)}{n - 1} = \frac{\sum x^2 - \dfrac{(\sum x)^2}{n}}{n - 1},$$

but rather write

$$\text{estimated variance } (\bar{x}) = (1 - f)\frac{s^2}{n}.$$

This estimated variance is used as before in obtaining an interval estimate for the population mean.

7.4.2 Stratified sample

In the case of a stratified sample, suppose we have selected random samples of $n_1, n_2, \ldots, n_k$ from population strata containing $N_1, N_2, \ldots, N_k$ individuals. We denote the sampling fractions by $f_1 = n_1/N_1, f_2 = n_2/N_2, \ldots, f_k = n_k/N_k$, and the total sample and population sizes by $n = \sum\limits_{i=1}^{k} n_i$ and $N = \sum\limits_{i=1}^{k} N_i$. We calculate the means $(\bar{x}_1, \bar{x}_2, \ldots, \bar{x}_k)$ and variance estimates $(s_1^2, s_2^2, \ldots, s_k^2)$ for the separate strata, where

$$\bar{x}_i = \frac{1}{n_i} \times (\text{total of the } n_i \text{ observations in stratum } i)$$

$$s_i^2 = \frac{1}{n_i - 1} \times (\text{s.s. of the observations in stratum } i).$$

The estimate $\hat{m}$ of the population mean is given by

$$\hat{m} = \frac{N_1}{N}\bar{x}_1 + \frac{N_2}{N}\bar{x}_2 + \cdots + \frac{N_k}{N}\bar{x}_k = \sum_{i=1}^{k}\left(\frac{N_i}{N}\right)\bar{x}_i$$

and the estimated variance of $\hat{m}$ is

$$\left(\frac{N_1}{N}\right)^2 (1 - f_1)\frac{s_1^2}{n_1} + \left(\frac{N_2}{N}\right)^2 (1 - f_2)\frac{s_2^2}{n_2} + \cdots + \left(\frac{N_k}{N}\right)^2 (1 - f_k)\frac{s_k^2}{n_k}$$

$$= \sum_{i=1}^{k}\left(\frac{N_i}{N}\right)^2 (1 - f_i)\frac{s_i^2}{n_i}.$$

It may be proved that, for a given total sample size n, this variance is made as small as possible if we take the sampling fraction for stratum i proportional to the standard deviation (s_i) for the stratum. In many cases we do not know these standard deviations until after we have carried out the survey, and so can make no use of this result, but if we do have some idea of them either from past observations or from a pilot survey, then the result is valuable. It means that if, for example, the standard deviation in stratum i is twice that in stratum j then the sampling fraction for i should also be twice that for j.

7.4.3 Two-stage sample

In defining a two-stage sample, suppose that the population is made up of K first-stage units containing $N_1, N_2, \ldots, N_K$ individuals. We will consider two of the many possible selection procedures.

In the first case, suppose that the first-stage units are all the same size, $N_1 = N_2 = \cdots = N_K = N^*$. Select k of these units at random, giving first-stage sampling fraction $f_1 = k/K$, and select n individuals at random from each of these chosen k units, to give second-stage sampling fraction $f_2 = n/N^*$. Let $\bar{y}_1, \bar{y}_2, \ldots, \bar{y}_k$ be the means of the observations on the chosen first-stage units, and let B and R be the between units and residual (within units) M.S.'s in the single-factor anovar of the total sample of nk individuals; the terms in this anovar are

Between chosen first-stage units	$k - 1$
Residual	$k(n - 1)$
Total	$kn - 1$.

Then the estimate, $\hat{m}$, of the population mean is

$$\hat{m} = \frac{\bar{y}_1 + \bar{y}_2 + \cdots + \bar{y}_k}{k} = \frac{1}{k} \sum_{i=1}^{k} \bar{y}_i$$

and the estimated variance of $\hat{m}$ is

$$\frac{1 - f_1}{kn} B + \frac{f_1(1 - f_2)}{kn} R.$$

In the second procedure, which can be used when the first-stage units are not all the same size, we select k first-stage units at random *with replacement* and *with probability proportional to size* of unit† and then take random samples all of size n from the k

---

† Suppose for example that we wish to choose $k = 2$ units from 8 units whose sizes are in the proportions 3:2:4:1:5:3:4:6. We assign the numbers 1, 2, 3, ... to these units as labels, giving the first three numbers $(1, 2, 3)$ to the first unit, the next 2

chosen units; the effect of this selection procedure is to give all the individuals in the population the same chance of being included in the final sample. As before let $\bar{y}_1, \bar{y}_2, \ldots, \bar{y}_k$ be the means of the observations in the k chosen units. Then the estimate, $\hat{m}$ of the population mean is

$$\hat{m} = \frac{\bar{y}_1 + \bar{y}_2 + \cdots + \bar{y}_k}{k} = \frac{1}{k} \sum_{i=1}^{k} \bar{y}_i$$

and the estimated variance of $\hat{m}$ is

$$\frac{\sum_{i=1}^{k} (\bar{y}_i - \hat{m})^2}{k(k-1)} = \frac{1}{k(k-1)} \left(\sum_{i=1}^{k} \bar{y}_i^2 - \frac{\left(\sum_{i=1}^{k} \bar{y}_i \right)^2}{k} \right),$$

that is the variance estimate for the mean of the $\bar{y}$'s calculated directly from the $\bar{y}$'s themselves.

The examples of procedure and calculation given in these sections are only a very small part of present knowledge of survey methods and anyone interested in undertaking a survey should consult one of the many specialized books available, for example Yates (1981), Sampford (1962), Cochran (1977).

7.5 Exercises

Exercises 1–5 require the use of single-factor anovar, discussed in 7.2 et sqq. In each case the analysis should begin with the drawing of histograms, and, when the anovar has been carried out, Bartlett's test and the Kolmogorov–Smirnov one-sample test should be used to check how far the distributional conditions required by the anovar are satisfied. Suggests for comparisons between levels are given, but the reader may well wish to consider others.

(4, 5) to the second unit, the next 4 (6, 7, 8, 9) to the third unit, and so on, giving each unit as many labels as its size proportion requires; the complete set of labels is thus

Unit	1	2	3	4	5	6	7	8
Size	3	2	4	1	5	3	4	6
Labels	1–3	4, 5	6–9	10	11–15	16–18	19–22	23–28

We now select at random two numbers from the range $1, \ldots, 28$ and take as our two units those for which the chosen numbers are labels; thus if we obtain 6 and 23 we take as chosen units the 3rd (corresponding to label 6) and 8th (corresponding to label 23). Sampling with replacement means that if the two numbers chosen happen to be labels for the same unit we use this unit twice in the sample.

1 Three dose levels of an appetite stimulant were compared with a control in an experiment using individually caged rats; the results give the total food consumptions (kg) for individual rats over the period of the experiment. Twelve rats were allocated to each treatment at random, with no attempt to achieve any kind of balance; the rats were then placed at random in the available cages. The values obtained were as follows:

| | Dose-level of appetite stimulant (per cent) | | |
Control	0.5	1	2
2.79	4.90	5.90	5.82
2.58	4.63	5.87	6.44
2.57	3.94	4.86	6.32
2.14	4.46	5.85	6.25
3.02	4.82	6.01	6.08
3.72	5.58	5.69	6.00
3.06	4.47	5.63	5.79
3.19	4.21	6.13	5.86
4.20	4.19	5.52	6.21
2.72	4.76	5.78	5.23
3.51	5.07	5.65	6.78
3.06	4.21	4.68	6.76

If the overall F-test in the anovar is significant the following comparisons between levels seem worth considering:

    (*a*) between the three dose levels of the stimulant,

    (*b*) between the control and the average of the three levels of the stimulant,

    (*c*) between levels 0.5 and 1 per cent, and between levels 1 and 2 per cent.

[Simple methods for evaluating the dose-response relationship are given in later exercises.]

2 The effect of selection for weight gain on the dressed weight of broiler chickens was determined after 5 generations of selection. Group *A* was bred by using only the heaviest 10 per cent in each generation: groups *B* and *C* were bred using respectively the heaviest 30 and 50 per cent: group *D* was obtained by crossing groups *A* and *C* of the previous generation. The dressed weights (kg) of 25 birds from each group were as follows:

A	B	C	D	A	B	C	D
1.66	1.50	1.47	1.60	1.65	1.67	1.50	1.61
1.63	1.65	1.37	1.64	1.39	1.45	1.55	1.76
1.59	1.54	1.47	1.40	1.47	1.48	1.53	1.41
1.69	1.58	1.42	1.53	1.59	1.47	1.51	1.35
1.49	1.55	1.30	1.58	1.61	1.40	1.58	1.59
1.66	1.47	1.69	1.57	1.49	1.46	1.56	1.47
1.51	1.43	1.41	1.64	1.48	1.66	1.47	1.41
1.57	1.42	1.40	1.50	1.69	1.59	1.63	1.47
1.50	1.67	1.40	1.45	1.61	1.32	1.59	1.61
1.45	1.41	1.38	1.51	1.49	1.44	1.50	1.47
1.53	1.52	1.51	1.53	1.54	1.81	1.49	1.59
1.69	1.48	1.60	1.71	1.43	1.48	1.51	1.39
1.45	1.45	1.31	1.57				

After the overall F-test the following comparisons could be helpful:

   (a) between A and B,
   (b) between B and C,
   (c) between B and D,
   (d) between $(A - B)$ and $(B - C)$; this contrast can be simplified as follows:

$$(A - B) - (B - C) = A - 2B + C.$$

This is an example of the general comparison with one degree of freedom, described in 7.2.2.3; if it is not significant it shows that the effect of the 20 per cent difference in selection between A and B can be regarded as the same as the effect of the 20 per cent difference in selection between B and C, so that the effect of selection is *linear* over this range. The contrast is therefore referred to as a test of *linearity*.

3 As part of a study of the newborn lamb, the level of immunoglobulin in serum (g/100 ml) was measured 24 hours *post partum* in individuals from 3 breeds, with the following results:

287

X	Y	Z
1.1	1.7	2.1
2.2	2.3	2.5
1.7	1.8	2.8
1.4	2.3	2.1
1.6	2.5	2.0
2.3	2.4	2.2
1.4	2.5	2.7
1.9	1.9	2.3
0.8	2.2	2.6
1.6	2.0	2.4
1.2	2.4	1.7
1.6	3.0	1.4
1.6	2.0	1.7
1.6		2.4
1.5		2.6
1.9		

Apart from the overall comparison between the three breeds, the only comparisons to consider are those between the breeds in pairs, X against Y, Y against Z, and X against Z.

4 The level of total protein (g/100 ml) was measured in the blood of cows receiving either a control ration or one of three rations supplemented with synthetic protein. The values obtained were as follows:

| Control | Synthetic supplement | | |
	P	Q	R
7.82	7.64	7.67	7.98
6.64	7.04	7.58	7.91
7.69	7.43	7.04	7.11
7.52	7.57	6.69	7.41
7.27	7.74	7.32	6.91
7.18	7.63	7.12	8.17
6.35	8.06	7.46	8.28
7.61	7.94	7.21	6.91
6.56	7.83	7.76	7.65
7.33	7.73	7.33	7.91
7.33		7.48	6.37
		7.63	7.21

Apart from the overall test, possible comparisons are:

(a) between the control group and the average of the other 3 groups, (*b*) between groups *P*, *Q* and *R*,

(*c*) between one group and another in any of the 6 possible pairings.

5 As part of a study of the effects of calcium level in the diet of ewes, the fresh weight (g) of the semitendinosus muscle was measured on carcasses from six treatment groups. The treatments were

A	no mineral supplement,
B, C, D	three supplements offered as licks,
E, F	two supplements added to the concentrate ration.

The measurements obtained were as follows:

A	*B*	*C*	*D*	*E*	*F*
61.8	65.4	65.7	20.8	51.0	43.3
46.0	53.0	48.1	54.5	58.5	54.2
21.7	50.9	64.2	41.0	58.3	51.5
25.2	28.6	67.9	51.1	57.7	77.3
48.8	81.5	42.3	42.5	71.3	70.1
47.7	44.7	62.0	35.5	73.6	69.5
39.0	64.5	44.3	78.1	67.6	63.1
37.2	38.0		57.3		83.2

In addition to the overall comparison, possible comparisons are

(*a*) between *B*, *C* and *D*,

(*b*) between *E* and *F*,

(*c*) between *A*, the average of *B*, *C*, *D* and the average of *E*, *F*.

When several comparisons are made within a set of treatments, as in the exercises above, there is often some difficulty in interpreting the results because the comparisons are not independent. One simple device which can be helpful in detecting this circumstance is to count the degrees of freedom of the comparisons. Thus in exercise 4, comparison (a) has 1 d.f., comparison (b) has 2 d.f. and the 6 comparisons in (c) have 1 d.f. each: these comparisons thus total 9 d.f. and clearly therefore cannot be independent, since the comparison between treatments in the main anovar has only 3 d.f., which is the maximum number of independent comparisons that are possible. In fact, in this exercise, comparisons (a) and (b) do provide a division of the 3 d.f. overall comparison into two independent terms – as is shown by the fact that the degrees of freedom and the sums of squares for these comparisons sum to give respectively the degrees of freedom and the sum of squares of the overall comparison – but the more general question of how to divide a comparison into independent terms is beyond the scope of this elementary discussion. If comparisons are found not to be independent, considerable care is needed in their interpretation; if at all possible, their use should be avoided.

Exercises 6–9, whilst arising from single-factor designs, involve some slightly more complex procedures than did the exercises above.

6 In a comparison between 4 wild species of potato one of the characters measured was the number of glandular hairs/mm^2 on the upper leaf surface. Twenty plants from each of the 4 species gave the following values:

A	B	C	D	A	B	C	D
2.3	8.1	4.8	11.6	1.1	3.7	4.9	6.4
1.7	8.2	7.9	9.7	1.8	4.8	9.4	15.2
1.7	6.4	3.1	8.3	1.0	1.6	2.9	10.5
1.8	3.3	12.5	11.9	1.5	3.1	5.4	11.2
0.7	3.2	8.1	12.6	1.3	5.1	3.8	8.4
1.2	5.2	7.2	5.5	0	5.2	6.1	6.5
2.1	7.4	6.0	11.2	1.7	7.5	5.9	7.4
2.7	5.0	7.9	4.2	1.3	6.5	5.4	14.6
0.6	9.9	6.1	6.1	1.6	6.7	6.8	7.8
1.1	3.7	8.9	8.1	1.7	6.9	5.9	11.1

This variable is not suitable for analysis because the variance increases with the mean, but this can be rectified, without unduly affecting the comparisons between individual species, by the transformation $x = \log(1 + \text{number})$; the discussion in exercise 6.8(18) is relevant.

After comparing the 4 species in the overall F-test, individual comparisons can be made between the species two at a time: as explained above, these comparisons are not independent.

7 In an experiment on the effect of progesterone on the length of the oestrous cycle in the Merino ewe, subcutaneous injections, using 4 dose levels, were given for 4 days starting on the day of oestrous. Eight ewes were tested at each dose level: the cycle lengths (days) were as follows:

Dose level (mg/day)			
0	10	25	40
18	15	11	9
14	14	13	10
18	17	11	12
18	14	11	10
18	12	12	11
18	13	11	11
18	12	11	10
19	13	12	11

Possible comparisons, after the overall F-test, are:

   (a) between levels 10, 25 and 40,

   (b) between level 0 and the average of 10, 25 and 40,

[These two comparisons are independent.]

   (c) level $10 - 2 \times$ level $25 +$ level 40,

[This is a test of linearity, as in exercise 2.]

   (d) level $40 -$ level 10.

[This is a test of average linear effect over the three treatment levels: comparisons (c) and (d) are independent and together are equivalent to (a). Thus (b), (c) and (d), each 1 d.f. comparisons, are one possible way of dividing the overall treatment difference, which has 3 d.f., into 3 independent comparisons.]

8 In another version of the experiment in exercise 7, the 4 dose levels were also used for injections starting on the day after oestrous; the treatments thus formed a two-factor table as follows:

Day on which injections commenced	Dose level (mg/day)			
	0	10	25	40
0				
1				

All the possible eight treatment combinations were used; the term *factorial* is applied for such an arrangement. By allocating 4 ewes to each treatment combination, the total number of ewes was the same as in the experiment of exercise 7; the cycle lengths were as follows:

	Dose level			
Commencement	0	10	25	40
0	17	15	12	8
	18	15	12	9
	17	14	11	11
	17	16	11	6
1	18	16	16	12
	20	14	14	13
	17	16	11	12
	14	16	14	12

The design is still of the single-factor type, with 8 treatment levels, the overall F-test indicating whether there are any differences between these levels. Since, however, these levels themselves form a two-factor arrangement, the comparisons between levels are more complex than in the earlier exercises, and will be discussed in greater detail. Denote the various totals of the observations by the following notation:

	Dose level				
Commencement	0	10	25	40	
0	$T_{0,0}$	$T_{0,10}$	$T_{0,25}$	$T_{0,40}$	R_0
1	$T_{1,0}$	$T_{1,10}$	$T_{1,25}$	$T_{1,40}$	R_1
	C_0	C_{10}	C_{25}	C_{40}	T

As a first step we will consider the 'main effect' comparisons:

(a) The effect of day of commencement is tested by comparing R_0 and R_1.

(b) The effect of dose level is tested by comparing C_0, C_{10}, C_{25} and C_{40}: the comparisons described in exercise 7 are relevant.

These contrasts have 1 d.f. (*a*) and 3 d.f. (*b*): the overall comparison between the eight T_{ij}'s has 7 d.f., the remaining 3 ($= 7 - 1 - 3$) being due to the interaction between dose level and day of commencement. We can calculate the s.s. for this term from the expression

$$\begin{array}{cccc} \text{interaction} = & \text{overall} & - \text{commencement} & - \text{dose level} \\ \text{s.s. (3 d.f.)} & \text{s.s. (7 d.f.)} & \text{s.s. (1 d.f.)} & \text{s.s. (3 d.f.)} \end{array}$$

The interaction can be divided in exactly the same way as the main effect for dose-level – if the other main effect (commencement) could be divided then the interaction could be divided to correspond to both main effect divisions at the same time. Since the commencement main effect has 1 d.f., the calculation of the various interaction terms is very simple, and we will consider only two examples (the reader may like to re-read the first half of 7.3.2): consider first contrast (*a*) from exercise 7, already used in the comparison of dose-level totals in (*b*) of this exercise. The calculation for the main effect contrast is

$$\frac{C_{10}^2}{8} + \frac{C_{25}^2}{8} + \frac{C_{40}^2}{8} - \frac{(C_{10} + C_{25} + C_{40})^2}{24}.$$

Remembering that $C_{10} = T_{0,10} + T_{1,10}$, etc., we can write this expression as

$$\frac{(T_{0,10} + T_{1,10})^2}{8} + \frac{(T_{0,25} + T_{1,25})^2}{8} + \frac{(T_{0,40} + T_{1,40})^2}{8}$$
$$- \frac{(T_{0,10} + T_{1,10} + T_{0,25} + T_{1,25} + T_{0,40} + T_{1,40})^2}{24}.$$

The interaction is approached by considering the differences

$$T_{0,10} - T_{1,10},$$
$$T_{0,25} - T_{1,25},$$
$$T_{0,40} - T_{1,40};$$

these are used exactly as the totals

$$T_{0,10} + T_{1,10},$$
$$T_{0,25} + T_{1,25},$$
$$T_{0,40} + T_{1,40}$$

in the main effect contrast.

293

Thus for the s.s. for 'commencement × contrast between levels 10, 25 and 40' we calculate

$$\frac{(T_{0,10} - T_{1,10})^2}{8} + \frac{(T_{0,25} - T_{1,25})^2}{8} + \frac{(T_{0,40} - T_{1,40})^2}{8}$$
$$- \frac{(T_{0,10} - T_{1,10} + T_{0,25} - T_{1,25} + T_{0,40} - T_{1,40})^2}{24}.$$

As a second example of the division of the interaction term, consider the linearity contrast used in (c) of exercise 7 and in (b) of this exercise.

The s.s. for the main effect contrast is

$$\frac{(C_{10} - 2C_{25} + C_{40})^2}{48},$$

the divisor being calculated as

$$1^2 \times 8 + (-2)^2 \times 8 + 1^2 \times 8$$

as in the general formula in 7.2.2.3.

Writing, as before, $C_{10} = T_{0,10} + T_{1,10}$, etc., gives

$$\frac{(T_{0,10} + T_{1,10} - 2(T_{0,25} + T_{1,25}) + T_{0,40} + T_{1,40})^2}{48}.$$

The interaction s.s., approached as before by using $T_{0,10} - T_{1,10}$, etc., is

$$\frac{(T_{0,10} - T_{1,10} - 2(T_{0,25} - T_{1,25}) + T_{0,40} - T_{1,40})^2}{48}.$$

The reader should consider how these calculations relate to the more general method of dividing an interaction described briefly towards the end of 7.3.3.

All the contrasts discussed in this exercise are parts of the overall 7 d.f. contrast between treatment combinations, and therefore are tested against the (within-treatment) residual used in testing that contrast.

9 The daily herbage dry matter consumption (kg) of grazing cattle was estimated by two methods; the cattle received in addition to grazing, one of three concentrate mixtures. The treatment combinations thus formed a 2 × 3 factorial arrangement, similar to that discussed in detail in exercise 8. Six cattle were used on each treatment combination; the results were as follows:

Method of estimating consumption	Concentrate mixture		
	R	S	T
A	8.3	9.9	10.2
	9.4	10.4	7.5
	9.4	8.6	7.8
	9.4	10.8	11.1
	9.4	10.0	8.0
	11.3	6.8	9.7
B	6.7	6.3	9.6
	8.7	8.6	4.9
	7.6	8.3	8.1
	7.4	7.4	7.3
	8.5	8.8	3.1
	10.6	10.2	8.1

The overall analysis is determined by the fact that the design is a single-factor arrangement, the overall comparison between treatment combinations having 5 d.f. This divides into

Mixtures	2
Methods	1
Mixtures × Methods	2

The interaction can be regarded either as measuring whether the effect of the mixtures on herbage consumption is estimated as the same by both methods, or as measuring whether the difference between methods is effectively the same with all mixtures. Possible individual comparisons within the mixtures main effect are between the mixtures two at a time: the interaction can be divided correspondingly.

Exercises 10–16 *use two-factor designs, discussed in 7.3 et sqq.; once again Bartlett's test and the Kolmogorov–Smirnov one-sample test can be used to check the distributional conditions. The idea of factorial arrangement of treatments, introduced above in exercises 8 and 9, is applied in some of the two-factor designs.*

10 To assess the usefulness of 3 new varieties (B, C, D) of daffodil for the cut-flower market, samples of each variety and of one standard variety (A) were grown àt each of 6 sites. The variable recorded was the date (days after 1 January) by which half the heads on a plot were ready for cutting. The results were as follows:

	Variety			
Site	A	B	C	D
1	43	34	36	37
2	46	30	35	43
3	46	33	28	40
4	42	25	27	33
5	40	32	29	39
6	37	22	28	35

After the overall between varieties F-test, the first individual comparisons to consider are of each of the new varieties against the standard variety.

[Each of these comparisons has 1 d.f., so between them they may be thought to account for the 3 d.f. between varieties; however, it will be found that the s.s. for these individual comparisons do not sum to that for the between varieties term, showing that these three comparisons are not independent – compare exercise 5 above.]

11 In an experiment on maize grown as a fodder crop, 5 plant densities were tested in a randomized block design with 5 blocks. The yields (kg of dry matter/plot) were as follows:

	Plant density (plants/unit area)				
Blocks	20	25	30	35	40
I	21.1	26.8	30.4	28.4	27.6
II	16.7	23.8	25.5	28.2	24.5
III	14.9	21.4	27.1	25.3	26.5
IV	15.5	22.6	26.3	26.5	27.0
V	19.7	23.6	26.6	32.6	30.1

The comparisons between treatments which are relevant in this case are those which help to make clear the form of the relationship between yield and plant density. Examples of this kind of comparison have been given above (exercises 2 and 7), testing the significance of an average linear relationship and of departures from linearity; the present example, however, involves more levels than those considered above and requires slightly more elaborate methods. A full discussion is beyond the scope of this book, but mention can be made of *orthogonal polynomials* described more fully in, for example, Fisher and Yates (1974). The object of this method is to provide a sequence of tests to determine whether

(*a*) the average linear response (i.e. the best straight line relating yield to density) accounts for a significant part of the between levels s.s.;

(*b*) the best quadratic relationship is a significant improvement over (*a*);

(*c*) the best cubic relationship is a significant improvement over (*b*);

and so on – Fisher and Yates (1974) consider relationships up to the quintic. The coefficients $a_1, a_2, \ldots, a_5$ for use, as in section 7.2.2.3, in the present case of equally spaced levels are:

Linear	-2	-1	0	1	2
Quadratic	2	-1	-2	-1	2
Cubic	-1	2	0	-2	1
Quartic	1	-4	6	-4	1

To illustrate the use of these coefficients, consider the calculation of the s.s. for the linear contrast. Denoting the totals of the observations for each plant density by $D_{20}, D_{25}, D_{30}, D_{35}, D_{40}$ – each a total of 5 observations – the s.s. is (by the formula in 7.2.2.3)

$$\frac{[(-2) \times D_{20} + (-1) \times D_{25} + 0 \times D_{30} + 1 \times D_{35} + 2 \times D_{40}]^2}{(-2)^2 \times 5 + (-1)^2 \times 5 + 0^2 \times 5 + 1^2 \times 5 + 2^2 \times 5}$$

$$= \frac{(-2D_{20} - D_{25} + D_{35} + 2D_{40})^2}{50}.$$

Each of these contrasts has 1 d.f. and it will be found that their s.s. sum to the 4 d.f. s.s. between plant densities, thus confirming that the contrasts are independent.

12 In an experiment to compare the yields of leys grown from 4 seed mixtures, a randomized block design was laid down with nine blocks; the dry matter yields (kg/plot) were as follows:

Block	Seed mixture			
	A	B	C	D
I	20.3	15.2	22.1	22.9
II	23.5	14.9	23.1	16.1
III	20.0	18.7	23.0	23.8
IV	22.7	12.4	17.8	18.9
V	19.1	16.8	19.9	19.5
VI	19.1	14.5	17.6	17.0
VII	19.7	15.0	26.0	17.2
VIII	19.3	15.1	21.9	20.4
IX	23.8	15.3	22.3	21.6

There are no groupings of the mixtures that can meaningfully be used, so that the only interesting comparisons, after the overall test, are between the mixtures two at a time. There are six such comparisons possible and, since each has 1 d.f. whilst the overall comparison has only 3 d.f., it is clear that the six comparisons are not independent: as is indicated in 7.2.2.5, one way of dealing with this situation is to use a multiple range test.

13 In a more complex version of the experiment in exercise 12, the 4 seed mixtures were combined in a factorial arrangement with 3 seed rates, giving 12 treatment combinations. These were arranged in 3 blocks of 12 plots, thus using the same total number of plots as in exercise 12. Since the blocks were larger in this experiment, we may expect it to prove slightly less sensitive, but this well may be compensated for by the fact that two main effects (seed rate as well as seed mixture) and the interaction between them are all estimable, whereas in the first experiment only one main effect could be estimated.

The yields (kg of dry matter/plot) were as follows:

	Seed rate (g/unit area)	Seed mixture			
		A	B	C	D
Block I	4	21.6	8.1	15.2	16.9
	6	19.4	15.3	19.4	20.9
	8	17.6	19.2	21.8	18.5
Block II	4	18.9	5.0	13.5	17.2
	6	19.4	11.1	14.5	16.2
	8	19.8	13.8	19.1	18.7
Block III	4	17.6	9.6	11.4	15.0
	6	19.4	10.2	17.6	20.0
	8	21.5	16.2	22.5	20.2

The overall anovar has the form

Treatments	11
Blocks	2
Residual	22
Total	35

showing that there is a small loss in residual d.f. by comparison
with the design in exercise 12; this illustrates further why the
present design may be expected to be slightly less sensitive. The
overall term for treatments can be divided into

Seed mixtures	3
Seed rates	2
Mixtures × rates interaction	6
Treatments	11

The s.s. are calculated by the methods described above. The seed
rates terms could be split into 'linear' and 'deviations from
linearity' terms, each with 1 d.f., as in earlier exercises, and this
same decomposition could be used in dividing the interaction.
All the contrasts would be tested against the residual M.S. because
this was used in testing the overall treatment term.

14 The effects of 4 cultivation sequences on the yield of sugar
from sugar-beet were tested in a randomized block design with
10 blocks. The yields (tonnes/ha) were as follows:

Blocks	Cultivation sequence			
	P	Q	R	S
I	3.04	5.04	5.20	8.89
II	4.43	7.12	6.98	9.79
III	6.47	7.52	7.73	8.26
IV	2.30	3.15	4.50	5.76
V	4.44	7.85	7.84	12.87
VI	1.83	7.82	7.59	7.68
VII	2.86	6.78	8.33	7.78
VIII	4.75	7.03	7.80	10.75
IX	6.90	6.26	8.09	9.89
X	1.51	8.23	6.16	7.86

As in exercise 12, the only individual comparisons that can be considered are between the sequences in pairs.

15 In a more elaborate version of the trial analysed in exercise 14, the four cultivation sequences were combined in a factorial arrangement with two systems of weed control, and the eight treatment combinations were arranged in a randomized block design in 5 blocks. The yields (tonnes/ha) were as follows:

	Weed control system	Cultivation sequence			
		P	Q	R	S
Block I	V	4.73	4.65	8.70	10.78
	W	4.36	6.02	9.29	13.13
Block II	V	3.23	6.29	3.69	6.95
	W	2.10	5.92	9.42	10.09
Block III	V	3.69	6.12	6.41	8.91
	W	6.99	8.33	8.52	11.21
Block IV	V	4.21	6.48	6.64	9.04
	W	5.78	9.52	10.11	11.53
Block V	V	2.92	7.42	6.16	6.16
	W	6.21	9.01	10.61	9.06

The treatment s.s. with 7 d.f. is divided into the two main effects (cultivation sequences with 3 d.f. and weed control systems with 1 d.f.) and the interaction between these factors (with 3 d.f.). The cultivation s.s. can be split to give comparisons between sequences in pairs, as in exercise 14, and the interaction can be split into corresponding terms using methods similar to those considered in exercise 8.

16 An experiment to compare the effects of four fertilizer treatments on the yield of strawberries was laid out in a randomized block design with 9 blocks. The treatment combinations were the factorial arrangement of two fertilizers, A and B, each used at two levels – in the jargon it is conventional to allow the term level to be used for the absence of a factor, so that the expression 'two levels' may mean either absence and presence or presence at two dose levels. The factorial combination of two factors each at two levels is referred to as a 2^2 *factorial*; this is the simplest case of the family of 2^n *factorial* arrangements, in which n factors, each at 2 levels, are used in all possible combinations. Two notations, both widely used to represent the treatment combinations in a 2^n factorial arrangement, are illustrated by application to the present experiment thus

Low level of A with low level of B	(1)	00
High level of A with low level of B	a	10
Low level of A with high level of B	b	01
High level of A with high level of B	ab	11

The fruit yields (kg/plot) were as follows:

	Treatment combination			
Block	00	01	10	11
I	11.9	10.6	13.3	14.4
II	9.2	9.9	10.2	13.0
III	8.9	9.8	11.5	14.0
IV	9.1	12.5	13.7	13.7
V	7.6	8.3	9.6	12.2
VI	12.6	9.5	14.2	16.6
VII	11.0	11.9	15.8	14.2
VIII	8.5	10.0	14.6	14.8
IX	10.4	9.8	11.0	16.0

The treatment main effects and interaction, each with 1 d.f., can be calculated by setting out the treatment totals T_{00}, T_{01}, T_{10},

T_{11} in a two-factor table, as in earlier exercises. However, it is worthwhile to use this example to illustrate the layout of calculations used for the 2^n factorial arrangement; this layout is made possible by the fact that all main effects and interactions in this arrangement have 1 d.f. each. The reader should check that this layout leads to the same contrasts as the general method used earlier. The coefficients a_1, a_2, a_3, a_4, used with the treatment totals in the manner of 7.2.2.3, are as follows:

	Treatment			
Contrast	00	01	10	11
Main effect of A	-1	-1	$+1$	$+1$
Main effect of B	-1	$+1$	-1	$+1$
Interaction	$+1$	-1	-1	$+1$

We note that the coefficients for the interaction are the products of the corresponding coefficients for the main effects; this relationship is typical of the 2^n factorial arrangement

Exercises 17–19 illustrate further some of the extensions of the randomized block design. Exercise 17 involves a randomized block design repeated at several centres; this is the design used in the first of the two examples in 7.3.3. Exercise 18 uses a randomized block design with split plots for a three-factor factorial arrangement of treatments; this extends the discussion in exercises 13, 15 and 16, where a randomized block design was used for a two-factor factorial arrangement. Exercise 19 is concerned with a randomized block design with replicated measurements on each plot; at first sight the analysis may appear similar to the single-factor design with factorial arrangement of treatments, used in exercises 8 and 9, but in fact there are important differences.

17 An experiment to measure the effect of a mineral supplement on the food conversion efficiency of bacon pigs was carried out at three centres. At each centre the design used was a randomized block, with 5 blocks and 4 treatments; the treatments were the control diet, and this diet supplemented with 10, 20 or 30 parts per million of the mineral supplement. Each plot of the design was a pen of 4 pigs; the observation for the plot was the average food conversion efficiency of the 4 pigs, calculated as

$$\frac{\text{total food consumed by the 4 pigs}}{\text{total gain of live weight by the 4 pigs}}.$$

The results of the experiment were as follows:

	Block	Supplement level (p.p.m.)			
		0	10	20	30
Centre A	I	3.54	3.49	3.37	2.79
	II	4.25	3.65	3.10	3.10
	III	4.16	3.57	3.31	3.20
	IV	4.02	3.80	3.39	3.08
	V	4.20	3.76	3.27	3.33
Centre B	I	4.33	3.62	2.96	3.55
	II	4.40	3.96	3.70	3.61
	III	4.33	4.12	3.39	3.44
	IV	3.98	3.63	3.53	3.29
	V	4.55	3.65	3.62	3.33
Centre C	I	3.14	2.66	2.38	2.46
	II	3.52	3.31	3.01	2.78
	III	3.86	3.68	3.42	3.24
	IV	3.22	3.06	2.63	2.18
	V	3.86	3.24	3.02	3.32

The analysis follows that of the first example in 7.3.3. As in that case, we undertake only those significance tests which are appropriate to evaluating the treatments when the centres can be regarded as a random selection from a wide possible choice. This circumstance, though useful, is not the only possible one but the proper consideration of the significance tests in other cases is beyond our present scope.

Possible comparisons between the treatments would be

(a) level 0 against mean of levels 10, 20 and 30;

(b) between levels 10, 20 and 30;

(c) the linear, quadratic and cubic contrasts between the four levels, obtained by the use of orthogonal polynomials, as in exercise 11. The appropriate coefficients for four equally spaced dose levels are:

Linear	-3	-1	1	3
Quadratic	1	-1	-1	1
Cubic	-1	3	-3	1

18 The investigation considered in exercises 14 and 15 was extended still further by the use of a second method for estimating the sugar content; two samples were taken from each plot and one method of estimation used on each. At first sight it may appear that the only effect of this change is to make the two-factor factorial arrangement used in exercise 15 (4 cultivation levels × 2 weed control levels) into a three-factor factorial arrangement (4 cultivation levels × 2 weed control levels × 2 estimation method levels), the design remaining as a randomized block in 5 blocks. In fact this is not correct because, whilst the different levels of the cultivation and weed control factors are tested on different plots, the different levels of the estimation method factor are tested on the same plot; this means that the comparisons of the methods levels are subject to less random variation – no additive plot-to-plot variation – than are the comparisons of the cultivation levels and the weed control levels. This is the first example we have had of the situation that characterizes the family of *split plot* designs. These designs occur whenever different factors are tested on different units of experimental material – in this case samples from the same plot for one factor and from different plots for the other factors – and therefore have different sources of variation and different standard deviations for the comparisons between levels. The larger units – plots in this case – are called *main plots* and the smaller units – samples within plots in this case – are called *subplots*: that part of the analysis dealing only with the factors of which different levels are used on different main plots is called the *main plot analysis*, whilst the rest of the analysis, which deals both with the factors of which the different levels are used on different subplots and the interaction of these factors with the main plot factors, is called the *subplot analysis*. There is a separate measure of random variation in each part of the analysis and each main effect and interaction can be tested only against its appropriate residual. This causes some difficulties, mentioned again later in this exercise, when comparisons are to be made between individual treatment combinations. The present example is a very simple illustration of a split plot design, and we shall not be able to attempt any general discussion, but it is worthwhile to consider even a simple case. It is also useful to be able to stress how a very small change in experimental procedure – in this case making two different determinations on one plot – can have a very substantial effect on the analysis.

The observations (sugar in tonnes/ha), including those already given in exercise 15, were as follows:

	Weed control system	Sugar estimation method	Cultivation sequence			
			P	*Q*	*R*	*S*
Block I	*V*	*M*	4.73	4.65	8.70	10.78
		N	0.61	6.31	3.45	8.49
	W	*M*	4.36	6.02	9.29	13.13
		N	5.65	9.74	7.63	12.00
Block II	*V*	*M*	3.23	6.29	3.69	6.95
		N	1.66	4.41	3.48	5.70
	W	*M*	2.10	5.92	9.42	10.09
		N	3.54	7.38	7.72	9.62
Block III	*V*	*M*	3.69	6.12	6.41	8.91
		N	4.84	0.78	7.03	5.96
	W	*M*	6.99	8.33	8.52	11.21
		N	4.04	8.64	8.10	7.88
Block IV	*V*	*M*	4.21	6.48	6.64	9.04
		N	4.53	3.53	6.59	7.93
	W	*M*	5.78	9.52	10.11	11.53
		N	5.28	8.95	10.08	11.46
Block V	*V*	*M*	2.92	7.42	6.16	6.16
		N	6.96	7.62	4.84	8.05
	W	*M*	6.21	9.01	10.61	9.06
		N	5.02	12.48	8.86	8.94

The main plot analysis follows exactly the same pattern as that in exercise 15 using, in place of the individual observations of that exercise, the totals of the *M* and *N* values for each plot. The only changes necessitated by using these totals are in the divisors in the s.s.; since each total contains twice as many observations as in exercise 15, the divisors are all twice as big. No other changes are necessary in obtaining the main plot analysis; in particular the degrees of freedom do not alter.

The subplot analysis is also easy in this example because the subplot factor – the method of estimating the sugar content – has only 2 levels and therefore only 1 d.f.; the method for the general case is a little more involved and will not be discussed here. In this 2-level case we first calculate the difference $(M - N)$ for every plot – the signs of these differences are important. There are 40 of these differences, each corresponding to a 1 d.f. comparison, so that the subplot analysis has a total of 40 d.f. One of these d.f. is used for the overall contrast between M and N: the s.s. is calculated as

$$\frac{[\sum (M - N)]^2}{80},$$

the sum being taken over all 40 plots. We note that the divisor follows the standard rule for a 1 d.f. contrast (7.2.2.3). A further 7 d.f. are associated with the two-factor table in which totals of $(M - N)$ are arranged according to the levels of the main plot factors. We calculate s.s.'s from this table exactly as usual for a two-factor table, obtaining terms for columns (cultivations) with 3 d.f., for rows (weed control) with 1 d.f. and for the column × row interaction with 3 d.f.; since, however, the $(M - N)$ differences are themselves measures of a main effect, these three terms in fact measure

cultivations × sugar estimation interaction	3
weed control × sugar estimation interaction	1
cultivations × weed control × sugar estimation interaction	3

In determining the divisors used in calculating these s.s.'s we need to remember that each $(M - N)$ difference involves *two* observations; the divisors are, in fact, exactly the same as those used in calculating the s.s. for the corresponding table in the main plot analysis, where the starting point was the $(M + N)$ totals, also involving two observations.

The terms so far considered in the subplot analysis have used $1 + 7 = 8$ of the 40 d.f. available for that analysis. The remaining 32 are used for the subplot residual. The s.s. is calculated by subtracting the s.s.'s so far considered from the 40 d.f. subplot total s.s., which is obtained as

$$\sum \frac{(M - N)^2}{2},$$

the $(M - N)$ values being those for the individual plots, and the sum being taken over all plots. The skeleton of the complete anovar is thus as follows:

Main plot analysis	Blocks	4
	Cultivations (C)	3
	Weed control (W)	1
	$C \times W$	3
	Main plot residual	28
	Main plot total	39
Subplot analysis	Sugar estimation (S)	1
	$C \times S$	3
	$W \times S$	1
	$C \times W \times S$	3
	Subplot residual	32
	Subplot total	40
	Total	79

The treatment comparisons, main effects and interactions, and any comparisons between level means, are tested against the residual for the section of the analysis in which they occur; the fact that the analysis is in two sections does not affect these comparisons. It does, however, substantially affect some of the comparisons between individual treatment combinations – this is one of the less welcome complications introduced by the split plot designs. Consider the layout of the table of treatment combination totals – which, of course, follows the layout given above for the individual blocks.

Weed control system	Sugar estimation method	Cultivation sequence			
		P	Q	R	S
V	M				
	N				
W	M				
	N				

Denote the main plot residual by s_a^2 and the subplot residual by s_b^2; for generality we will also write n for the number of levels of the subplot factor (in this case $n = 2$) and r for the number of replications (in this case $r = 5$).

We then distinguish two cases:

(a) Comparisons between the totals of two treatment combinations which have the same levels of the main plot factors but different levels of the subplot factors – in this case this means treatment combinations which differ only in M, N – require

$$\text{var(difference of totals)} = 2rs_b^2.$$

(b) Comparisons between the totals of two treatment combinations other than those covered by case (a) require

$$\text{var(difference of totals)} = 2r\left[\left(1 - \frac{1}{n}\right)s_b^2 + \frac{1}{n}s_a^2\right].$$

Significance tests involving the use of this expression present some difficulty, because it is not at once clear what degrees of freedom to associate with this variance. An approximate rule is to use the degrees of freedom of the main plot residual (28 in this case); a more precise rule is to use as the value of t required for significance (at some specified level) the results of the following calculations:

$$t = \frac{t_a s_a^2 + (n - 1)t_b s_b^2}{s_a^2 + (n - 1)s_b^2},$$

where t_a and t_b are the values of t required for significance (at the specified level) when using respectively the degrees of freedom of the main plot residual and the subplot residual. Thus in the present case t_a is taken with 28 d.f. and t_b with 40 d.f.

19 Three new varieties of grapefruit (R, S, T) were compared with a standard variety (X) in an experiment carried out at 7 farms. The experimental area on each farm was divided into 4 plots and one plot assigned at random to each variety; six trees, of the same variety, were grown on a plot. One of the characters measured was the skin thickness (mm) of the fruit; the table gives the mean thickness estimated for each tree.

	Variety					Variety			
	X	R	S	T		X	R	S	T
Farm 1	4.0	7.9	8.8	11.2	Farm 5	2.0	6.7	4.0	10.7
	4.0	8.8	7.7	11.3		6.8	7.9	7.0	7.0
	4.7	5.6	10.2	10.8		0.8	3.7	10.3	8.1
	9.6	5.1	9.1	11.9		2.3	5.1	9.6	13.2
	7.0	5.7	10.9	11.5		8.5	4.9	8.1	6.3
	7.0	4.7	9.9	10.1		0.9	6.8	7.6	5.7
Farm 2	7.0	3.6	4.9	11.0	Farm 6	6.8	8.6	5.2	13.2
	6.7	2.9	9.9	10.9		5.2	10.8	10.1	9.0
	3.0	3.5	10.8	6.9		5.0	5.5	8.3	8.1
	5.5	2.8	5.3	6.4		6.8	4.7	9.3	13.3
	5.6	6.1	10.7	7.6		5.5	6.4	8.4	14.6
	3.5	7.2	11.5	7.7		3.6	7.0	10.7	9.8
Farm 3	3.0	4.4	8.3	9.6	Farm 7	4.5	5.7	7.1	8.6
	4.6	5.6	8.3	11.8		5.6	4.4	8.5	10.5
	5.1	4.3	5.9	12.3		4.9	7.1	7.3	12.0
	2.8	8.6	6.5	8.6		8.2	5.9	11.1	6.9
	3.3	4.8	7.8	11.8		8.7	6.8	9.4	11.6
	3.0	7.1	8.4	10.1		7.3	4.6	9.2	10.3
Farm 4	4.7	4.8	8.9	10.1					
	4.6	5.0	6.0	8.3					
	3.4	4.5	9.1	11.3					
	2.2	2.5	7.1	9.5					
	3.8	4.6	8.2	8.7					
	6.6	5.9	8.1	12.1					

The design of this experiment is a randomized block with 7 blocks (farms) and 4 treatments (varieties); there are, however, 6 observations on each plot, and the differences within plots between these observations require an extra term in the anovar. Each of the 28 plots contributes $6 - 1 = 5$ d.f. to this term, so that it has in all $28 \times 5 = 140$ d.f. The skeleton anovar is thus as follows:

Farms	6
Varieties	3
Farms × varieties interaction	18
Variation within plots	140
Total	167

The s.s.'s for farms and varieties are calculated in the standard 'main effect' pattern, remembering in determining the divisors that each farm total, for example, is a sum not of 4 observations (4 varieties) but rather 24 (4 varieties and 6 observations – trees – on each). The total s.s. follows the standard pattern. It remains therefore to consider the calculations either of the interaction or the within-plots term; whichever one is calculated directly the other can be obtained by subtraction. Although the interaction is a familiar expression by now, it is probably easier to calculate the within-plots term because this follows the well established pattern for the residual in a single-factor design.

When the s.s.'s and M.s.'s are obtained, it is necessary to consider the significance tests. Heretofore in a randomized block analysis we have always used the blocks × treatment interactions as residual, but in the present case we must ask ourselves whether this is still appropriate when we have the within-plots term available. At first sight the analysis is like those in exercises 8 and 9, which took the form

Main effect (M_1)
Main effect (M_2)
Interaction $(M_1 \times M_2)$
Residual within units

and thus appear very similar to the present case. In those exercises every effect was tested against the within-units term and there was no mention of the possibility of using the interaction as residual; it appears therefore that there is some difference between the designs in exercises 8 and 9 and that in the present case.

This difference can be understood by considering the nature of the factors in the various exercises; for convenience these factors are listed below:

Exercise 8	Exercise 9	Exercise 19
Dose of progesterone	Concentrate mixture	Farms
Starting day of injections	Method of estimating consumption	Varieties of grapefruit

Examining these we find that all except 'farms' in exercise 19 have in common the property that the particular levels considered are essential to the experiment, whereas this is not true of 'farms'. Thus different doses of progesterone or different starting days for the injections would have made exercise 8 refer to a different experiment, answering different questions: different concentrate mixtures or methods of estimating consumption would have had the same effect in exercise 9; so would using different varieties in exercise 19; but using different farms in exercise 19 would not, because we want our conclusions about the varieties to apply to 'all' farms and we are prepared to regard the set of farms used as a random selection, representative of the wider set. We are thus aware of the distinction between *fixed factors*, of which treatments are the commonest example, whose precise levels are essential to the investigation, and *random factors*, such as blocks, whose levels can be regarded as representative of a large set of possible levels. In exercises 8 and 9, where there was no consideration of using the interaction as residual, this interaction did not contain a random factor, but in the present exercise it does. This is the first point in deciding whether an interaction should be used as residual; if it contains only fixed factors the decision is 'only in very rare cases' – which we shall not discuss – but if it contains one or more random factors the decision is 'possibly, but it depends on the step-by-step consideration of the analysis'. The general rules for this consideration are beyond our scope here, but for the present simple case the procedure is as follows:

(*a*) test the interaction against the within-plots variation;

(*b*) if it is significantly larger then use it as residual for testing the main effects;

(c) if it is not significantly larger then use the within plots term as residual.

The application of even this simple rule can on occasion lead to misgivings – for example, are we happy to follow it strictly when the interaction is just not significantly large – but at least the rule offers some guidance. Sources for further reading are considered later.

Comparisons between the levels of the varieties factor follow the patterns considered in earlier examples.

Exercises 20–4 are concerned with surveys, discussed in 7.4 et sqq. The first two exercises contain observations for analysis, one from a small survey, the other from a rather larger one, whilst the remaining exercises merely present for the reader's consideration circumstances in which a survey would be an appropriate means of collecting information.

20 A survey of potato yield (quintals/ha) in the Patna region of India was designed as a two-stage sample. The first-stage units were villages, and 20 were selected at random with replacement with probability proportional to number of holdings: second-stage units were holdings, 10 being selected at random from each of the chosen first-stage units. The yields for individual holdings are given below; estimate the mean yield for the region, and give 95 per cent confidence limits for this mean.

Village	1	2	3	4	5	6	7	8	9	10
	150	163	140	151	172	180	147	190	173	191
	141	182	145	169	167	172	152	173	152	167
	153	166	129	165	145	174	166	173	172	167
	178	154	158	171	149	173	147	159	164	170
	177	163	158	152	156	178	168	168	185	167
	169	160	159	145	148	169	157	192	176	155
	148	148	152	163	150	179	158	170	176	167
	150	182	150	166	157	168	146	169	100	179
	163	162	155	141	154	169	130	184	170	186
	157	162	140	153	147	177	158	175	181	173

Village	11	12	13	14	15	16	17	18	19	20
	178	176	157	156	165	162	146	170	167	142
	181	146	177	179	144	169	147	155	182	150
	153	169	158	163	148	165	146	136	169	188
	167	157	176	168	159	166	163	177	172	159
	170	160	160	167	154	154	169	172	169	157
	171	159	171	179	171	170	145	148	179	162
	157	166	161	164	141	144	143	168	181	145
	158	171	174	181	149	161	154	156	170	159
	182	158	172	162	155	141	156	151	174	153
	176	158	179	168	147	162	152	142	185	154

21 A survey of milk production performance was carried out in one region of Great Britain. The region was divided into 2 areas which were expected, on the basis of a pilot survey, to prove relatively homogeneous. Three breeds of cow were included in the survey and the 6 area–breed combinations were used as strata in the design. The approximate numbers of milking cows in each stratum, and the corresponding numbers in the sample were as follows:

Area	Numbers	Breed		
		A	*B*	*C*
1	Stratum	3000	6800	4400
	Sample	120	270	175
2	Stratum	2000	2000	2500
	Sample	80	80	100

The sampling fraction was approximately the same for all strata because the pilot survey had shown no clear evidence of differences of variability between strata.

The samples within strata were selected by a two-stage process. First-stage units – herds – were selected at random with replacement with probability proportional to size and equal numbers of individuals – cows – were selected at random from each chosen herd. As explained in 7.4.3 this method of selection ensured that every cow in the stratum had the same chance of being chosen. The numbers of herds and of cows per herd in each stratum were as follows:

Area	Breed					
	A		*B*		*C*	
	Herds	Cows	Herds	Cows	Herds	Cows
1	6	20	9	30	7	25
2	4	20	4	20	5	20

The sample thus consisted of 825 cows from 35 herds. The milk yields of the individual cows are given in Table A23, which for convenience of layout is included in the Appendix.

The estimation procedures for this survey require some extensions of the formulae in 7.4.2 and 7.4.3. Let:

N = number of cows in the population sampled,

N_i = number of cows in the ith stratum,

k_i = number of herds selected in the ith stratum,

n_i = number of cows selected from each of the k_i chosen herds in the ith stratum,

$\bar{y}_{ij}$ = mean yield of the n_i cows selected in herd j in the ith stratum,

$\hat{m}_i$ = estimated mean for the ith stratum.

Then

$$\hat{m}_i = \frac{1}{k_i} \sum_{j=1}^{k_i} \bar{y}_{ij}.$$

The estimated variance of this estimate is

$$\text{var}(\hat{m}_i) = \frac{\sum_{j=1}^{k_i} (\bar{y}_{ij} - \hat{m}_i)^2}{k_i(k_i - 1)}.$$

Then the estimate of the mean yield for the whole population is

$$\hat{m} = \sum_{i=1}^{6} \frac{N_i}{N} \hat{m}_i,$$

which is of similar form to the corresponding expression in 7.4.2. However, the variance of $\hat{m}_i$ is not the same form as the variance of $\bar{x}_i$ in 7.4.2, and so the estimated variance of $\hat{m}$ is also of a different form from that in that section, namely

314

$$\text{var}(\hat{m}) = \sum_{i=1}^{6} \left(\frac{N_i}{N}\right)^2 \text{var}(\hat{m}_i)$$

$$= \sum_{i=1}^{6} \left(\frac{N_i}{N}\right)^2 \frac{\sum_{j=1}^{k_i} (\bar{y}_{ij} - \hat{m}_i)^2}{k_i(k_i - 1)}.$$

22 Plan, carry out and analyse a survey amongst your friends/neighbours/classmates to study the distances that they travel from home/digs/hostel to work/college.

You will need to define your target population, and to decide the type of sample and to select the individuals in the sample, to decide the information needed, both background and direct, to choose the method of collecting the information – interview or questionnaire – to obtain the information, and to analyse and present it.

23 Investigate, as in exercise 22, the food energy consumption (calories/day) of your target population.

Similar considerations apply in this case, but the required measurement is much more difficult to obtain and this will affect the sample size you can undertake.

24 Investigate, as in exercise 22, the opinions of your target population on some topical issue.

This exercise will illustrate difficulties of other kinds that may be encountered in collecting information.

8 ASSOCIATED NORMAL VARIABLES

8.1 Correlation and regression

In 4.3.1 we introduced Kendall's rank correlation coefficient to measure the association between two measured or ranked variables, without requiring any assumption about the form of the distributions appropriate to the variables. In this chapter we will consider what methods are appropriate when one or both of the variables follows a normal distribution. We consider first another kind of correlation analysis, which is essentially a test of whether there is any association between the variables, and secondly the methods of regression analysis, in which we investigate the form of the relationship; we shall give only the simplest procedures of both techniques.

8.2 Covariance

In measuring the association between two normal variables, it is convenient first to generalize the variance estimate used for measuring the variability of a single variable. Suppose then that we have observed the variables x and y on n individuals, obtaining values x_1, y_1; x_2, y_2; $\ldots$; x_n, y_n. The variances of x and y are estimated by

$$s_x^2 = \frac{1}{n-1} \sum_{i=1}^{n} (x_i - \bar{x})^2,$$

$$s_y^2 = \frac{1}{n-1} \sum_{i=1}^{n} (y_i - \bar{y})^2.$$

A simple first measure of the association between x and y is the *covariance* which is estimated by

$$C_{xy} = \frac{1}{n-1} \sum_{i=1}^{n} (x_i - \bar{x})(y_i - \bar{y}).$$

This expression can be simplified to give

$$C_{xy} = \frac{1}{n-1}\left\{ \sum_{i=1}^{n} x_i y_i - \frac{\left(\sum_{i=1}^{n} x_i\right)\left(\sum_{i=1}^{n} y_i\right)}{n} \right\};$$

the calculation is illustrated in Table 49. The expression in the curly braces in the last equation is known as the *corrected sum of products about the sample means* (or the *sum of products*), by analogy with the sum of squares for a single variable (6.3.1). It is convenient to represent the sums of squares and products (s.s.p.) by the following notation:

$$S_{xx} = \sum_{i=1}^{n} (x_i - \bar{x})^2, \quad S_{yy} = \sum_{i=1}^{n} (y_i - \bar{y})^2,$$

$$S_{xy} = \sum_{i=1}^{n} (x_i - \bar{x})(y_i - \bar{y}).$$

The point estimates of variance and covariance can then be written

$$s_x^2 = \frac{S_{xx}}{n-1}, \quad s_y^2 = \frac{S_{yy}}{n-1}, \quad C_{xy} = \frac{S_{xy}}{n-1}.$$

The interval estimate for variance was given in 6.4.1, but the interval estimate for covariance is beyond the scope of this book.

We note that, whilst s.s.'s and variance estimates are always positive, s.p.'s and covariance estimates can be negative; a positive covariance indicates that x and y tend to increase together, whilst a negative covariance shows that when one variable increases the other tends to decrease.

8.2.1 Product-moment correlation

The covariance estimate is a simple measure of association, but it is not easy to interpret because it depends on the scales on which the variables are measured; just as we found it useful to standardize deviation (6.2) so it is useful to standardize association. The quantity used for this purpose is the *coefficient of product-moment correlation*,† which is denoted by ρ: it is

† This is usually referred to simply as the *correlation coefficient*, but it is sometimes convenient to use the more explicit term.

estimated by

$$r = \frac{C_{xy}}{s_x s_y},$$

that is the sample covariance divided by the product of the sample standard deviations. It is easy to show that this formula may be simplified to the form

$$r = \frac{S_{xy}}{\sqrt{(S_{xx} S_{yy})}}.$$

Table 49. *An example of the calculations of* s.s.p. *and the estimation of covariance and product-moment correlation (see 8.2, 8.2.1).*

Observations

x	2.1	2.4	3.6	3.7	4.3	5.1	5.5	5.8
y	4.1	6.0	5.5	8.2	7.5	12.6	8.1	10.8
x	5.9	6.6	7.4	8.2	8.8	9.0	9.1	9.8
y	7.2	13.1	11.3	15.6	13.4	19.0	15.8	14.6

Calculations for s.s.p.

$$n = 16 \qquad\qquad \sum x = 97.3 \qquad\qquad \sum y = 172.8$$

$$\sum x^2 = 682.87 \qquad \sum y^2 = 2145.42 \qquad \sum xy = 1193.11$$

$$\frac{(\sum x)^2}{n} = 591.71 \qquad \frac{(\sum y)^2}{n} = 1866.24 \qquad \frac{(\sum x)(\sum y)}{n} = 1050.84$$

$$S_{xx} = 91.16 \qquad\qquad S_{yy} = 279.18 \qquad\qquad S_{xy} = 142.27$$

Calculations of s_x^2, s_y^2, C_{xy}, r

$$S_x^2 = \frac{91.16}{15} \qquad\qquad s_y^2 = \frac{279.18}{15} \qquad\qquad C_{xy} = \frac{142.27}{15}$$

$$= 6.077 \qquad\qquad = 18.612 \qquad\qquad = 9.485$$

$$s_x = 2.465 \qquad\qquad s_y = 4.314$$

$$r = \frac{9.485}{2.465 \times 4.314} = \frac{9.485}{10.634} = 0.8920$$

Alternatively

$$r = \frac{S_{xy}}{\sqrt{(S_{xx} S_{yy})}} = \frac{142.27}{\sqrt{25,450.05}} = \frac{142.27}{159.53} = 0.8918$$

The calculation is illustrated in Table 49. It may be shown that the correlation coefficient never exceeds 1 in numerical value, i.e. $-1 \leqslant r \leqslant 1$. When $r = 1$ the sample relationship is a straight line, with x and y increasing together, and when $r = -1$ the relationship is again a straight line, y now decreasing when x

increases; these relationships are shown in Fig. 23a, b. Inter-
mediate values of r are not so easy to interpret, however;
thus the relationships shown in Fig. 23c, d both give $r = 0$. In
general we may say that values of r near $+1$ or -1 indicate a
close relationship, with little departure from linearity, but that

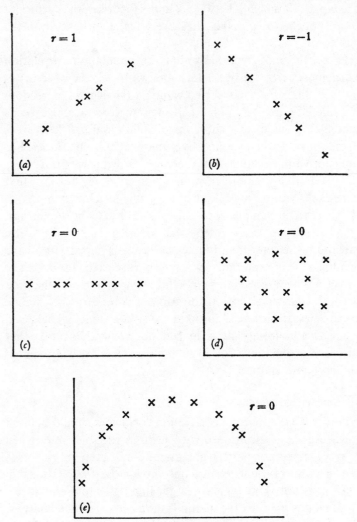

Fig. 23. Scatter diagrams illustrating various values of the sample product-moment
correlation; see 8.2.1.

319

other values do not necessarily indicate the absence of a relationship; thus in Fig. 23e we see a close sample relationship but in this case $r = 0$. Because of this difficulty, the correlation coefficient should be used only in conjunction with a scatter diagram; if, in particular, this diagram gives any evidence that the average relationship between x and y is curved, the coefficient should not be used as a measure of the closeness of the relationship.

To test the null hypothesis that the population correlation coefficient is zero, we use Table A18 with degrees of freedom $v = n - 2$; this rather unexpected value for the degrees of freedom is considered more fully in 8.3.1. If the sample value of r exceeds the tabulated value for a chosen probability and $(n - 2)$ degrees of freedom, we reject the null hypothesis at this chosen level of significance and conclude that there is clear evidence for a linear average relationship between x and y, remembering however that this test tells us nothing about other forms of relationship.

Even when a sample correlation is found to be significant, implying that we must accept that there is some population correlation, the interpretation of the result is subject to all the difficulties of interpreting association discussed in 4.4. The product-moment correlation coefficient was one of the first statistical measures of association to reach the public consciousness, and was misused so often that it achieved considerable notoriety. It is in fact no worse than any other measure of association, but all such measures must be presented with very great caution.

8.3 Linear regression

We noted in 8.2.1 that the correlation coefficient should be used to measure the association of two normal variables only when the average relationship is a straight line. Even in this case however the coefficient tells us only how close the relationship is, and does not show which straight line best represents it. To determine this we need the methods of *linear regression* analysis, which is the only example we consider of a very general and powerful set of procedures.

We have already seen (2.4.2) how Galton introduced the term regression, and we noted in the same section that the two variables involved in a regression relationship were not regarded equally, changes in one (the independent variable) being thought of as causing changes in the other (the dependent variable). This distinction is fundamental to all regression procedures, and these methods cannot be expected to give useful and reliable results unless the distinction is justified in the relationship under study.

We consider first how to represent a straight line relationship in algebraic symbols. We conventionally denote the independent variable by x and the dependent variable by y; we will show that any straight line can be represented by an equation of the form

$$y = \alpha + \beta x.$$

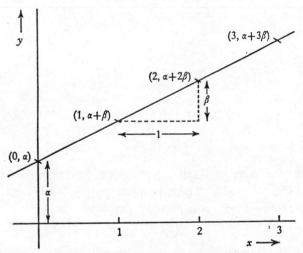

Fig. 24. Diagram illustrating the interpretation of the constants α, β in the equation $y = \alpha + \beta x$; see 8.3.

This can be seen most easily by calculating the value of y for a few values of x, and plotting the points so obtained. Thus if $x = 0$ we have $y = \alpha$, so that the point $(0, \alpha)$ lies on our line. If $x = 1$ we obtain $y = \alpha + \beta$, so that $(1, \alpha + \beta)$ is a second point: similarly $x = 2$ and $x = 3$ lead to $(2, \alpha + 2\beta)$ and $(3, \alpha + 3\beta)$. When these points are plotted (Fig. 24) we find that they lie on a straight line, that this line cuts the y-axis at a height α above the origin, and that the gradient of the line is β – in the sense that when x increases by 1 then y increases by β. The reader should experiment with various values of α and β to illustrate that any line can be specified by the appropriate choice of these constants; he will find, for example, that negative values of α correspond to lines which cut the y-axis below the origin and that negative values of β correspond to relationships in which one variable decreases when the other increases.

When we undertake a regression analysis we do not know the values of α and β for our relationship, and we cannot determine them simply by plotting points corresponding to our observations because these will deviate randomly from their straight line average. We have therefore to obtain estimates of the parameters of our line using the information given by the n pairs of observations

$$x_1, y_1; \; x_2, y_2; \; \ldots; \; x_n, y_n.$$

This is more easily done if we rewrite the equation of the line in the form

$$y = \gamma + \beta(x - \bar{x}),$$

where

$$\bar{x} = \frac{1}{n} \sum_{i=1}^{n} x_i$$

and is thus the average of our sample values of the independent variable. This change – which is made purely for reasons of algebraic convenience – is achieved by writing

$$\gamma = \alpha + \beta \bar{x},$$

that is γ is the y-value of the point on the line corresponding to $x = \bar{x}$. It may then be shown that the point estimates of β and γ are

$$b = \text{estimate}(\beta) = S_{xy}/S_{xx}$$

and

$$c = \text{estimate}(\gamma) = \bar{y}.$$

The s.s.p.'s S_{xy} and S_{xx} were defined in 8.2.

To obtain the interval estimates of β and γ, we need first an estimate of the random deviation of the points from the straight line average relationship. This is measured by a variance which is estimated by

$$s^2 = \frac{1}{n-2}\left\{S_{yy} - \frac{S_{xy}^2}{S_{xx}}\right\};$$

this form, which is quite unlike anything we have met in earlier sections, is discussed more fully in 8.3.1. The variances of b and c are now estimated by

$$\text{estimate var}(b) = s^2/S_{xx}$$

and

$$\text{estimate var}(c) = s^2/n;$$

these lead to interval estimates for β and γ in the usual way, namely

$$\text{interval estimate for } \beta = \frac{S_{xy}}{S_{xx}} \pm t\sqrt{\frac{s^2}{S_{xx}}},$$

$$\text{interval estimate for } \gamma = \bar{y} \pm t\sqrt{\frac{s^2}{n}}.$$

Here t is taken with $(n-2)$ degrees of freedom – corresponding to the estimate s^2 – and the probability level of our choice. The forms of these interval estimates should be compared with those given for other parameters in earlier sections (6.4, 6.6.1, 7.2.2, 7.3.1); in all cases we have the appropriate point estimate $\pm\, t \times$ (the estimated standard deviation of this estimate).

As usual we can derive significance tests from these interval estimates. Particularly useful is the test for the null hypothesis $\beta = 0$, because this gives an answer to the question 'Is there any linear relationship between the variables x and y?; if we assume that there is no relationship, we imply symbolically that $\beta = 0$. The test for this hypothesis is to note whether or not the value 0 is enclosed within the interval estimate for β. If this test is the only thing that concerns us, we may, as usual, reduce the calculation somewhat by working out an observed value of t, namely

$$\text{observed } t = \frac{S_{xy}/S_{xx}}{\sqrt{(s^2/S_{xx})}},$$

and comparing this value with the value tabulated for significance at our chosen level with $(n-2)$ degrees of freedom; we reject the

323

null hypothesis if our observed value of t exceeds the chosen tabulated value.

The calculations described in this section are illustrated in Table 50, using the means and s.s.p.'s obtained in Table 49.

Table 50. *The estimation of the linear regression line relating the observations in Table* 49 (*see* 8.3).

$$\text{Point estimate of } \beta = \frac{S_{xy}}{S_{xx}} = \frac{142.27}{91.16} = 1.561$$

$$\text{Point estimate of } \gamma = \frac{\sum y}{n} = \frac{172.8}{16} = 10.80$$

$$\text{Point estimate of residual variance } s^2 = \frac{1}{n-2}\left\{S_{yy} - \frac{S_{xy}^2}{S_{xx}}\right\} = \frac{1}{14}(279.18 - 222.04)$$

$$= \frac{1}{14} \times 57.14 = 4.081$$

$$\text{Estimate of variance of } b = \frac{s^2}{S_{xx}} = \frac{4.081}{91.16} = 0.04477$$

$$\text{Estimate of variance of } c = \frac{s^2}{n} = \frac{4.081}{16} = 0.25506$$

$$\text{Interval estimate of } \beta = \frac{S_{xy}}{S_{xx}} \pm t\sqrt{\frac{s^2}{S_{xx}}}$$

$$= 1.561 \pm 2.145 \times 0.21158$$

$$= 1.561 \pm 0.454$$

$$= 1.107, 2.015$$

$$\text{Interval estimate of } \gamma = \bar{y} \pm t\sqrt{\frac{s^2}{n}}$$

$$= 10.80 \pm 2.145 \times 0.505$$

$$= 10.80 \pm 1.08$$

$$= 9.72, 11.88$$

Significance test of $\beta = 0$,

$$\text{observed } t_{14} = \frac{S_{xy}/S_{xx}}{\sqrt{(s^2/S_{xx})}} = \frac{1.561}{0.21158} = 7.378***$$

8.3.1* Residual variance

In this section we consider more fully the estimation of the variance of the deviations of the observed y-values from the regression line (8.3). It is convenient to denote the values calculated from the regression equation by Y, so that

$$Y_i = \bar{y} + b(x_i - \bar{x}),$$

where $b = S_{xy}/S_{xx}$. The deviations of the observed y-values from the line are then represented by

$$y_i - Y_i = y_i - \bar{y} - b(x_i - \bar{x});$$

in this expression x_i is the observed x-value corresponding to y_i, the observed y-value.

The total of these deviations is

$$\sum_{i=1}^{n} (y_i - Y_i) = \sum_{i=1}^{n} (y_i - \bar{y}) - b \sum_{i=1}^{n} (x_i - \bar{x}) = 0,$$

and so the s.s. of the deviations is simply $\sum_{i=1}^{n} (y_i - Y_i)^2$, without any subtracted term. Thus

s.s. of deviations of observed points from the regression line

$$= \sum_{i=1}^{n} (y_i - Y_i)^2$$

$$= \sum_{i=1}^{n} \{(y_i - \bar{y}) - b(x_i - \bar{x})\}^2$$

$$= \sum_{i=1}^{n} \{(y_i - \bar{y})^2 - 2b(x_i - \bar{x})(y_i - \bar{y}) + b^2(x_i - \bar{x})^2\}$$

$$= \sum_{i=1}^{n} (y_i - \bar{y})^2 - 2b \sum_{i=1}^{n} (x_i - \bar{x})(y_i - \bar{y}) + b^2 \sum_{i=1}^{n} (x_i - \bar{x})^2$$

$$= S_{yy} - 2bS_{xy} + b^2 S_{xx}$$

$$= S_{yy} - 2\frac{S_{xy}}{S_{xx}} S_{xy} + \frac{S_{xy}^2}{S_{xx}^2} S_{xx}$$

$$= S_{yy} - \frac{S_{xy}^2}{S_{xx}}.$$

This s.s. has $(n-2)$ degrees of freedom, rather than $(n-1)$ as has a s.s. of the form $\sum (y - \bar{y})^2$, because it involves two estimated constants, β and γ, rather than one (the population mean) as in the simpler $\sum (y - \bar{y})^2$ form. Thus the estimate of residual variance is

$$s^2 = \frac{1}{n-2} \left\{ S_{yy} - \frac{S_{xy}^2}{S_{xx}} \right\}.$$

8.3.2 Regression anovar

A convenient means of bringing together two important parts of the calculations described in 8.3 is provided by the anovar of linear regression, which gives the estimate of variance as residual M.S. and gives an F-test equivalent to the t-test of the null hypothesis $\beta = 0$. The anovar also enables us to answer the question 'how much of the variation of the dependent variable y, is due to the independent variable, x, and how much to random causes?' The reverse question, with the variables interchanged, is not sensible because we have from the start assumed a one-way relationship, with changes in x causing changes in y.

The variation of the dependent variable is measured by S_{yy} and this is the total s.s. in the anovar, with $(n-1)$ degrees of freedom. The amount of this variation accounted for by the independent variable, the regression s.s., is S_{xy}^2/S_{xx} with 1 degree of freedom; the residual s.s. is obtained by subtraction and has therefore $(n-2)$ degrees of freedom. These formulae are brought together in Table 51, and the calculation is illustrated there using the s.s.p.'s from Tables 49 and 50.

Table 51. *The anovar of linear regression (see 8.3.2). We note that F (54.41) is, to within rounding errors, equal to the square of the observed t (7.378) calculated in Table 50 (7.378² = 54.43).*

Source of variation	Anovar of linear regression			
	S.S.	d.f.	M.S.	F
Regression	S_{xy}^2/S_{xx}	1	s.s./d.f.	$\dfrac{\text{Reg. M.S.}}{\text{Res. M.S.}}$
Residual	$S_{yy} - S_{xy}^2/S_{xx}$	$n-2$	s.s./d.f. $(= s^2)$	
Total	S_{yy}	$n-1$		
	Worked on the observations in Table 49			
Regression	222.04	1	222.04	54.41
Residual	57.14	14	4.081	
Total	279.18	15		

To verify that the F-test of the anovar is equivalent to the t-test for $\beta = 0$ given in 8.3 we note first that

$$\text{Residual m.s.} = \frac{1}{n-2} \{\text{Total s.s.} - \text{Regression s.s.}\}$$

$$= \frac{1}{n-2} \left\{ S_{yy} - \frac{S_{xy}^2}{S_{xx}} \right\}$$

$$= s^2,$$

so that
$$F_{1,n-2} = \frac{\text{Regression m.s.}}{\text{Residual m.s.}}$$

$$= \frac{S_{xy}^2/S_{xx}}{s^2}$$

$$= \frac{(S_{xy}/S_{xx})^2}{s^2/S_{xx}}$$

$$= \left\{ \frac{S_{xy}/S_{xx}}{\sqrt{(s^2/S_{xx})}} \right\}^2$$

$$= (\text{observed } t_{n-2})^2.$$

The F- and t-tests are thus mathematically equivalent; this is a further example of the relation between these tests that was noted in 7.2.2.2.

8.3.3 The model

The model for the analysis of linear regression is easily obtained from the equation of the regression line used in 8.3. This line represented the average relationship between x and y, and all that we need to complete the specification of the relationship is to include a term for the random deviations from the average. Thus the model is

$$y_i = \gamma + \beta(x_i - \bar{x}) + r_i,$$

where the r_i's are independent, normal, with zero mean and variance σ^2. The reader will note the exact analogy between these conditions and those for other anovars (7.2.2, 7.3.1). In the present case a convenient way of emphasizing the assumption that all the residuals have the same variance σ^2 is to say that the variance of the residuals is the same for all x_i. This means that a scatter of points such as is shown in Fig. 25 cannot properly be analysed by the methods of linear regression, even though the average relationship is a straight line, because the deviation from the line clearly increases in variance as x increases; this form of invalidity is quite common. The assumptions of normality and constant residual variance can be tested by the use of the Kolmogorov–Smirnov one-sample test and of Bartlett's test, as in the earlier examples of anovar (7.2.3.1, 7.2.3.2).

The point and interval estimates of β and γ have already been given (8.3); the point estimate of σ^2 is s^2 (8.3, 8.3.1) and the corresponding interval estimate is obtained by the method usually used for variances (6.4.1).

We note that the assumptions in the model do not specify in any way the distributional properties of the independent variable x. All the conditions relate to the properties of the residuals,

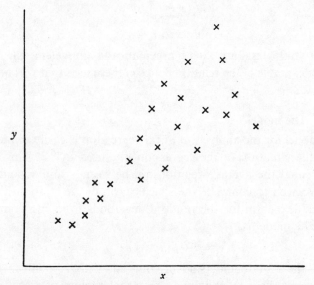

Fig. 25. Scatter diagram illustrating one common form of invalidity affecting regression analysis; see 8.3.3.

which measure how the observed values y_i of the dependent variable deviate from their average values, which in turn are determined by the x_i's and the constants β and γ. In other words all the assumptions are expressed in terms of the conditional distribution of y_i given x_i; this is a further example of the one-way relationship between the variables.

One implication of the assumptions, which needs careful attention in practical application of the technique, is that there is no random element in x; this means that the use of linear regression methods may give unreliable results if there is any appreciable 'experimental error' in the measurements of x.

8.3.4   The connection between correlation and regression

The techniques described above for use with product-moment correlation and linear regression are obviously very similar, so that it is necessary to make clear the relationship between the two types of analysis. Both procedures, in the simple forms we have described, can be used only when the average relationship between the variables is linear; regression methods can however be extended to allow for curved relationships, whilst correlation is of limited value in this case. In correlation the variables are on an equal footing, and the normal assumption is required for each, whilst in regression we assume that the conditional distribution of y given x is normal, but impose no conditions on x.† In correlation we measure the closeness of the linear relation without particularizing it at all, but in regression the essential point is to learn all we can about the constants of the straight line. The significance tests for $\rho = 0$ and $\beta = 0$ are, by mathematical accident, equivalent in this simple case, but in more complex analyses they may be different.

The concept of 'closeness of the relationship' can be applied in regression analysis by considering what proportion of the total variation in y is accounted for by regression on x. This

---

† In fact in correlation a stronger assumption, that of bivariate normality, is required; this specifies that the *joint frequency function*, $f(x, y)$, is of the form

$$f(x, y) = \frac{1}{2\pi\sigma_1\sigma_2\sqrt{(1 - \rho^2)}}$$
$$\times \exp\left[-\frac{1}{2(1 - \rho^2)} \left\{ \frac{(x - \mu_1)^2}{\sigma_1^2} - \frac{2\rho(x - \mu_1)(y - \mu_2)}{\sigma_1\sigma_2} + \frac{(y - \mu_2)^2}{\sigma_2^2} \right\} \right].$$

We shall not attempt here to interpret this expression.

proportion, measured on a percentage scale, can be expressed in terms of the s.s.'s as follows.

Percentage of variation in y due to linear regression on x

$$= 100 \times \frac{\text{Regression s.s.}}{\text{Total s.s.}}$$

$$= 100 \times \frac{S_{xy}^2/S_{xx}}{S_{yy}}$$

$$= 100 \left\{ \frac{S_{xy}}{\sqrt{(S_{xx}S_{yy})}} \right\}^2$$

$$= 100r^2.$$

This quantity, sometimes known as the *coefficient of determination*, is a further example of the close relationship between the two types of analysis.

Both procedures can be generalized to cases where more than two variables are involved. We can consider the *multiple regression* of a dependent variable on several independent variables, and can measure the closeness of the relationship by the *coefficient of multiple correlation*. These techniques are very valuable, but their concepts and calculations are too complex to consider here.

8.3.5(c) Linear regression calculations using Minitab, SPSS-X or Genstat

We will analyse the observations in Table 52 to illustrate the applications of the three packages in some of the simplest regression calculations, but we will not attempt to show the full range of output that is available. The commands and data for all three packages are shown in CTable 29.

The data layout used for SPSS-X and Genstat is not acceptable in Minitab; using read rather than set, the layout would have to be

```
read c1 c2
0.253 3.85
0.291 4.40
......
```

with each of the 22 cases on a separate row – although the use of 0 before the decimal point is optional and the space left between the c1 and c2 entries need not be the same in every case. On the other hand the layout used for Minitab is available in Genstat but not in SPSS-X; in Genstat the read statement would require an option as follows

$$\text{read [serial = yes]}$$

and the values for each variable would require a colon as terminator.

CTable 29. *Commands and data for the use of Minitab, SPSS-X or Genstat to carry out simple regression calculations on the observations in Table 52.*

Minitab

```
set c1
.253 .291 .3 .407 .427 .502
.511 .544 .595 .619 .623 .734
.752 .768 .789 .803 .818
.832 .854 .864 .868 .95
set c2
3.85 4.4 5.6 7.5 6.89 8.53
8.38 10.43 10.95 10.52 10.03 12.89
12.1 15.83 15.41 14.31 14.42
15.32 14 17.93 18.11 17.8
let c3 = c1 - mean(c1)
plot c2 c1
brief
regress c2 on 1 in c3
stop
```

SPSS-X

```
data list free / abs conc
variable labels abs 'Absorptiometer reading'
               conc 'Concentration of spermatozoa'
begin data
0.253 3.85   0.511  8.38   0.752 12.10   0.832 15.32
0.291 4.40   0.544 10.43   0.768 15.83   0.854 14.00
0.300 5.60   0.595 10.95   0.789 15.41   0.864 17.93
0.407 7.50   0.619 10.52   0.803 14.31   0.868 18.11
0.427 6.89   0.623 10.03   0.818 14.42   0.950 17.80
0.502 8.53   0.734 12.89
end data
regression variables - abs conc / dependent - conc / enter
```

Table 29 (*continued*)

```
units [nvalues = 22]
read           abs, conc
0.253 3.85    0.511   8.38    0.752 12.10    0.832 15.32
0.291 4.40    0.544 10.43    0.768 15.83    0.854 14.00
0.300 5.60    0.595 10.95    0.789 15.41    0.864 17.93
0.407 7.50    0.619 10.52    0.803 14.31    0.868 18.11
0.427 6.89    0.623 10.03    0.818 14.42    0.950 17.80
0.502 8.53    0.734 12.89  :
model          conc
terms [print = correlations]   abs, conc
fit            abs
rkeep          fittedvalues = pred
graph [xtitle = 'absorptiometer reading'; ytitle = 'concentration  \
    of spermatozoa'; xlower = 0; xupper = 1.0; ylower = 0;   \
    yupper = 20; ncolumns = 50; nrows = 30]   pred, conc;   \
                              abs; method = line, point
```

The use in Minitab of the command

$$\text{let } c3 = c1 - \text{mean}(c1)$$

has the effect that the regression equation is fitted in the form

$$y = c + b(x - \bar{x})$$

rather than

$$y = a + bx.$$

The use of the command brief determines the amount of output; the manual lists seven levels of the command in the form

brief K

with $K = 1, \dots, 7$. The form used is equivalent to brief 2. The command

regress c2 on 1 in c3

specifies that the variable in c2 is to be used as the dependent variable in regression on one independent variable which is in c3.

In the SPSS-X version, the regression command requires three subcommands separated from each other by /:

variables = {varlist}

specifies all the variables which are to be used in the regression;

    dependent = varlist

specifies the dependent variable, whilst the final subcommand influences the calculation only when we are fitting a multiple regression. Since, however, something must be stated, the form

      enter

has been chosen in the present case, giving *forced entry*.

In Genstat, the terms statement sets out all the variates used in the regression, in the same way as does the subcommand variables in SPSS-X; the option [print = correlations] calls for the outputting of the array of all the correlations between the variates, referred to as the *correlation matrix*. The model statement defines the dependent variable, whilst the fit statement specifies the independent variable. The command

    rkeep   fittedvalues = pred

requires that the fitted values – the values on the fitted regression line – shall be stored in a variate to be known as pred.

The commands for Minitab and for Genstat include the printing of the scatter diagram relating the two variables; in SPSS-X this could be achieved by the use of the command scattergram, which was introduced in 2.4.2(c) and is not illustrated again here. In Minitab the command is

    plot c2 c1

The first named variable is plotted on the y-axis and the second named on the x-axis. In the Genstat version two plots are shown on the same diagram, one for the fitted values and the other for the observed; this is controlled by the parameters to the graph command, as follows:

- the first parameter is the identifiers of the variates to be plotted on the y-axis, in this case pred and conc;
- the second parameter is the identifiers of the variates to be plotted on the x-axis, in this case abs. When, as for these two parameters, lists of different lengths are given the shorter lists

are recycled until the longest list is exhausted; thus, in this case pred is plotted against abs and then conc is plotted against abs. This order of plotting is chosen so that in the event of an observed point (the second plot) falling close to the fitted line (the first plot) the observed point will overwrite the fitted line instead of being obscured by it;

- the third parameter is the method of plotting, the two values in the string (i.e. line, point) referring to the two plots; the default setting is point;
- the fourth parameter, symbols, is not set and so takes the default values which are a mixture of . . . and ' ' ' for lines and * for points with multiple points indicated by integers.

We note that the parameters are separated by semi-colons, as we have seen above to be the case for options. For the graph

CTable 30a. *Output from the Minitab commands and data in CTable* 29; *the plot is shown in CFig.* 4a.

```
- - regress c2 on 1 in c3

THE REGRESSION EQUATION IS
Y -   11.6 +  20.2 X1

                                ST. DEV.    T-RATIO =
          COLUMN   COEFFICIENT   OF COEFF.   COEFF/S.D.
          --        11.6000       0.2273      51.04
    X1    C3        20.161        1.103       18.27

THE ST. DEV. OF Y ABOUT REGRESSION LINE IS

S = 1.066
WITH (  22 - 2) = 20  DEGREES OF FREEDOM

R-SQUARED = 94.3 PERCENT
R-SQUARED - 94.1 PERCENT, ADJUSTED FOR D.F.

ANALYSIS OF VARIANCE

    DUE TO       DF      SS        MS= SS/DF
    REGRESSION   1    379.490      379.490
    RESIDUAL     20    22.726        1.136
    TOTAL        21   402.217

- - stop
```

statement the options specify
- the titles to be used on the two axes;
- the plotting area limits on the two axes;
- the size of the plotting area, columns being plotted by –'s and rows by I's.

The output from the Minitab commands is given in CTable 30*a*, apart from the lineprinter plot which is reproduced in CFig. 4*a*; the SPSS-X output is in CTable 30*b* and that for the Genstat commands is in CTable 30*c* with the lineprinter plot in CFig. 4*b*. Many features are common to all three outputs. All give the fitted constants appropriate to the model used – which was explained above to have been chosen to be different in the Minitab

CTable 30*b*. *Output from the SPSS-X commands and data in CTable 29.*

```
* * * * * * * * * * * * * * * * * * * * * * * * * * * *
```

Listwise Deletion of Missing Data

Equation Number 1   Dependent Variable.. CONC   Concentration
    of spermatozoa

Beginning Block Number 1.   Method: Enter

Variable(s) entered on Step Number 1..   ABS   Absorptiometer
    reading

Multiple R	.97134
R Square	.94350
Adjusted R Square	.94067
Standard Error	1.06598

Analysis of variance

	DF	Sum of Squares	Mean Square
Regression	1	379.49322	379.49322
Residual	20	22.72638	1.13632

F = 333.96174   Signif F = .0000

--------------- Variables in the Equation ----------------------------

Variable	B	SE B	Beta	T	Sig T
ABS	20.161487	1.103242	.971338	18.275	.0000
(Constant)	-1.325346	.742895		-1.784	.0896

End Block Number   1   All requested variables entered

CTable 30c. *Output from the Genstat commands and data in CTable 29; the plot is shown in CFig. 4b.*

*** Degrees of freedom ***

Correlations:　　20

*** Correlation matrix ***

abs	1	1.0000	
conc	2	0.9713	1.0000
		1	2

***** Regression Analysis *****

Response variate: conc
Fitted terms: Constant, abs

*** Summary of analysis ***

	d.f.	s.s.	m.s.
Regression	1	379.49	379.493
Residual	20	22.73	1.136
Total	21	402.22	19.153
Change	-1	-379.49	379.493

Percentage variance accounted for 94.1

* MESSAGE: The following units have high leverage:
　　　　　　　　1　　　　　0.207

*** Estimates of regression coefficients ***

	estimate	s.e.	t
Constant	-1.325	0.743	-1.78
abs	20.16	1.10	18.27

analysis – with the standard deviation and *t*-test for each constant; the SPSS-X analysis also gives the significance level for each test. All three also give the analysis of variance but only SPSS-X includes the value of F with its probability level – the option [fprob = yes] is not available with these Genstat statements.

The most obvious difference between the three outputs is in the information given about the degree of correlation between the variables. The only quantity common to all three is actually

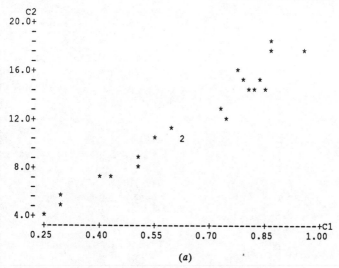

(a)

CFig. 4*a*. Scatter diagram of the observations in Table 52, drawn in response to the Minitab commands in CTable 29; the remainder of the output is in CTable 30*a*.

named differently in each and is moreover a measure not mentioned earlier in this chapter, namely

$$1 - \frac{\text{Residual M.S.}}{\text{Total M.S.}};$$

in the Minitab output this (expressed as a percentage) is described as 'R^2 adjusted for d.f.', in SPSS-X as 'adjusted R^2' and in Genstat (again as a percentage) as 'percentage variance accounted for'. The purpose of this measure is to express closeness of relationship in a way that is less affected by the number of independent variables than is the direct multivariate extension of the coefficient of determination mentioned in 8.3.4; in the present case, with only one independent variable, the distinction between the two measures is less important. However, Minitab and SPSS-X both give R^2 also – which is the same as r^2 in the present *univariate regression* case.

Further differences between the three outputs are:

(*a*) Minitab gives the regression equation explicitly and also

337

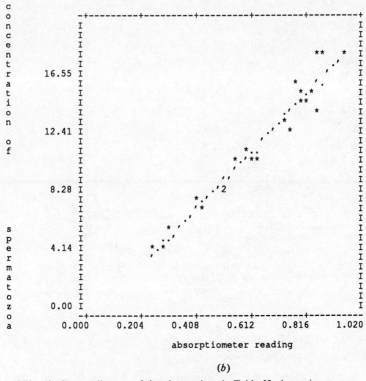

CFig. 4*b*. Scatter diagram of the observations in Table 52, drawn in response to the Genstat commands in CTable 29; the remainder of the output is in CTable 30*c*.

the value of s, the square root of the residual mean square (see 8.3.1);

(*b*) SPSS-X states how missing observations would be dealt with, describes the procedure chosen for the analysis, gives the value of R as well as that of R^2 and also the value of s;

(*c*) Genstat gives the correlation matrix – which of course shows the value of r – because this was specified in the option to the terms statement. The output also contains the following warning:

* MESSAGE: The following units have high leverage:
                    1   0.207

The purpose of this notification is to show us that unit 1 – with values abs = 0.253, conc = 3.85, exerts a non-negligible degree of influence on the position of the fitted line. The formal definition of *leverage* need not concern us here but one helpful approximate interpretation is to note that the reciprocal of leverage for any unit is more or less equivalent to the number of observations entering into the determination of the predicted y-value corresponding to the x-value of the specified unit. Thus a unit with leverage = 1 effectively determines the position of the fitted line near it, one with leverage = 0.5 has 50 per cent influence. This type of warning can be useful if, for example, we know that the observations on some units are less reliable than those on others. A fuller discussion of the subject can be found in, for example, the article by Velleman and Welsch (1981).

In all three languages further output is available to give other results derivable from the regression calculation.

8.4 Accuracy of prediction from linear regression

When we have fitted a regression line – that is, have estimated the constants β and γ – we frequently want to use it to predict the value of y appropriate to some particular x. For example, in determining the concentration of spermatozoa in a suspension it is convenient to use the property that the amount of light from a constant source absorbed by the suspension – this can be measured by an absorptiometer – is related to the required concentration. To use this property we need to calibrate the absorptiometer by determining for several samples the light absorption, x, read from the absorptiometer, and the concentration of spermatozoa, y, determined from haemocytometer counts. We fit the regression line of y on x and thereafter, when determining the concentration in a new sample, reads its x value from the absorptiometer and predict its y value from the equation

$$Y = \bar{y} + b(x - \bar{x}).$$

Here $\bar{x}$, $\bar{y}$ and b are determined from the samples used in the calibration, x is measured for the new sample and Y is the estimate

339

we make of the concentration in this sample; it is conventional to write Y for a predicted value and y for an observed value. An example of such a calibration is given in Table 52 and Fig. 26.

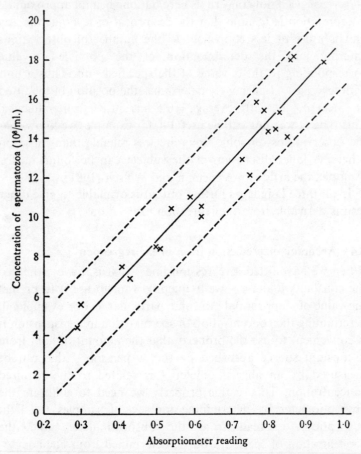

Fig. 26. Scatter diagram illustrating the calibration of an absorptiometer for use in estimating the concentration of spermatozoa in a suspension; see 8.4 and Table 52. The solid line is the fitted linear regression of y on x; the broken lines give limits to the predictions that can be made using this line—they are both slightly curved, diverging from the regression line.

The value of Y is a point estimate and we shall clearly need the interval estimate corresponding to it. Determining this interval requires some care, because we need to be very clear in

Table 52. *An example of the calculations in the calibration of an absorptiometer for determining the concentration of spermatozoa in a suspension. The interpretation of the limits is discussed in 8.4; the data and limits are plotted in Fig. 26.*

Observations of absorptiometer reading (x) and concentration of spermatozoa (y) (in units of 10^8/ml)

x	y	x	y	x	y	x	y
0.253	3.85	0.511	8.38	0.752	12.10	0.832	15.32
0.291	4.40	0.544	10.43	0.768	15.83	0.854	14.00
0.300	5.60	0.595	10.95	0.789	15.41	0.864	17.93
0.407	7.50	0.619	10.52	0.803	14.31	0.868	18.11
0.427	6.89	0.623	10.03	0.818	14.42	0.950	17.80
0.502	8.53	0.734	12.89				

The preliminary calculations give

$$n = 22 \qquad \sum x = 14.104 \qquad \sum y = 255.20$$
$$\bar{x} = 0.6411 \qquad \bar{y} = 11.600$$

$$\sum x^2 = 9.975542 \qquad \sum xy = 182.42908 \qquad \sum y^2 = 3{,}362.5396$$

$$\frac{(\sum x)^2}{n} = 9.041946 \qquad \frac{(\sum x)(\sum y)}{n} = 163.60640 \qquad \frac{(\sum y)^2}{n} = 2{,}960.3200$$

$$S_{xx} = 0.933596 \qquad S_{xy} = 18.82268 \qquad S_{yy} = 402.2196$$

$$b = S_{xy}/S_{xx} = 20.1615$$

$$S_{xy}^2/S_{xx} = 379.4931 \qquad s^2 = \frac{1}{n-2}\{S_{yy} - S_{xy}^2/S_{xx}\} = 1.1363$$

$$t_{20}\,(5\text{ per cent}) = 2.086 \qquad s = 1.0660$$

Hence the limits within which we would expect a single observed y to fall are calculated as

$$Y \pm ts\left\{1 + \frac{1}{n} + \frac{(x-\bar{x})^2}{S_{xx}}\right\}^{\frac{1}{2}} = Y \pm 2.086 \times 1.0660\left\{1 + \frac{1}{22} + \frac{(x-0.641)^2}{0.933596}\right\}^{\frac{1}{2}}$$
$$= Y \pm \{5.169 + 5.296(x-0.641)^2\}^{\frac{1}{2}},$$

where Y is given by

$$Y = \bar{y} + b(x-\bar{x}) = 11.600 + 20.1615(x - 0.641).$$

For convenient values of x these formulae give

$$V = 5.169 +$$

x	$x - 0.641$	Y	$5.296(x-0.641)^2$	$V^{\frac{1}{2}}$	$Y \pm V^{\frac{1}{2}}$	
0.241	−0.4	3.535	6.016	2.453	1.082,	5.988
0.341	−0.3	5.552	5.646	2.376	3.176,	7.928
0.441	−0.2	7.568	5.381	2.320	5.248,	9.888
0.541	−0.1	9.584	5.222	2.285	7.299,	11.869
0.641	0 0	11.600	5.169	2.274	9.326,	13.874
0.741	0.1	13.616	5.222	2.285	11.331,	15.901
0.841	0.2	15.632	5.381	2.320	13.312,	17.952
0.941	0.3	17.648	5.646	2.376	15.272,	20.024

our minds how we are going to interpret the interval. There are two quite different interpretations possible and we will consider these before giving the formulae. In the first approach we look for limits which are relevant to our prediction of the observed value of y we would obtain for the test sample; in determining these limits we take account not only of the variance of Y as an estimated point on the regression line but also of the variance of an observed y about this line. In the second approach we seek limits for the average of the y-values we would obtain if we took many samples all having the same value of x; in this calculation we include the variance of Y but regard the variance of the y-values about the line as unimportant because we have averaged over many samples. In the author's opinion the first approach is usually of more practical meaning, but the distinction between the approaches is often not made clear and formulae are sometimes given only for the second method. The limits relevant to the two approaches are as follows:

limits within which we expect a single observed y to fall

$$= Y \pm ts \left\{ 1 + \frac{1}{n} + \frac{(x - \bar{x})^2}{S_{xx}} \right\}^{\frac{1}{2}};$$

limits within which we expect an average of many observed values of y, all with the same x, to fall

$$= Y \pm ts \left\{ \frac{1}{n} + \frac{(x - \bar{x})^2}{S_{xx}} \right\}^{\frac{1}{2}}.$$

Here t is taken at the chosen probability level with $(n - 2)$ degrees of freedom; s^2, n, $\bar{x}$ and S_{xx} refer to the calibration samples; x refers to the new sample. The variances used in determining these limits are discussed in the next section.

Values of the limits for several values of x, using the first formula, are calculated in Table 52 and plotted in Fig. 26. We see at once that the limits are not all equally close to the regression line, but, because of the term $(x - \bar{x})^2/S_{xx}$, diverge from the line as x diverges from $\bar{x}$. This is a reasonable result, for it means simply that as we move away from the average of our calibration samples we become less precise in our predictions.

342

8.4.1* The derivation of two variances

In this section we derive the variances used in the interval estimates of prediction in 8.4. We note first that all the predictions are given for a specified value of x, so that in the calculations we can both treat x as a constant, and can use the variance of y conditional on x. We then have

$$\text{var}(Y) = \text{var}\{\bar{y} + b(x - \bar{x})\}$$
$$= \text{var}(\bar{y}) + (x - \bar{x})^2 \text{var}(b);$$

this follows from the formula in the footnote on page 174 since $\bar{y}$ and b may be shown to be independent. Furthermore

$$\text{var}(\bar{y} - Y) = \text{var}(\bar{y}) + \text{var}(Y),$$

because y and Y are based on different sets of independent observations; also, as in 8.3,

$$\text{estimated var}(\bar{y}) = \frac{s^2}{n},$$

$$\text{estimated var}(b) = \frac{s^2}{S_{xx}}.$$

Combining these expressions we have

$$\text{estimated var}(y - Y) = s^2\left\{1 + \frac{1}{n} + \frac{(x - \bar{x})^2}{S_{xx}}\right\},$$

which is used in the first pair of limits, and

$$\text{estimated var}(Y) = s^2\left\{\frac{1}{n} + \frac{(x - \bar{x})^2}{S_{xx}}\right\},$$

which is used in the second pair.

8.5 Exercises

These exercises, like the whole of the preceding chapter, deal only with the most elementary aspects of linear regression analysis. In each exercise the scatter diagram relating y to x should first be plotted and then the regression line estimated – if necessary, after the transformation of one of the variables. Confidence limits should be obtained for the parameters of the regression line and the anovar computed. In some cases, confidence limits are also needed for some predicted values of y.

343

1 As part of a study of the effect of housing conditions on the health of fattening pigs, a single animal was exposed to a range of relative humidities (%) and the rectal temperature (°C) noted. Estimate the regression of temperature on humidity from the following observations:

Temp.	R.H.	Temp.	R.H.	Temp.	R.H.
39.10	48.1	39.08	47.8	39.74	69.8
39.18	44.0	39.24	55.5	39.06	32.6
39.45	49.6	39.61	61.6	39.40	57.5
39.57	64.5	39.58	67.7	39.28	57.3

2 The following observations of body weight (kg) and food consumption (calories/day) were collected in a study on obesity in adolescent girls:

kg	cal/day	kg	cal/day	kg	cal/day
60.4	2680	71.6	2710	94.3	3390
81.1	3280	64.6	2600	61.8	2700
94.9	3890	75.1	2880	78.1	3090
86.4	3170	89.6	3430	74.8	3020
90.3	3390	84.4	3160	59.0	2410
60.4	2670	93.0	3330	69.2	2830
77.8	2770	61.3	2360	67.1	2620
85.0	3330	74.9	3030	82.4	2820

Estimate the regression line, and obtain point and interval estimates for

(a) the weight of a girl whose daily consumption was 3000 calories,

(b) the mean weight of a group of girls, all of whom consumed 3000 calories per day.

3 In a greenhouse experiment on the growth of a grass mixture, known amounts of available nitrogen (g/pot) were added to the compost in which the grasses were grown, and the amounts of nitrogen (g/pot) in the mature grasses were recorded. Use the following observations to estimate the regression of nitrogen in the grasses on nitrogen added to the compost.

Grass	Compost	Grass	Compost	Grass	Compost
0.12	0	0.26	0.4	0.63	0.8
0.23	0.1	0.52	0.5	0.69	0.9
0.32	0.2	0.53	0.6	0.73	1.0
0.25	0.3	0.63	0.7		

Estimate the regression line and use the result

$$\text{var}(\bar{y} - b\bar{x}) = \text{var}(\bar{y}) + \bar{x}^2\,\text{var}(b)$$

in testing whether the regression line passes through the origin, that is whether $\alpha = 0$: comment on the result of this test.

4 Bull spermatozoa were washed clean of seminal plasma and resuspended in buffered media containing known concentrations (mg/100 ml) of fructose. The suspensions were incubated and the fructose utilizations in the first hour of incubation were recorded, with the following results:

Utilization	Initial	Utilization	Initial	Utilization	Initial
13.8	100	66.7	180	123.4	260
57.5	120	84.9	200	129.1	280
63.3	140	89.8	220	134.6	300
59.2	160	121.4	240		

Estimate, with confidence limits, the mean utilization to be expected from suspensions initially containing 190 mg/100 ml.

5 The milk yield (kg/24 hours) of a lactating ewe was determined from weighings of her lamb and after suckling. The following observations were taken at 5-day intervals:

Yield	Day	Yield	Day	Yield	Day
1.84	14	1.60	39	1.25	64
1.86	19	1.62	44	1.30	69
1.75	24	1.55	49	1.26	74
1.68	29	1.51	54	1.25	79
1.50	34	1.34	59	1.15	84

345

Estimate, with confidence limits, the yield that would have been recorded had weighings been carried out on day 67.

6 The dependence of daily rate of development of maize seedlings on soil temperature (°C at a depth of 5 cm) was investigated by taking observations on 18 plots, with the following results:

Development	Temp.	Development	Temp.	Development	Temp.
0.069	13.1	0.019	9.8	0.074	15.3
0.080	11.2	0.098	16.7	0.071	14.9
0.034	10.3	0.068	12.6	0.022	10.1
0.028	12.2	0.051	10.1	0.111	16.0
0.027	9.9	0.086	16.1	0.057	11.3
0.054	13.6	0.043	12.8	0.102	14.5

Estimate, with confidence limits, the mean rate of development expected from seedlings grown in soil of temperature 15°C.

7 Investigate the regression of y on x in exercise 2.6(20).

8 Investigate the relationships between the bacon measurements given in exercise 2.6(24).

This exercise may help to make clear the limitations of regression methods given in this chapter when more than two variables have been recorded, that is when multiple regression methods are needed. For example, try to use the regressions of A on C and C on H to predict the regression of A on H, and compare the relationship thus obtained with the regression of A on H estimated directly.

9 The dependence of yield (kg) from a strawberry plant on the number of flower heads produced in the fruiting season was estimated from measurements on 37 plants:

Yield	Heads	Yield	Heads	Yield	Heads
0.69	5	0.75	5	1.06	10
1.39	15	1.05	10	0.71	9
1.09	12	1.44	14	1.19	10
0.76	9	0.76	6	0.83	7
0.86	12	0.72	5	0.49	6
1.04	10	1.29	10	1.24	15
1.10	13	0.71	6	1.22	9

Yield	Heads	Yield	Heads	Yield	Heads
1.02	10	0.86	8	0.98	8
1.10	10	0.93	10	0.71	6
1.07	9	0.91	6	1.60	14
0.94	7	0.70	8	1.21	13
0.98	14	1.56	15	1.05	7
0.71	5				

Use the result $\text{var}(\bar{y} - b\bar{x}) = \text{var}(\bar{y}) + \bar{x}^2 \, \text{var}(b)$ in testing whether the regression line passes through the origin, that is whether $\alpha = 0$; comment on the result of this test.

10 In a class experiment to determine the specific latent heat of fusion of ice, each member of the class used the same volume of water, but added to it a convenient weight of ice (g), noting the fall in temperature (°C) produced. The individual observations were as follows:

Fall	Ice	Fall	Ice	Fall	Ice
19	13.4	16	10.5	29	19.3
19	12.8	17	11.8	21	13.9
22	15.1	16	10.2	24	15.6
30	19.7	20	13.6	16	10.7
27	17.5	29	18.1	28	18.5
29	19.3	27	17.8	25	17.2
18	12.5	28	18.1	22	15.0
18	11.5	23	14.2	20	14.2
21	14.2	25	15.9	28	19.0
17	11.2	27	18.3	26	17.2

In this analysis the calculations given in the text need to be modified to make use of the prior information that if no ice is added ($x = 0$) no fall in temperature can result ($y = 0$), that is we known in advance that the regression line passes through the origin. We cannot use this knowledge simply by fitting the line as before and then testing (as in exercises 3 and 9) whether it deviates from the origin; rather we have to change the calculation, basing it on the model

$$y_i = \beta' x_i + r_i.$$

347

ASSOCIATED NORMAL VARIABLES

The point estimate of β' may be shown to be

$$b' = \frac{\sum\limits_{i=1}^{n} x_i y_i}{\sum\limits_{i=1}^{n} x_i^2}.$$

The residual variance is estimated by

$$s'^2 = \frac{1}{n-1}\left\{\sum_{i=1}^{n} y_i^2 - \frac{\left[\sum\limits_{i=1}^{n} x_i y_i\right]^2}{\sum\limits_{i=1}^{n} x_i^2}\right\}$$

with $(n-1)$ degrees of freedom. The estimated variance of b' is

$$\text{estimated var}(b') = \frac{s'^2}{\sum\limits_{i=1}^{n} x_i^2},$$

which leads as usual to an interval estimate for β'. The t-test for the null hypothesis $\beta' = 0$ is

$$\text{observed } t = \frac{b'}{\sqrt{\left(s'^2 \bigg/ \sum\limits_{i=1}^{n} x_i^2\right)}},$$

with $(n-1)$ degrees of freedom. The anovar form of the calculation is

Source of variation	S.S.	d.f.	M.S.	F
Regression	$\dfrac{\left(\sum\limits_{i=1}^{n} x_i y_i\right)^2}{\sum\limits_{i=1}^{n} x_i^2}$	1	s.s./d.f.	$\dfrac{\text{Reg. M.S.}}{\text{Res. M.S.}}$
Residual	Subtraction	$n-1$	s.s./d.f. $(= s'^2)$	
Total	$\sum\limits_{i=1}^{n} y_i^2$	n		

As before, the residual M.S. estimates the residual variance and the F-value is the square of the t used to test $\beta' = 0$.

The volume of water used by each member of the class was 55 cm^3. Estimate, with confidence limits, the latent heat of fusion of ice.

11 The concentration (mg/100 ml) of a certain substrate in a cell suspension was measured at ten-minute intervals with the following results:

Conc.	Time	Conc.	Time	Conc.	Time
436	10	144	60	74	110
258	20	121	70	84	120
230	30	120	80	77	130
212	40	97	90	62	140
144	50	82	100	40	150

The plot of these observations shows that the relationship is not linear; indeed it is to be expected that the form would be exponential, so that the plot of log(concentration) against time should be considered. This will be found to be approximately linear, so we analyse the regression of $y = \log(\text{concentration})$ on $x = \text{time}$.

The discussion of exercise 6.8(18) is relevant here. In the present analysis the logarithmic transformation achieves purpose (*b*) of that discussion, in that it makes the regression relationship approximately linear – and so easier to analyse – and also makes the variation more nearly uniform along the regression line, as may be seen by comparing the two plots using concentration and log(concentration); the transformation has no effect as far as purpose (*a*) of 6.8(18) is concerned.

12 The growth of yam tubers over the period July–December was studied by recording the dry weight (g) at weekly intervals, with the following results:

Weight	Week	Weight	Week	Weight	Week
20.5	1	67.1	8	230	15
21.8	2	59.6	9	225	16
29.3	3	80.0	10	240	17
30.5	4	104	11	333	18
37.8	5	114	12	395	19
41.2	6	167	13	406	20
50.8	7	174	14	492	21

The relationship between dry weight and week will be found to be approximately exponential, so that the discussion of the previous exercise is relevant.

Estimate, with confidence limits, the mean weight of tubers harvested at $17\frac{1}{2}$ weeks – this requires the point and interval estimates to be calculated from the regression of log(weight) on week, and then to be converted back to the weight scale; it is not justifiable to carry out any calculations directly with dry weights.

9 SOME NON-NORMAL DISTRIBUTIONS

9.1 Non-normal distributions

In the preceding chapters we have considered either methods applicable to all types of distribution or, in chapters 6, 7 and 8, methods applicable to one particular family of distributions, the normal family. The purpose of the present chapter is to discuss certain cases where the distribution of the variable under study is not normal but is nevertheless specified in a way which makes invalid the use of the non-parametric methods considered earlier.

Non-normal families of distributions can occur in many different forms, each characterized by particular circumstances and conditions of experimentation; in the present chapter we shall give most attention to two of the more important, the *binomial* and *Poisson* families. The binomial distribution has already been mentioned briefly in the discussion of the sign test (3.4.1.2); the Poisson distribution is a special case of it. We shall consider first the conditions under which these distributions occur.

9.2 The binomial and Poisson families

A distribution belonging to one of these families may occur when the experimental situation obeys the following three conditions:

(*a*) that for any one observation only two results are possible;

(*b*) that the chances of these two results do not vary from one observation to the next;

(*c*) that successive observations are independent in the sense of 4.2.3.

Condition (*a*) limits the relevance of these distributions to the cases where taking an observation can have only an 'all-or-nothing' result; thus a child who is exposed to the risk of measles either does or does not contract the disease; a cow which is inseminated either does or does not conceive; a rat challenged with a poison either does or does not die. The two possible results to any one trial are conventionally known as *success* and *failure*.

Condition (*b*) requires that the chances of success and failure, conventionally represented by p and q, must be the same for all trials. Thus all children of the population we consider must have the same chance of contracting measles; all inseminated cows must have the same chance of conceiving; all challenged rats must have the same chance of dying. This restriction clearly restricts considerably the experiments to which these families of distributions apply; removing the condition leads to families of distributions which are more widely applicable, but which are too complex to consider here. The condition means in particular that the binomial and Poisson families do not apply to situations where the population under investigation is known to divide into sub-populations which differ in the chance of success (in the sense here used) of their members. Thus, if we believed that boys were more liable to measles than girls we should violate condition (*b*) if we included children of both sexes in our sample; the presence of completely sterile cows in the group to be inseminated would similarly violate the condition, as would the inclusion of some rats which had been desensitized by a course of gradually increasing challenge doses.

Condition (*c*) requires that the result of any particular trial shall not have any effect on the result of any other trial. In considering the incidence of measles, the inclusion in the sample of two children in the same household might be thought to violate this condition: presumably if one child became infected this might increase the chance of infection in the other. If two cows were inseminated from the same pipette then venereal infection might be transferred from the one to the other, and reduce the chance of conception of the second; similar considerations might apply to administering the challenge dose to rats.

These examples show that the conditions are relevant to the way in which the experiments are conducted, and to the populations under investigation. In other words the conclusion that we are dealing with a variable belonging to the binomial or Poisson family should be based on consideration of the experimental procedure, and not on inspection of the results; such inspection provides, at best, very imperfect grounds for such a conclusion.

9.2.1 The binomial distribution

Under the conditions (a), (b) and (c) discussed in 9.2, the binomial distribution is relevant when we consider the question 'What is the probability, P_r, of obtaining r successes in a given number, n, of trials?' The formula answering this question is

$$P_r = \frac{n(n-1)(n-2)\ldots(n-r+1)}{1 \cdot 2 \cdot 3 \ldots r} p^r q^{n-r};$$

this is discussed more fully in 9.2.1.1, but a brief introduction is appropriate here. If we write $1 \cdot 2 \cdot 3 \ldots r = r!$, called r *factorial*, and similarly $1 \cdot 2 \cdot 3 \ldots n = n!$, then the coefficient in this expression can be put in the form

$$\frac{n(n-1)\ldots(n-r+1)(n-r)(n-r-1)\ldots 3 \cdot 2 \cdot 1}{1 \cdot 2 \cdot 3 \ldots r \cdot (n-r)(n-r-1)\ldots 3 \cdot 2 \cdot 1} = \frac{n!}{r!\,(n-r)!},$$

so that

$$P_r = \frac{n!}{r!\,(n-r)!} p^r q^{n-r}.$$

The form of the distribution depends on the values of the parameters n (the number of trials) and p (the chance of success). The dependence on p is illustrated in Fig. 27, which shows the values of P_r for $n = 12$ and $p = 0.1, 0.5$ and 0.8. The distribution for $p = 0.5$ is symmetric, but the others are skew, that for $p = 0.1$ in the positive sense and that for $p = 0.8$ in the negative sense (see 2.1.1). If we had considered the case $p = 0.9$ we would have found that it was identical in form with $p = 0.1$, but reflected in the line $r = 6$; similarly $p = 0.2$ would give the reflexion of $p = 0.8$ in this same line. The dependence of the distribution on n is illustrated in Fig. 28, which shows the values of P_r for $p = 0.7$ and $n = 9, 29$ and 49; we note that as n increases the skewness of the distribution decreases. This effect is more obvious if we consider the distribution not of r itself (the number of successes) but of the *binomial proportion* r/n, which measures the proportion

353

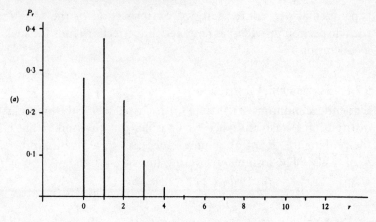

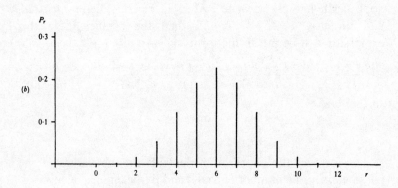

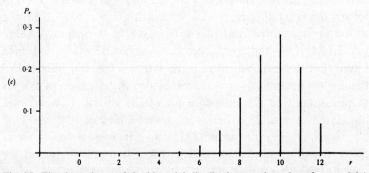

Fig. 27. The dependence of the binomial distribution on the value of p: see 9.2.1. For all three figures $n = 12$ whilst the values of p in (a), (b) and (c) are 0.1, 0.5 and 0.8.

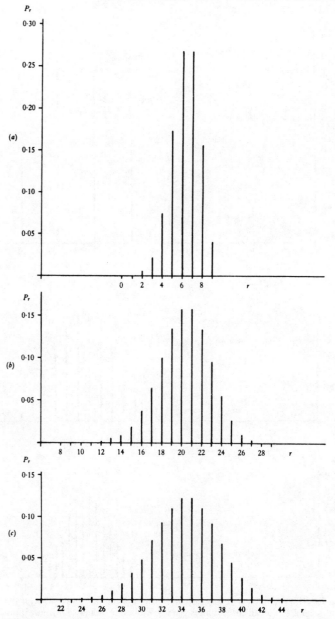

Fig. 28. The dependence of the binomial distribution on the value of n: see 9.2.1. For all three figures $p = 0.7$ whilst the values of n in (a), (b) and (c) are 9, 29 and 49.

355

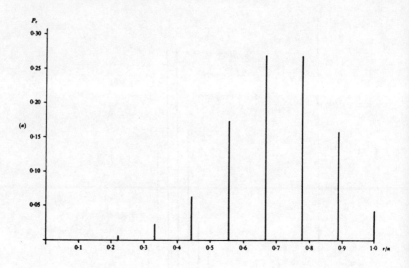

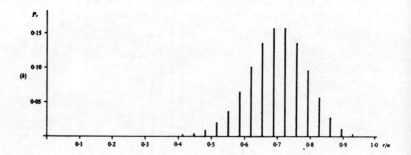

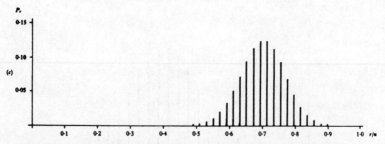

Fig. 29. The dependence of the distribution of a binomial proportion on the value of n: see 9.2.1. For all three figures $p = 0.7$ whilst the values of n in (a), (b) and (c) are 9, 29 and 49. These distributions are the same as those in Fig. 28, but are transformed to have the same horizontal scale.

of trials which lead to successes; the distributions of r/n are illustrated in Fig. 29.

These figures show that both the average (mean) value of the distribution and its spread (deviation) depend on the values of n and p; formulae expressing this dependence are given in 9.2.1.1. This implies that both the binomial variable, r, and the binomial proportion, r/n, fail to satisfy either the basic conditions required for the validity of anovar (see, for example, 7.2.2) or even those required for most of the *non-parametric* methods (see, for example, 3.2) because, if the mean of the distribution varies from level to level of a factor – as it may be expected to do – so also will the variance and skewness. Thus, even though some examples of the binomial distribution look somewhat 'normal' in form and might therefore tempt us to appeal to the robustness of anovar (see 7.2.3) as a justification for using this technique to analyse binomial variables, this procedure is not in fact valid and can lead to very unreliable conclusions.

9.2.1.1* Formulae relating to the binomial distribution

In this section we prove the formula for P_r when r is binomial, and discuss the mean and variance of r.

We consider a series of n independent trials with chance of success p at each trial and obtain the probability of r successes occurring in these n trials. One particular sequence leading to this result is for the r successes to occur consecutively in the first r trials and for the $(n - r)$ failures to occur in the remaining $(n - r)$ trials. The problem of obtaining the probability of this particular sequence is in essence just the same as that of determining the probability that a head is thrown in each of two successive tosses of a fair coin; this probability is clearly $\frac{1}{4}$, and is the product of the probability of a head at the first throw $(\frac{1}{2})$ multiplied by the probability of a head at the second throw $(\frac{1}{2})$. This reasoning applies exactly to a longer sequence of tosses of the coin and, similarly, to our sequence of binomial *trials* with constant chance of success p.† Thus the probability of the sequence defined

---

† This is an example of the *product law of probability*, which states that in a series of independent trials the probability of any particular sequence of results is the product of the probabilities of the individual results in the sequence.

above is

$$p \times p \times \cdots \times \underbrace{p \times q \times q \times \cdots \times q}_{} = p^r q^{n-r}.$$
$$\underbrace{}_{r \text{ terms}} \quad \underbrace{}_{(n-r) \text{ terms}}$$

If we consider any other possible sequence – i.e. any other sequence containing r successes and $(n-r)$ failures – we find that the same final probability is obtained, even though the order of the terms in the detailed version of the expression is not the same. Thus the probability of any one acceptable sequence is $p^r q^{n-r}$.

This, however, does not answer our question because we require not the probability of some one particular sequence but that of any one or other from the full set of acceptable sequences. To obtain this probability we use the summation law of probability (5.4, footnote), which tells us to add the probabilities of all the acceptable sequences. Since these probabilities are all equal this summation is equivalent to multiplying the probability for any one sequence ($p^r q^{n-r}$) by the number of acceptable sequences. We require therefore to determine this number, and to do this we consider how acceptable sequences are generated. Any sequence is acceptable if we start by postulating r successes anywhere among the n trials and then fill all the remaining $(n-r)$ places with failures; the number of acceptable sequences is thus the number of different ways in which we can choose sets of r positions in the sequence of n. This number is known as the *number of combinations of n things chosen r at a time* and is written $\binom{n}{r}$; it is proved in books on elementary algebra that

$$\binom{n}{r} = \frac{n(n-1)(n-2)\ldots(n-r+1)}{1 \cdot 2 \cdot 3 \ldots r} = \frac{n!}{r!\,(n-r)!}.$$

We thus have, as stated in 9.2.1,

$$P_r = \frac{n!}{r!\,(n-r)!}\,p^r q^{n-r}$$

which is written alternatively as

$$P_r = \binom{n}{r} p^r q^{n-r}.$$

To calculate the average of r over the population, we use, as explained in the second footnote to 5.3.2, the concept of expectation, which provides a population quantity analogous to the familiar sample average. The expectation of r is given by the formula

$$E(r) = \sum rP_r,$$

the summation being taken over the full range of values of the variable r; thus in the present case we have

$$E(r) = \sum_{r=0}^{n} r \binom{n}{r} p^r q^{n-r}$$

$$= \sum_{r=0}^{n} r \frac{n(n-1)(n-2) \ldots (n-r+1)}{1.2.3 \ldots r} p^r q^{n-r}$$

$$= \sum_{r=1}^{n} \frac{n(n-1)(n-2) \ldots (n-r+1)}{1.2.3 \ldots (r-1)} p^r q^{n-r}.$$

To evaluate this sum we use the *binomial theorem*, discussed in books on elementary algebra, which states that

$$(a+b)^m = \sum_{s=0}^{m} \frac{m(m-1)(m-2) \ldots (m-s+1)}{1.2.3 \ldots s} a^s b^{m-s}.$$

The expression for $E(r)$ is not exactly in this form but if we remove np as a factor common to all terms of the sum, writing therefore

$$E(r) = np \sum_{r=1}^{n} \frac{(n-1)(n-2) \ldots (n-r+1)}{1.2.3 \ldots (r-1)} p^{r-1} q^{n-r}$$

and then compare $n-1$ with m, $r-1$ with s, p with a and q with b we obtain

$$E(r) = np(p+q)^{n-1}.$$

Since p and q refer to success and failure, the only two possible results to our trials, we can write $p+q=1$ (cf. 5.4, footnote) so that we obtain

$$E(r) = np.$$

This is the population average of the binomial variable r, exactly analogous to the sample average obtained by totalling all the observations and dividing by the number of observations.

If we calculate this for the examples of Fig. 27 we obtain $E(r) = 1.2$, 6 and 9.6 for the three diagrams; it is instructive to note where these values fall in the distributions. Similarly for Fig. 28 we obtain $E(r) = 6.3$, 20.3 and 34.3, whilst in considering Fig. 29 it is easy to show that

$$E\left(\frac{r}{n}\right) = \frac{1}{n} E(r) = \frac{1}{n} \cdot np = p,$$

so that the mean is at 0.7 for all three diagrams.

To measure the variability shown by a binomial variable we use, as in the normal case (6.2) the standard deviation and its square, the variance. As was explained in that section, the definitions of these quantities which we have used for the normal distribution do not apply to other distributions, such as the binomial, and the general definition involves concepts beyond our present scope. It may be shown, however, that for a binomial variable

$$\text{variance } (r) = npq,$$

$$\text{standard deviation } (r) = \sqrt{(npq)},$$

$$\text{standard deviation } (r/n) = \sqrt{(pq/n)}.$$

Thus in the examples of Figs. 27 and 28 the standard deviations of r are 1.04, 1.73, 1.39 and 1.38, 2.47, 3.21; in the case of Fig. 29 the standard deviations of r/n are 0.153, 0.085, 0.065. Comparing these values with the diagrams we see that the standard deviation is larger for the more widely spread distributions and less for the more compact cases, as we could naturally require of any reasonable measure of variation. However, it is not possible to relate multiples of the standard deviation to probability, as can be done for the normal distribution. In fact, in the binomial case there is no simple interpretation of the standard deviation, nor is there any other measure of deviation as convenient as is the standard deviation in the normal case.

9.2.2 The Poisson distribution

This distribution occurs under the same conditions as the binomial $\{(a), (b) \text{ and } (c) \text{ of } 9.2\}$ when the event considered is 'rare'; we shall first give an example to illustrate the meaning of this term. Suppose that we are interested in the metabolism of a certain substance and are studying this by the use of a labelled form of the substance; we thus make counts of the number of radioactive emissions from an animal treated with the labelled substance. The number of emissions in a given time is of course a variable quantity, and we need to know the distribution of this variation in order to interpret the counts we observe. Suppose that each count is taken over a fairly short period, say of length t seconds, so that the average rate of emission can be regarded as constant. We can gain some understanding of the distribution of the number of counts by imagining the period t to be divided up into a large number of short (equal) subintervals and defining as a 'trial' the observation of the counter for a single subinterval; such trials satisfy the conditions of 9.2 because

(a) if the subintervals are short we can effectively ignore the chance of 2 or more emissions occurring so that the only possible results are 0 or 1 emission;

(b) t is short enough to justify our treating the chance of success (emission) as the same for all the subintervals of t;

(c) emissions are independent, so that our observations on successive subintervals are also.

We can sensibly suppose that the chance of emission in any one subinterval is proportional to the length of the subinterval so that, if we denote the number of subintervals by n, so that the length of each is t/n seconds, we can represent the chance of success by

$$p = \lambda \frac{t}{n},$$

where λ is the constant of proportionality. We therefore have, as a first step to representing the distribution of the number of counts, a series of n binomial trials, each with chance of success $\lambda t/n$. Thus the probability of r successes, that is of r emissions, in the period t is to a first approximation

$$\frac{n(n-1)\ldots(n-r+1)}{1.2.3\ldots r}\left(\frac{\lambda t}{n}\right)^r\left(1-\frac{\lambda t}{n}\right)^{n-r}.$$

This formula is, however, no more than an approximation because time does not really proceed as a series of discrete subintervals; the approximation improves if we make our imaginary subintervals shorter, that is make n larger, and we should therefore consider how the above formula changes if n becomes very larger. It may be shown that the expression comes nearer and nearer to the value given by the formula

$$P_r = \frac{e^{-\lambda t}(\lambda t)^r}{r!};$$

the proof of this result is given in 9.2.2.1. The distribution defined by this formula is known as the *Poisson distribution*; it was described above as relevant to 'rare' events because, just as in this example an emission occurred at only a small proportion of the moments in the observation time t seconds, so in the general case a success occurs in only a small proportion of conceptual 'trials'.

Arguments similar to the above can be used to explain both superficial and spatial occurrences of the distribution. Thus in counting cells on a microscope slide we can imagine the field of view divided by a fine grid, treating a 'trial' as the observation of a single square of this grid and a 'success' as the occurrence of a cell in this square; again, in counting small organisms in soil we can regard the soil sample as divided up by a fine lattice and define a 'trial' as the observation of a single cube in this lattice and a 'success' as the occurrence of an organism in this cube. In these, as in all cases, the essential prerequisite for the validity of the Poisson distribution is that uniformity conditions (*b*) and (*c*) of 9.2 hold for the model we use to represent the variability; thus in a microscope count we require the cells to be equally likely to settle at any point on the slide, with no tendency to cluster, and in considering the soil counts we must satisfy ourselves that there is no factor which may lead to variation in the chance that an organism is found at any particular point. The requirement that the chance of an event occurring must be constant over time or space does not, of course, mean that the observed distribution of the event will be uniform, but it does imply that we can validly consider the Poisson distribution as a description of the observed variable.

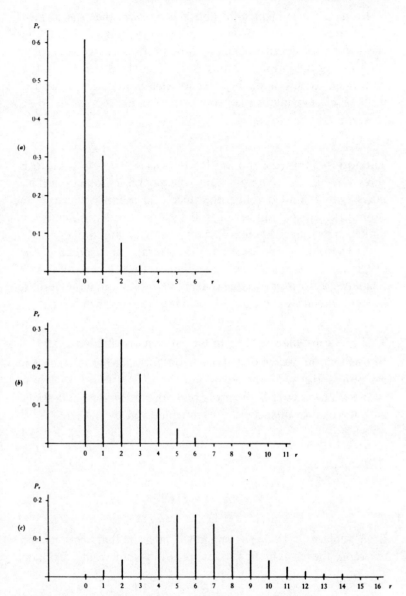

Fig. 30. The dependence of the Poisson distribution on the value of m: see 9.2.2. The values of m in (a), (b) and (c) are 0.5, 2 and 6. Although in theory very large values of r can occur the values of P_r are too small to plot; the sums of the unplotted values of P_r in the three figures are 0.001, 0.002 and 0.000.

From the argument used above it appears that the form of the distribution depends on two parameters, t and λ, but in fact these quantities occur together as the product λt and it is solely the value of this product which determines the distribution. It is convenient to use a single letter for this parameter, so we can write $m = \lambda t$, giving as the probability formula

$$P_r = \frac{e^{-m}m^r}{r!}.$$

The form of this distribution for the cases $m = 0.5$, 2 and 6 is shown in Fig. 30. We note that, whereas the binomial variable takes only a limited range of values (the number of successes cannot exceed the number of trials), the Poisson variable is not limited in its range although P_r is very small for large values of r.

As in the binomial case both the average and spread of the distribution depend on the value of m, so that the use of anovar to analyse a Poisson variable is likely to lead to unreliable results. Formulae relating average and spread to m are given in 9.2.2.1.

9.2.2.1* Formulae relating to the Poisson distribution

In this section we consider further the formula for P_r when r is Poisson, and discuss the mean and variance of r.

In 9.2.2 we saw that the probability of r 'successes' in n 'trials' with the chance of success at each trial given by

$$p = \lambda \frac{t}{n}$$

was expressed as

$$P_r = \frac{n(n-1)\ldots(n-r+1)}{1 \cdot 2 \cdot 3 \ldots r}\left(\frac{\lambda t}{n}\right)^r\left(1-\frac{\lambda t}{n}\right)^{n-r}.$$

If we replace λt by m, as suggested later in that section, and rearrange the terms in the above expressions somewhat, we obtain

$$P_r = \frac{m^r}{r!}\frac{n(n-1)\ldots(n-r+1)}{n^r}\frac{\left(1-\dfrac{m}{n}\right)^n}{\left(1-\dfrac{m}{n}\right)^r}.$$

We now wish to determine how this formula is modified as n becomes large; in more formal language we want to determine its limit as n tends to infinity. The first term on the right-hand side of the equation $(m^r/r!)$ does not depend on n at all, and is thus unchanged in the limit. The second term may be rewritten thus

$$\frac{n(n-1)\ldots(n-r+1)}{n^r} = 1\left(1-\frac{1}{n}\right)\cdots\left(1-\frac{r-1}{n}\right);$$

each of the factors on the right-hand side becomes near to 1 as n becomes large so that, since the number of factors, r, does not change, the whole term tends to the limit 1. The same reasoning applies to the denominator of the third term, i.e.

$$\left(1-\frac{m}{n}\right)^r \to 1 \quad \text{as } n \to \infty.$$

The numerator of the third term, $(1 - (m/n))^n$, is shown in books on elementary algebra to have the limit e^{-m}. Thus we have in the limit

$$P_r = \frac{m^r}{r!} \cdot 1 \cdot \frac{e^{-m}}{1} = \frac{e^{-m}m^r}{r!},$$

as stated in 9.2.2.

To obtain the mean of r we evaluate, as in the binomial case,

$$E(r) = \sum r P_r$$

$$= \sum_{r=0}^{\infty} r\, \frac{e^{-m}m^r}{1\,.\,2\,.\,3\ldots r}$$

$$= e^{-m} \sum_{r=1}^{\infty} \frac{m^r}{1\,.\,2\,.\,3\ldots(r-1)}.$$

To evaluate this sum we use the *exponential series*

$$e^x = \sum_{s=0}^{\infty} \frac{x^s}{s!}.$$

We first remove a factor m from every term in the expression for $E(r)$, thus obtaining

$$E(r) = m\,e^{-m} \sum_{r=1}^{\infty} \frac{m^{r-1}}{1\,.\,2\,.\,3\ldots(r-1)};$$

comparing m with x and $(r-1)$ with s we obtain

$$E(r) = m\, e^{-m} \cdot e^{m} = m.$$

We note that this result can also be obtained from the expression for $E(r)$ for the binomial, namely $E(r) = np$; in this we substitute $p = \lambda t/n = m/n$. This leads at once to $E(r) = m$, but there are in fact certain analytical difficulties in this form of argument, so that the approach given above is preferable.

For the distribution shown in Fig. 30 the means are thus 0.5, 2 and 6; it is instructive to note where the means fall on the distributions.

As in the case of the binomial, we do not consider the formal definition of the variance; it may be shown that

$$\text{var}(r) = m, \quad \text{s.d.}(r) = \sqrt{m},$$

but, as in the binomial case, there is no simple interpretation of the standard deviation in relation to probability.

9.2.3 Large sample forms

If we take large numbers of observations in either a binomial or a Poisson situation we find that the distributions become more and more nearly normal as the sample size increases; we have, indeed, already commented on the approximately normal form of some of the distributions plotted above.† The degree of closeness to normality depends on the sample size and, in the binomial case, on the chance of success; if this chance is close to 0 or 1 the sample size needs to be larger to give any required closeness to normality than if the chance is near 0.5. It is not easy to specify just what values of sample size and chance of success give distributions which can be regarded, for practical purposes, as normal. A reasonable guide is to require, in the binomial case, that the mean frequency of the less common event (success or failure) or, in the Poisson case, that the mean frequency of the event recorded, shall not be less than 10.

The fact that the distributions are close to normal does not, of course, mean that normal methods can be used uncritically,

† It was the consideration of the binomial distribution for large values of n which led to the first discovery of the normal distribution.

because all but the simplest normal methods require the equality of two or more variances. This condition is most unlikely to be satisfied for binomial and Poisson variables, because, for both distributions, the value of the variance is related to that of the mean. Thus in any analysis in which we are interested in the possibility of differences of means, the use of normal methods to analyse binomial or Poisson variables is clearly inappropriate.

9.3 Estimation and significance testing

In practical problems we are often reasonably confident that the conditions, discussed above, which lead to the binomial and Poisson distributions are satisfied, but we do not know the values of the parameters. In the binomial case the only unknown parameter is p, the chance of success in each trial, because n, the number of trials, is known; in the Poisson case the mean m is unknown. If we want values for these parameters we have to estimate them from the samples available; as usual we consider first point and then interval estimates.

9.3.1 Estimating p for a binomial distribution

Suppose we have a single sample of binomial observations in which we have obtained r successes in n trials. The estimate we use for the chance of success, p, is the proportion of successes in the sample, that is r/n.

To obtain the interval estimate for p we use one of the charts in Fig. A1, selecting the chart according to the degree of confidence (either 95 or 99 per cent) that we wish to have in our estimate. We enter the chart by first locating the proportion of successes, r/n, on one of the horizontal axes; the bottom axis contains values from 0 to 0.50, and the top axis contains values from 0.50 to 1. We then note where the ordinate through the value of r/n intersects the two curves corresponding to the sample size n; the chart shows only the curves for $n = 8, 10, 12, 16, 20, 24, 30, 40, 60, 100, 200, 400$ and 1000, but the points of intersection for other values of n can be found visually with sufficient accuracy. The confidence limits for p are then read on the appropriate ordinate scale as the values corresponding to the points of intersection just located; if r/n is read on the bottom horizontal scale we use the left-hand vertical scale, whilst if r/n is on the upper horizontal scale we use the right-hand vertical scale.

As an example, suppose we have observed 6 successes in 20 binomial trials, i.e. $r = 6$, $n = 20$. The point estimate of p is thus $r/n = 0.3$, whilst the 95 per cent confidence limits for p, read from the first of the two charts, are approximately 0.12 and 0.54. We note that these limits are not symmetrically placed about the point estimate; this is as we would expect because, except when $p = 0.5$, the binomial is a skew distribution.

9.3.2 Estimating the mean of a Poisson distribution

Suppose that in a single period of observation of a Poisson variable the event under study has occurred r times; we wish to estimate the value of the parameter m in the probability formula

$$P_r = \frac{e^{-m}m^r}{r!}$$

and to obtain confidence limits for this estimate. The point estimate of m is the observed value of r; the confidence limits are obtained from Table A19 which gives, for values of r up to 50, lower and upper confidence limits for m for the confidence levels 95 and 99 per cent. We note that, as in the binomial case, these limits are not symmetrical about the point estimate r; this is as we would expect, because the Poisson distribution is skew.

9.3.3 Two binomial samples

Consider the situation where in one set of n_1 binomial trials we have observed r_1 successes whilst in another set of n_2 trials there have been r_2 successes; these samples can be considered separately by the methods of 9.3.1, but we may well wish to go further and to learn what we can about the difference between the chances of success, p_1 and p_2, in the two populations. As in the previous discussions of differences (3.2, 6.6) we need to consider the point and interval estimates of the difference and also the significance test form of the comparison.

As usual, the point estimate of the difference between the populations is the difference between the point estimates for the two populations, i.e.

$$\text{point estimate } (p_1 - p_2) = \frac{r_1}{n_1} - \frac{r_2}{n_2}.$$

No simple interval estimate is available for this difference but a significance test of the null hypothesis $p_1 = p_2$ can be made using the χ^2-test for a 2×2 contingency table (4.2.1) with columns of the table corresponding to the two samples and rows corresponding to successes and failures.

If the null hypothesis is accepted then we may require an estimate of the (common) chance of success based on the two samples together. This is obtained by using the methods of 9.3.1 with number of successes $(r_1 + r_2)$ and number of trials $(n_1 + n_2)$.

9.3.4 Two Poisson samples

If we have observed a Poisson type of event to occur r_1 and r_2 times, and denote the means of the respective populations by m_1 and m_2, we have

$$\text{point estimate } (m_1 - m_2) = r_1 - r_2.$$

As in the binomial case, it is not easy to give confidence limits for this difference, but there is an approximate significance test for the null hypothesis $m_1 = m_2$, using Table A20. This table gives, for approximate significance levels of 5 and 1 per cent and for values of $(r_1 + r_2)$ from 1 to 80, values v such that if the smaller of r_1 and r_2 is less than or equal to v then the null hypothesis $m_1 = m_2$ should be rejected, at the stated level, in a two-sided test; in a one-sided test the significance levels would be half those given, i.e. 2.5 and 0.5 per cent. If, for example, we observe $r_1 = 18$, $r_2 = 6$ then $r_1 + r_2 = 24$, and the critical values for two-sided 5 and 1 per cent significance are 7 and 5 so that we reject the null hypothesis at the 5 per cent level, since the smaller of r_1, r_2 is less than the 5 per cent value but greater than the 1 per cent value.

This test is approximate because the calculation of probabilities depends on the value of $m_1 = m_2$, which is, of course, not known. Table A20 is constructed so that the true probability is always less than the value given – by an amount depending on the value of $m_1 = m_2$ – so that the chance of concluding that the populations differ when they are in fact the same is less than the stated significance level.

If the null hypothesis is accepted then we can say that $(r_1 + r_2)$ is a Poisson variable with mean $2m$, where m is the common mean of the two distributions under investigation. We can use the procedure of 9.3.2 to obtain point and interval estimates for $2m$; values applicable to m are obtained by halving these values for $2m$.

9.3.5 More than two samples

If we have binomial or Poisson samples from more than two populations we can estimate the values of p or m for individual populations by the methods used above (9.3.1, 9.3.2) but we need to extend the significance tests used for comparing values from different populations. In this section we consider the case where the population factors form a single-factor arrangement; more complex designs require the methods discussed in 9.4.

In the binomial case the appropriate significance test is that for a $2 \times k$ contingency table, the rows of the table corresponding to successes and failures and columns corresponding to the k levels of the (single) population factor. The computations for this test have been described in 4.2.2; the test is the direct extension of the two-level case considered in 9.3.3. An example of the use of this test is given in 9.3.5.1.

In the Poisson case we use another form of χ^2-test. Suppose that we have k Poisson observations $r_1, r_2, \ldots, r_k$ and wish to test the null hypothesis $m_1 = m_2 = \cdots = m_k$. To do this we calculate first

$$\bar{r} = \frac{1}{k} \sum_{i=1}^{k} r_i,$$

and then use as test statistic

$$\chi^2 = \frac{1}{\bar{r}} \sum_{i=1}^{k} (r_i - \bar{r})^2,$$

determining its significance from Table A5 with degrees of

freedom $v = k - 1$.† The sum of squares $\sum (r - \bar{r})^2$ can of course be calculated from the expression

$$\sum_{i=1}^{k} (r_i - \bar{r})^2 = \sum_{i=1}^{k} r_i^2 - \frac{\left(\sum_{i=1}^{k} r_i\right)^2}{k}$$

given in 6.3.1.

9.3.5.1 An example of the comparison of binomial distributions

As part of a study of the disease scrapie in sheep it was wished to determine whether susceptibility in the ewe was associated with susceptibility in the lamb. Sheep from a susceptible breed were tested, the ewes representing two strains (A and B) bred in different parts of the United Kingdom. In what follows an animal is referred to as ($+$) if it developed scrapie after challenge with infected brain and ($-$) if it did not.

For strain A the results were as follows:

4 ($+$) ewes had single ($+$) lambs,
1 ($+$) ewe had twins, one of which was ($+$) and one ($-$),
3 ($+$) ewes had only ($-$) lambs,
3 ($-$) ewes had single ($+$) lambs,
2 ($-$) ewes had twins, one of which was ($+$) and one ($-$),
1 ($-$) ewe had triplets, two of which were ($+$) and one was ($-$),
13 ($-$) ewes had only ($-$) lambs.

The numbers in these groups are too small to permit reliable use of the contingency table test (4.2.3) and so groups were combined by taking together all ($+$) ewes which had any ($+$) lambs and similarly all ($-$) ewes which had any ($+$) lambs. This reduced

---

† This test is a simple example of the use of the χ^2 distribution in a *test of goodness of fit*, that is a test for comparing an observed distribution with one calculated on theoretical grounds. We have already used the Kolmogorov–Smirnov one-sample test for such a purpose (7.2.3.1), but in the present case the χ^2-test is more commonly used. The comparison is between the observed frequencies $r_1, r_2, \ldots, r_k$ and the value $\bar{r}$ which is the best estimate of the frequency we would expect to get in every sample if the null hypothesis $m_1 = m_2 = \cdots = m_k$ were true; the calculation takes the form

$$\sum \{(\text{obs. freq.} - \text{exp. freq.})^2/\text{exp. freq.}\}$$

used for χ^2 in 4.2.2.

the data to the following form:

  5 (+) ewes had at least one (+) lamb,
  3 (+) ewes had only (−) lambs,
  6 (−) ewes had at least one (+) lamb,
13 (−) ewes had only (−) lambs.

The corresponding frequencies for strain B ewes were 18, 8, 15 and 42; the next step was to test whether the two strains could reasonably be regarded as the same in susceptibility, for if they could, then clearly the best test of maternal influence on susceptibility would be obtained by combining the results from the two strains. The test to compare strains required the calculation of χ^2 for the 2×4 contingency table shown in Table 53. This table also shows the steps of the calculation, which for purposes of illustration has been performed by two different methods, (a) using the general formula given in 4.2.2 and (b) a shorter method applicable only to the $2 \times k$ case. The resulting value of $\chi^2 - 0.351$ with 3 d.f. − is very far from significance at the conventional levels so that we are clearly entitled to claim that the two strains are the same in susceptibility and therefore to use the total frequencies for the two strains. This leads to the 2×2 contingency table shown in Table 54, for which the value of χ^2 is

$$\frac{(55 \times 23 - 21 \times 11)^2 \times 110}{34 \times 76 \times 44 \times 66} = 15.67.$$

This value is highly significant ($P < 0.1$ per cent with 1 d.f.) and shows that the data do not allow us to maintain that susceptibility in the lamb is not associated with susceptibility in the ewe: we are not, of course, able to say from these data what causes the association because the ewes both bore and suckled the lambs.

9.4 Transformations

We saw above that, except for large samples, the binomial and Poisson distributions differ appreciably from the normal and that, even in large samples, normal methods are further precluded because the variance of the binomial or Poisson is related to the

Table 53. *Contingency table showing numbers of ewes of two strains classified according to their own and their lambs' response to challenge with scrapie-infected brain, together with two methods of calculating the χ^2 which tests association between strain and susceptibility (see 9.3.5.1).*

Ewe susceptibility	+		−		Totals
Lamb susceptibility	+	−	+	−	
Strain A	5	3	6	13	27
Strain B	18	8	15	42	83
Totals	23	11	21	55	110

(a) Calculation of χ^2 by the method of 4.2.2
   (i) Expected frequencies

5.645	2.700	5.155	13.500	27
17.355	8.300	15.845	41.500	83
23	11	21	55	110

   (ii) (obs. freq. − exp. freq.)^2/exp. freq.

0.074	0.033	0.139	0.019	
0.024	0.011	0.045	0.006	$\chi_3^2 = 0.351$

(b) Calculation of χ^2 by a short-cut method applicable to the $2 \times k$ contingency table
   (i) In the notation of 4.2.2 calculate a row of values $p_j = O_{ij}/C_j$ choosing $i =$ either 1 or 2, and, for the same value of i, calculate $\bar{p} = R_i/N$. In the present case, taking $i = 1$ gives

$$p_1 = 5/23 \qquad p_2 = 3/11 \qquad p_3 = 6/21 \qquad p_4 = 13/55 \qquad \bar{p} = 27/110$$
$$ = 0.21739 \quad\ = 0.27273 \quad\ = 0.28571 \quad\ = 0.23636 \quad\ = 0.24545$$

   (ii) Calculate

$$\chi_{k-1}^2 = \frac{\sum\limits_{j=1}^{k} O_{ij} p_j - R_i \bar{p}}{\bar{p}(1 - \bar{p})},$$

where i has the same value as was chosen in (i). Then

$$\chi_3^2 = \frac{0.06493}{0.18520} = 0.351.$$

mean and so is not constant throughout a set of data. Nevertheless, there are some methods of analysis applicable to binomial and Poisson variables which are closely related to the normal methods discussed in chapters 6–8; these methods all involve the use of some transformation of the binomial or Poisson variable, that is the analysis is carried out not on the variable observed, but on some other quantity calculated from it. The theory underlying these procedures, both in regard to the choice

Table 54. *Contingency table showing numbers of ewes in the four classes defined by ewe/lamb susceptibility/non-susceptibility to challenge with scrapie (see 9.3.5.1).*

		Ewe		
		(+)	(−)	
Lamb	(+)	23	21	44
	(−)	11	55	66
		34	76	110

of transformation and to the methods used for analysing the transformed variable will not be discussed here; it can be approached from several points of view and a balanced description requires concepts that are too advanced for this book.

9.4.1 The angular transformation

This transformation is applied to binomial variables, usually with the object of making the variance independent of the mean. The transformation converts the binomial proportion r/n to the angle θ defined by the relations

$$\frac{r}{n} = \sin^2 \theta \quad \text{or} \quad \theta = \sin^{-1} \sqrt{\frac{r}{n}}.$$

The angle is sometimes measured in degrees and sometimes in radians; a table of the transformation in the Appendix (Table A21) gives the angle in degrees. It may be proved that

$$\text{var}(\theta) \approx \frac{820.7}{n} \quad \text{if } \theta \text{ is measured in degrees,}$$

$$\approx \frac{1}{4n} \quad \text{if } \theta \text{ is measured in radians;}$$

in neither case does the variance depend on the mean of θ, and if n, the number of trials, is the same for all samples then the variance of θ may be taken as approximately constant, so that one obstacle to the use of normal methods is removed. This is particularly convenient when the population factors influencing the chances of success of the binomial distribution do not form a single-factor arrangement, because in these cases we cannot use the χ^2 procedures discussed in 9.3.3, 9.3.5. There is, however, no reason to assume that, just because the variance of θ is constant, we can use normal methods to analyse values of θ, because we need also to consider the form of the distribution of θ; this is not exactly normal, but is often sufficiently close to normality to ensure that reliable results can be obtained by using normal methods.†

If n is not the same for all samples the variance of θ is not constant and so normal methods as described earlier cannot be used. It is, however, possible still to use these methods provided that we modify the formulae which lead to the sums of squares: in most cases these modifications involve algebra and computing procedures beyond our present scope, but in a few simple cases – single-factor designs and regression – the only modification needed is to use *weighted* sums of squares in place of the forms given earlier. The idea underlying this procedure is that since different values of θ are not all equally accurate it is necessary to attach more importance (more weight) to the more accurate values; this is achieved by associating with each value of θ a weight, denoted conventionally by w, which is in fact the number of trials n, and by altering the previous formulae for sums of squares as follows:

(a) replace the number of observations by $\sum w$;
(b) replace $\sum \theta$ by $\sum w\theta$;
(c) replace $\sum \theta^2$ by $\sum w\theta^2$.

Thus a sum of squares that we would expect to calculate as

$$\sum \theta^2 - \frac{(\sum \theta)^2}{n}$$

(where, of course, n is the number of values of θ) should be calculated as

$$\sum w\theta^2 - \frac{(\sum w\theta)^2}{\sum w}.$$

The degrees of freedom of the s.s. do not change.

---

† More exact methods, which do not depend on this approximation, are available when maximum accuracy is needed; see, for example, the Introduction to Fisher and Yates (1974). These methods are to be preferred to the use of the 'continuity correction' recommended by some authors.

As an illustration of these procedures we will consider the analysis of the observations reported by Gordon (1957). With the object of separating spermatozoa carrying the Y chromosome from those carrying the X, samples of rabbit semen were electrophoresized; spermatozoa collected from the cathode and anode were used separately for artificial insemination. The numbers of males obtained in each litter are shown in Table 55

Table 55. *Observations of offspring produced by insemination with spermatozoa collected after electrophoresis of rabbit semen* (Gordon, 1957) (*see* 9.4.1).

Spermatozoa attracted to cathode			Spermatozoa attracted to anode		
Number of males (r)	Litter size $(n = w)$	Angular transform $\left(\theta = \sin^{-1}\sqrt{\dfrac{r}{n}}\right)$	Number of males (r)	Litter size $(n = w)$	Angular transform $\left(\theta = \sin^{-1}\sqrt{\dfrac{r}{n}}\right)$
1	2	45.0	2	3	54.7
3	5	50.8	1	4	30.0
4	4	90.0	1	4	30.0
6	8	60.0	2	5	39.2
1	2	45.0	0	4	0
3	8	37.8	0	7	0
3	8	37.8	1	1	90.0
3	7	40.9	1	8	20.7
4	7	49.1	4	10	39.2
3	7	40.9	2	5	39.2
			5	8	52.2
			2	7	32.3
			1	6	24.2
$\sum r = 31$	$\sum w = 58$ $\sum w\theta = 2795.1$ $\sum w\theta^2 = 145\,359.65$		$\sum r = 22$	$\sum w = 72$ $\sum w\theta = 2232.6$ $\sum w\theta^2 = 91\,052.58$	

– the samples which Gordon described as 'not blind tested' have been omitted. This table also gives the angular transforms in degrees of the proportions of males in the litters and the values of $\sum w\theta$ and $\sum w\theta^2$. The anovar comparison between the two samples now proceeds as in 7.2.1; Table 56 is analogous to Table 34, containing the preliminary calculations, and Table 57 is analogous to Table 35, giving the anovar itself. The difference between the means of θ in the two groups is significant at the 5 per cent level, showing that we cannot reasonably regard the mean proportion of male offspring as the same in litters produced by the use of the two types of semen.

Table 56. *Preliminary calculations for the weighted anovar to compare the two groups in Table 55; the layout of the table is based on that of Table 34.*

| | Levels | | |
	1 (cathode)	2 (anode)	Total
Number of samples	10	13	23
$\sum w$	58	72	$\sum (\sum w) = 130$
$\sum w\theta$	2795.1	2232.6	$\sum (\sum w\theta) = 5027.7$
$\bar{\theta} = \dfrac{\sum w\theta}{\sum w}$	48.19	31.01	
$\sum w\theta^2$	145 359.65	91 052.58	$\sum (\sum w\theta^2) = 236\,412.23$
$\dfrac{(\sum w\theta)^2}{\sum w}$	134 699.72	69 229.21	$\sum \dfrac{(\sum w\theta)^2}{\sum w} = 203\,928.93$
			$\dfrac{[\sum (\sum w\theta)]^2}{\sum [\sum w]} = 194\,444.36$

The residual mean square provides a measure of the variation in the proportion male between litters and may be compared with the value of 820.7 (see above) which would be expected if the binomial conditions held both between and within litters. This comparison is made by a two-sided F-test (6.6.4); we have $s^2 = 1546.8$ with 21 d.f. and $\sigma^2 = 820.7$ with ∞ d.f. (as it is a *known* value). The value of $F_{21,\infty}$ is thus 1.88; from Table A15

Table 57. *Weighted single-factor anovar to compare the two groups in Table 55; the preliminary calculations are given in Table 56 and the procedure is based on that in Table 35.*

Source of variation	S.S.	d.f.	M.S.	F
Between levels (cathode v. anode)	$203\,928.93 - 194\,444.36 = 9484.57$	$2-1 = 1$	9484.57	6.13*
Residual	$41\,967.87 - 9484.57 = 32\,483.30$	$23-2 = 21$	1546.82	
Total	$236\,412.23 - 194\,444.36 = 41\,967.87$	$23-1 = 22$		

the acceptable limits (5 per cent level) for $F_{12,\infty}$ are 0.364, 1.94 and for $F_{24,\infty}$ are 0.515, 1.64. It is clear, even with very rough interpolation, that our observed value is outside the acceptable limits and so we cannot reasonably maintain that the binomial conditions hold.

This, in turn, means that we are not justified in simply taking totals over litters and comparing the results in a 2×2 contingency table (i.e. a table with columns corresponding to cathode and anode counts and rows corresponding to male and female counts). This procedure is at first sight attractive because of its simplicity, but is shown not to be valid by the F-test just described; as a general rule it is preferable not to use a theoretical value, nor any procedure based upon it (as the χ^2-test for the 2×2 table is based on the theoretical argument justifying the use of the binomial), if an empirical value is available. We may, of course, reasonably ask whether the use of the angular transformation is justified if the binomial conditions do not hold; the best answer we can give is that it probably provides a nearer approximation to meeting anovar conditions than does the analysis of the untransformed proportions.

9.4.1(c)   Weighted anovar in Genstat

The commands and data for the Genstat analysis of the observations in Table 55 are shown in CTable 31; the commands follow closely the pattern of earlier examples with only two

CTable 31. *Commands and data for the Genstat analysis of the observations in Table 55.*

```
units [nvalues = 23]
factor [labels = !t(Cath, An)]   electrode
read           electrode, male, litter
1 1 2   1 3 5   1 4 4   1 6 8   1 1 2   1 3 8   1 3 8
1 3 7   1 4 7   1 3 7   2 2 3   2 1 4   2 1 4   2 2 5
2 0 4   2 0 7   2 1 1   2 1 8   2 4 10  2 2 5   2 5 8
2 2 7   2 1 6 :
treatments   electrode
calculate     transform = ang(100*male/litter)
anova [weight = litter; fprob = yes]   transform
```

features requiring comment. The first of these is the use of a calculate statement to obtain the angular transforms of the proportions of males; three of the many functions available in calculations are shown, namely

*   indicating multiplication,
/   indicating division, and
ang   indicating the angular transform (in degrees) of a percentage.

The second new feature is the use of the option [wt = litter] in the anova statement; this has the effect of modifying the anovar computations in the manner described in 9.4.1.

The output of this analysis is given in CTable 32; the reader will notice small divergences from the results given in Table 57, occurring because the angles in the present analysis are calculated more precisely than the values quoted in Table 55.

The calculations of means differ from the familiar pattern in exactly the same way as do the calculations of sums of squares,

CTable 32. *Output of the Genstat analysis commanded in CTable 31.*

************* Analysis of variance *************

Variate: transfor

Source of variation	d.f.	s.s.	m.s.	v.r.	F pr.
electrod	1	9461.	9461.	6.11	0.022
Residual	21	32533.	1549.		
Total	22	41994.			

************* Tables of means *************

Variate: transfor

Grand mean   38.7

electrod	Cath	An
	48.2	31.0
wt.rep.	58.00	72.00

********** Standard errors of differences of means **********

Table	electrod
wt.rep.	unequal
s.e.d.	6.94

applying the rules for altering formulae explained in 9.4.1; thus, the grand mean value of 38.7 (degrees) can be verified as the ratio

$$\frac{\sum (\sum w\theta)}{\sum (\sum w)} = \frac{5027.7}{130} = 38.67,$$

where the numerator and denominator in this fraction are to be found in the Total column of Table 56. The values given for Cath and An relate similarly to entries in the cathode and anode columns of the same table. The values of wt.rep. in the tables of means section are the totals of the weights $(\sum w)$ in the two groups; this form of (abbreviated) title is appropriate because one of the more common uses of weighting is – as here – to take account of inequalities in replication.

The information provided in the standard errors of differences of means section takes due account of the unequal replication in the treatment groups (as was the case in CTable 20), but in the present table only one s.e.d. is given (whereas there were three in the earlier table), since with only two groups to compare no useful purpose would be served by quoting a selection of values. In addition, the actual calculation of the s.e.d. has to be modified to take account of the use of weighting in the analysis; this is achieved by the use of the expression

$$\frac{s^2}{\sum w}$$

for the variance of a mean rather than the familiar s^2/n; thus, in the present case the variance of the difference between the two means is estimated by

$$\frac{1549}{58} + \frac{1549}{72} = 48.22,$$

giving

$$\text{s.e.d.} = 6.944.$$

9.4.2 The square root transformation

This transformation is used with Poisson variables in the same way as the angular transformation is used with binomial variables; in place of the observed Poisson count r the variable used in analysis is x where

$$x = \sqrt{r}.$$

It may be shown that

$$\mathrm{var}(x) \fallingdotseq 0.25$$

whatever the mean of the parent Poisson distribution. The distribution of x is often sufficiently close to normal to justify the use of normal methods: since $\mathrm{var}(x)$ does not depend on the mean of x, the need for weighting, as described for the angular transformation, does not arise.

9.4.3 The probit transformation

This transformation is often used in the analysis of dose-response relationships when the response variable is a binomial proportion, r/n. It is useful when, for example, the threshold of response is normally distributed in the test population because in this case it reduces the dose-response relationship to a straight line. The probit corresponding to r/n is defined as $(x + 5)$ where x is the standardized normal deviate such that $\Phi(x) = r/n$ (see 6.2.1); the constant 5 is added so that the probit is positive in all practical cases. Thus if $r = 7$, $n = 8$ we have $r/n = 0.875$ and the value of x given by Table A11 such that $\Phi(x) = 0.875$ is $x = 1.15$, so that the probit corresponding to r/n is 6.15. Tables giving the probit transformation directly are available in, for example, Fisher and Yates (1974).

The variance of the probit depends both on n and on the population proportion of which r/n is an estimate, so that weights must be used in all analyses of probits; the computations tend therefore to be lengthy and, since the usage is a little specialized, will not be discussed here. A suitable book with which to begin further reading is Finney (1971).

9.4.4 The effect of transformations on interactions

One difficulty found with the use of any transformation is that it affects the apparent importance of interactions; in some cases it can make us think that an interaction exists when the untransformed data show no evidence of it. Suppose, for example, that in studying the effect of two factors, A at 4 levels and B at 3 levels, on a Poisson variable we obtained the counts shown in section (a) of Table 58. These counts show no interaction at all; the difference between any two levels of A is the same for all levels of B (and, of course, vice versa). When, however, we consider the counts transformed to square roots, given in section (b) of the table, we find a noticeable interaction. Thus the difference between levels 1 and 2 of A is 0.54 at level 1 of B, 0.45 at level 2 and 0.42 at level 3; the difference between levels 3 and 4 of A, which, in the untransformed counts, was the same as that between levels 1 and 2, is 0.39 at level 1 of B, 0.35 at level 2 and 0.34 at level 3. These changes in the differences between levels must clearly make it very difficult to interpret the result of any analysis of the transformed counts.

Table 58. *The effect of a transformation in producing an apparent interaction between two factors (see 9.4.4).*

	Levels of factor A			
Levels of factor B	A_1	A_2	A_3	A_4
(a) Untransformed values				
B_1	12	16	24	28
B_2	18	22	30	34
B_3	21	25	33	37
(b) Transformed values				
B_1	3.46	4.00	4.90	5.29
B_2	4.24	4.69	5.48	5.83
B_3	4.58	5.00	5.74	6.08

Similar difficulties arise in the use of a transformation in the analysis of any design involving an interaction. Moreover in regression analysis, although in simple cases no interactions are involved, the use of a transformation will alter the shape of the regression relationship and thus may make interpretation difficult – although in some cases, such as probit analysis, the transformation may actually make the relationship easier to interpret. In most analyses where an interaction is relevant or where a regression relationship is made more difficult to interpret, it is probably better to work with the variables observed rather than with transformations of them. In many cases the means will be large enough for the variables to be approximately normal (see 9.2.3); it may then be justifiable to use *non-orthogonal* weighted anovar methods to take account of differences of variance. These procedures are computationally elaborate, and will not be discussed here, but they can provide conclusions relating to meaningful variables: there is no sense in using a transformation if we thereby obtain a valid and simple analysis of a meaningless variable.

9.5 The multinomial distribution

In this section we will consider rather briefly an extension of the binomial model. Suppose that conditions (*a*), (*b*) and (*c*) of 9.2 are replaced by the following:

(*a*) that for any one observation only k results are possible;

(*b*) that the chances of these k results, denoted by $p_1, p_2, \ldots, p_k$ with

$$\sum_{i=1}^{k} p_i = 1,$$

do not vary from one observation to the next;

(*c*) that successive observations are independent in the sense of 4.2.3.

We now extend the question considered in 9.2.1 to the form 'What is the probability, $P(r_1, r_2, \ldots, r_k)$, that in a given number, n, of trials we obtain the first possible result r_1 times, the second r_2 times, $\ldots$, the kth r_k times?' We require of course

$$\sum_{i=1}^{k} r_i = n.$$

The answer to this question is the *multinomial distribution*, specified by

$$P(r_1, r_2, \ldots, r_k) = \frac{n!}{r_1! \, r_2! \ldots r_k!} \, p_1^{r_1} p_2^{r_2} \ldots p_k^{r_k}.$$

If $k = 2$ this formula is easily shown to reduce to the binomial probability given in 9.2.1; the proof of the formula is very similar to that for the binomial given in 9.2.1.1. To summarize this distribution it is necessary to consider the mean and variance of each separate variable (i.e. of $r_1, r_2, \ldots, r_k$); it may be shown that

$$E(r_i) = np_i,$$

$$\text{var}(r_i) = np_i(1 - p_i).$$

The expression for $E(r_i)$ is derived by an argument similar to that used in 9.2.1.1 for the binomial but, as in that section, the derivation of the variance requires concepts beyond our present scope. In addition, as we have several variables, we require the population form of the covariance, mentioned in 8.2;† as in the case of variance, we will not give the formal definition, but it may be shown that

$$\text{cov}(r_i, r_j) = -np_i p_j.$$

The analysis of multinomial observations involves all the difficulties discussed above in considering the analysis of binomial observations and, in addition, since there are several related variables, may require methods of *multivariate analysis*; we will give in the next section one example which does not require new methods.

9.5.1 An example of a test of goodness of fit

Suppose that in a simple genetic experiment we consider alleles A and a at one locus and alleles B and b at a second locus: suppose that A is dominant over a and B over b and that we wish to test whether the two loci are linked. If they are not linked then in the progeny of matings between double heterozygotes, AB/ab, the phenotypes AB, Ab, aB, ab will occur in the proportions $9:3:3:1$. We thus have a multinomial situation; if r_1, r_2, r_3, r_4 denote the numbers of offspring of the four phenotypes then, under the null hypothesis of no linkage, we have $p_1 = 9/16$, $p_2 = 3/16$, $p_3 = 3/16$, $p_4 = 1/16$. Suppose that among 64 offspring ($n = 64$) we find

$$r_1 = 41, \qquad r_2 = 9, \qquad r_3 = 8, \qquad r_4 = 6;$$

are these numbers compatible with the null hypothesis?

---

† In the binomial case we could at first sight consider the covariance between the number of successes and the number of failures, but this is not interesting because, for a given n, these two numbers determine each other.

For this comparison we use the χ^2-test of goodness of fit, described in the footnote to 9.3.5. The observed frequencies are the values of r_1, r_2, r_3, r_4 given above and the expected frequencies are the means of r_1, r_2, r_3, r_4, namely, np_1, np_2, np_3, np_4, as given in 9.5; we thus have expected frequencies

$$64 \times \tfrac{9}{16} = 36 \quad \text{corresponding to } r_1,$$
$$64 \times \tfrac{3}{16} = 12 \quad \text{corresponding to both } r_2 \text{ and } r_3$$

and

$$64 \times \tfrac{1}{16} = 4 \quad \text{corresponding to } r_4.$$

Then

$$\chi^2 = \frac{(41-36)^2}{36} + \frac{(9-12)^2}{12} + \frac{(8-12)^2}{12} + \frac{(6-4)^2}{4}$$
$$= 0.693 + 0.750 + 1.333 + 1.000$$
$$= 3.78.$$

We test this with degrees of freedom

$$v = \text{number of classes} - 1$$
$$= 3,$$

using Table A5. We find the probability corresponding to this value of χ^2 with 3 d.f. is greater than 10 per cent, and so we are entitled to accept the null hypothesis of no linkage.

The rule just used for finding the degrees of freedom in a test of goodness of fit is adequate provided that, as in the present case, the null hypothesis is enough to enable us to calculate how the total frequency n may be expected to divide among the classes. If this is not the case we have to estimate the undefined parameters of the problem from the observed frequencies, and lose one degree of freedom for each parameter estimated. Thus if, in this example, we had not been able to assume complete dominance of A and B, the null hypothesis of no linkage would have left us needing to estimate the degree of dominance for each locus, so that the number of degrees of freedom would have been reduced to 1.

9.6 Exercises

Exercises 1–7 give the opportunity to study a few members of the binomial and Poisson families using the formulae in 9.2 et sqq. In each case:

(*a*) *evaluate the probabilities;*

(*b*) *plot the distribution;*

(*c*) *calculate the mean directly, using the formula mean* $= \sum rP_r$ *with the values of* P_r *obtained in* (*a*);

(*d*) *calculate the mean, variance and standard deviation by using the formulae given in 9.2.1.1 or 9.2.2.1: compare the mean with the value obtained in* (*c*);

(*e*) *mark the mean on the plot of the distribution: note where it lies relative to the mode* (*the value of r for which* P_r *is a maximum*);

(*f*) *calculate the limits* (*mean* $\pm$ 1.645 *s.d.*) *and mark them on the plot of the distribution: in the normal case these limits would enclose 90 per cent of the distribution, excluding 5 per cent at each end, and it is instructive to see how far this property holds in the present cases.*

1 Consider the binomial with $n = 12$, $p = 0.3$.

2 Consider the binomial with $n = 12$, $p = 0.5$.

3 Consider the binomial with $n = 18$, $p = 0.2$; compare it with the distribution in exercise 1.

4 Consider the binomial with $n = 30$, $p = 0.2$; compare it with the distributions in exercises 2 and 3.

5 Consider the Poisson with $m = 6$; compare it with the distributions in exercises 2 and 4.

6 Consider the Poisson with $m = 1$.

7 Consider the Poisson with $m = 10$.

Exercises 8–15 involve the methods of estimation and significance testing described in 9.3 et sqq. for use with one or two binomial or Poisson samples.

8 In an influenza epidemic 14 children out of a class of 38 contracted the infection. Estimate, with confidence limits, the probability of infection.

9 A small survey showed that out of 43 men aged 25 or less, 12 had false teeth. Test whether this proportion differed significantly from the value 0.37 for this proportion obtained in a large survey reported by another worker.

10 Jessica and Anne were twin sisters: Jessica had blue eyes and Anne had brown. Out of the 23 children and grandchildren born to Jessica before her death, 6 had blue eyes: out of the 31 born to Anne by the same date, 14 had blue eyes. Test whether these proportions differ significantly, and give confidence limits either for the separate proportions or, if they can justifiably be combined, the overall proportion.

11 A smallholder kept a few chickens of each of two breeds. In a certain period, out of the 95 eggs he obtained from birds of one breed, 20 were brown whilst out of 87 eggs from birds of the other breed, 34 eggs were brown. Test whether these proportions differed significantly.

12 In a six-month period the number of sensitive reactions to tetanus antigen treated in a small hospital was 14. Obtain confidence limits for the number to be expected in the following six months.

13 A greengrocer found 8 damaged tomatoes in a crate from a new supplier. Estimate the limits within which he expected to find the number damaged in the next crate.

14 In the month before a hospital theatre was closed for sterilization, 14 patients contracted post-operative staphylococcic infections; in the month after sterilization the number was 4. Test whether this difference was significant.

15 Two seed merchants each supplied a sample of a certain variety of wheat, and one half-acre plot was grown from each sample. One plot contained 16 plants with a certain genetic defect, whilst the other contained 24 plants. Test whether these frequencies differed significantly and give confidence limits either for the separate population frequencies or, if they can justifiably be combined, for the overall frequency.

Exercises 16–25 illustrate the use of the angular and square root transformations discussed in 9.4 et sqq. The experimental designs used have been limited to those of the single-factor family because, as explained in 9.4.1 and 9.4.4, two-factor designs involve complexities of analysis that we do not consider.

16 In an experiment to compare the resistances to rust of 5 varieties of wheat, 30 batches of 10 randomly selected ears were taken from single demonstration plots of the 5 varieties; the numbers of infected ears in the individual batches were as follows:

Variety						Variety				
A	B	C	D	E		A	B	C	D	E
2	1	3	4	3		5	0	5	1	3
4	2	6	1	5		1	3	3	0	3
7	1	2	3	2		2	3	4	2	2
4	3	4	5	4		7	4	6	4	4
5	2	5	2	2		5	3	7	4	6
3	4	7	2	3		2	3	4	1	3
4	2	4	4	4		4	2	8	2	4
0	4	6	4	4		2	2	5	2	4
1	0	4	3	6		6	1	7	1	4
1	2	7	2	2		3	1	5	1	1
2	3	7	2	2		4	1	3	1	2
2	4	3	3	3		1	5	5	0	4
3	3	6	0	4		0	5	3	0	4
6	4	5	0	4		0	2	5	2	4
4	2	8	1	4		1	2	2	3	2

Transform the proportions of infected ears to angles and compare the angular values for the 5 varieties in a single-factor anovar; weights are not needed since each proportion is based on the same number of ears (10). Comparisons between the varieties can be made by the methods discussed in chapter 7. The residual mean square can be compared with the value (82.07) expected under binomial conditions, to serve as an indication of how far these conditions hold.

The design of this investigation can be criticized on the ground that since the different varieties were grown on different plots, the residual used to test differences between varieties should include the between-plots random variation, whilst that obtained by this

*design – one demonstration plot for each variety – did not do so.
This error is very easy to slip into; it can be avoided by ensuring
that there is replication at all relevant levels of the design process,
in this case by growing several plots of each variety in a properly
randomized arrangement. An example of such an experiment is
given in the next exercise.*

17 Thirty plots were used, 6 for each variety, and 5 batches of
10 ears were taken at random from each plot, thus using the
same total number of batches as in exercise 16; the numbers of
infected ears were as follows:

Variety	Plot	Numbers of infected ears				
A	1	4	4	5	4	5
	2	2	3	3	2	2
	3	1	2	2	8	2
	4	2	4	7	4	4
	5	3	1	1	4	4
	6	3	4	3	6	4
B	7	3	1	2	2	1
	8	6	3	3	1	2
	9	3	5	1	4	4
	10	1	1	3	2	3
	11	5	3	5	4	5
	12	1	2	1	1	2
C	13	5	8	6	4	4
	14	6	4	8	6	6
	15	8	4	4	7	4
	16	2	4	6	1	5
	17	6	4	7	5	5
	18	2	5	5	4	3
D	19	4	5	3	3	3
	20	2	2	1	1	5
	21	3	3	2	3	3
	22	0	3	1	5	3
	23	2	2	0	2	2
	24	4	2	2	4	0
E	25	2	5	4	3	4
	26	2	2	2	1	5
	27	5	4	3	4	1
	28	0	5	5	2	3
	29	4	2	3	2	5
	30	0	4	3	2	2

Once again the variable analysed is the angular transformation of the proportion of infected ears, without the use of weights. The anovar, however, requires an extension of the single-factor type used above: it is best approached by considering first the anovar for a single variety: this is

Plots within the variety	5
Batches within the plots	24
Variety total	29

This follows the familiar single-factor pattern, both in degrees of freedom and in the calculation of the s.s.'s. When we consider the overall analysis, including all 5 varieties, the total number of observations is 150, so that the d.f. for the total s.s. are 149. The 5 single-variety analyses account for $5 \times 29 = 145$ of these, and the remaining 4 relate to the between varieties term; the anovar thus takes the form

Varieties	4	
Plots within varieties	25 –	5 from each variety
Batches within plots	120 –	24 from each variety
Total	149	

The s.s.'s for varieties and total are obtained in the normal manner; the s.s.'s for plots within varieties and batches within plots are obtained, as are their degrees of freedom, by totalling the appropriate terms from the individual variety analyses.

This type of analysis, characterized by the sequence of sources of variation each one within the one above – batches *within* plots, plots *within* varieties – is known as a *nested* or *hierarchical* anovar. We note especially that it does not contain any interaction terms, because the between batches variation is a single factor, albeit of a more complex type than we have met before. It is, however, possible to find a two-factor design in which one – or indeed both – of the factors forms a nested arrangement, and in the analysis of which there will therefore be interaction between (one level of) one nested factor and (one level of) the other nested factor, but never between two levels of one factor: such analyses are beyond our scope here.

The general treatment of significance tests in nested designs is complex, but a satisfactory procedure for the present case is to test (*a*) the plots within varieties M.S. against the batches within plots M.S., thus testing the null hypothesis that all the plots of a single variety have the same mean rate of infection, and (*b*) the varieties M.S. against the plots within varieties M.S., testing the null hypothesis that the variety means are all the same. Some authors argue that the procedure for testing this latter null hypothesis should depend on the result of test (*a*) but in this example, with adequate degrees of freedom for the plots within varieties term, there is little need for this refinement.

18 In an experiment to compare the effects on conception rate of 4 semen diluents for use in artificial insemination in cattle, 96 semen collections from a group of bulls were assigned randomly 24 to each diluent, the results for individual collections being recorded separately as follows (*r* conceptions out of *n* inseminations):

			Diluent				
P		*Q*		*R*		*S*	
r	*n*	*r*	*n*	*r*	*n*	*r*	*n*
33	57	20	38	16	35	32	43
28	49	36	66	38	53	33	51
12	21	41	58	30	44	42	63
34	59	23	40	47	74	20	25
29	58	23	36	54	77	14	20
47	74	34	54	29	42	33	49
43	61	20	38	42	57	41	60
24	52	35	53	33	48	53	80
25	38	35	55	39	53	28	47
26	43	32	52	42	65	23	30
9	19	25	44	43	64	28	45
30	58	33	42	27	42	26	41
27	47	20	33	23	36	38	61
37	65	24	37	26	41	44	61
21	35	31	54	54	76	52	78
40	64	29	53	31	38	41	64
20	33	36	59	30	48	30	49
29	53	43	62	37	51	24	36

	Diluent							
P		Q		R		S	
r	n	r	n	r	n	r	n
28	44	38	57	37	55	33	47
29	43	20	34	46	69	15	27
22	35	19	32	38	63	32	50
34	51	47	81	39	59	32	48
23	48	33	50	29	45	35	53
31	50	23	37	41	59	33	59

The angular transformation of the proportion of conceptions is the appropriate variable, but the s.s.'s have to be calculated using weights, because the number of inseminations is not the same for all collections. In all other respects the terms in the anovar follow the usual single-factor pattern.

The design is better than that in exercise 16 because the use of different bulls, as well as of collections within bulls, was randomized over collections. However this use was not balanced – that is no steps were taken to ensure that each bull contributed the same number of collections for use with each diluent – and this makes the tests somewhat less reliable. An experiment using a balanced design is described in the next exercise.

19 In this design, 24 bulls were allocated randomly into 4 groups of 6, one group for each diluent, and 4 collections were used from each bull; the experiment thus used the same number of collections as that described in exercise 18, but their allocation was balanced. The design was therefore nested, with terms for diluents, bulls within diluents and collections within bulls. The observations (*r* conceptions out of *n* inseminations) were as follows:

	Bull	r	n	r	n	r	n	r	n
Diluent P	1	18	38	20	45	27	59	6	11
	2	17	40	30	51	19	34	11	23
	3	51	61	29	42	35	47	15	20
	4	21	31	36	64	24	44	41	61

	Bull	r	n	r	n	r	n	r	n
Diluent P	5	53	79	33	57	27	46	49	77
	6	26	53	12	40	32	67	22	45
Diluent Q	7	31	59	31	54	46	71	35	62
	8	29	48	35	47	45	69	38	62
	9	28	65	24	48	4	18	13	46
	10	24	43	33	50	39	65	34	59
	11	8	13	21	35	30	60	20	45
	12	25	49	20	45	31	52	16	30
Diluent R	13	24	31	29	53	27	45	45	69
	14	37	72	19	32	25	47	23	33
	15	38	47	40	52	42	58	53	71
	16	22	37	42	76	58	82	28	40
	17	28	40	38	51	36	43	37	45
	18	46	57	51	68	38	48	38	46
Diluent S	19	25	42	15	30	32	46	33	52
	20	39	56	39	58	49	61	48	67
	21	34	44	52	74	20	29	41	62
	22	27	38	27	37	25	40	33	47
	23	27	52	22	49	39	52	22	43
	24	49	60	42	71	42	53	35	56

The analysis follows the same general pattern as that for the nested design in exercise 17, except that the s.s.'s are calculated using weights. The fact that the number of inseminations cannot be fixed introduces a second source of lack of balance, but there is no practical means of overcoming this and at this level of complexity its effect can only be ignored and more caution than usual exercised in attaching importance to results that are only marginally significant.

The experiment could of course have been set up with a two-factor design, using one collection from each bull for each diluent, but this would have required methods of analysis outside the scope of this book.

20 In an experiment on the survival time of a cell suspension, 5 subsamples were taken daily on 6 successive days, and a count made on each sample of the number of dead cells in 100 cells. The results were as follows:

		Day			
0	1	2	3	4	5
24	26	19	30	46	59
18	27	19	29	37	62
20	16	21	24	38	67
21	25	16	31	47	65
20	24	21	30	38	54

The design is a single-factor, with sufficient replication at the subsample within days level to ensure that the correct residual is estimated for use in comparing day means. The angular transformation is appropriate but there is no need to use weights in calculating the s.s.'s. The choice of comparisons between levels – between days – presents some difficulties because of the form of the relationship with time. Orthogonal polynomials can be used, the appropriate coefficients for this case being

Linear	−5	−3	−1	1	3	5
Quadratic	5	−1	−4	−4	−1	5
Cubic	−5	7	4	−4	−7	5
Quartic	1	−3	2	2	−3	1
Quintic	−1	5	−10	10	−5	1

but, in general, it will be found that polynomials do not provide a satisfactory approximation to relationships which begin with very gradual change but eventually become very rapid. On the other hand, an exponential relationship cannot sensibly be considered because the variable is bounded – the angular transformation of the proportion of dead cells cannot exceed 90.0. It is probably best to compare successive times and to consider such contrasts as

mean of days 0 and 1 compared with day 2;

the best choice of contrasts is usually that which leads to the simplest description of the results.

21 In studying the effect of soil temperature on germination rate in maize, seed boxes containing 20 seeds were placed in a range of environments. The soil temperature ($x°C$) was recorded at a depth of 5 cm and the number of seeds germinating (r) was also noted: the results were as follows:

r	x	r	x	r	x	r	x
16	16.7	13	11.8	12	14.8	11	12.5
13	13.4	15	11.4	15	13.3	10	10.4
7	10.1	9	9.9	13	16.3	7	11.7
11	11.1	7	10.5	9	10.4	10	10.0
12	15.7	17	16.1	16	17.0	13	14.5

The angular transformation of $r/20$ is used as dependent variable in regression on x.

22 Six types of insect trap were tested under uniform conditions over 20 one-hour periods: the numbers of a certain species captured in each trap in each period were as follows:

		Type of trap			
F	G	H	J	K	L
11	10	24	18	18	24
11	20	20	16	13	18
10	14	19	17	17	24
11	13	20	18	15	15
6	9	12	13	15	25
10	19	8	15	7	22
5	10	20	15	12	23
8	10	19	20	18	20
7	22	20	12	10	29
5	10	19	19	8	16
9	12	24	14	7	13
9	16	25	6	16	17
11	12	23	21	15	15
12	12	15	15	6	21
12	10	10	8	11	19
9	6	18	16	11	23
12	11	15	13	13	25
11	6	16	14	13	24
13	9	20	18	10	18
12	10	20	12	14	17

The square-root transformation is the most likely to be successful in this case.

23 In an investigation of colour blindness, groups of 200 children were examined in 4 urban districts. The numbers of children per group with a certain vision defect were as follows:

	District		
Q	R	S	T
2	4	9	10
4	9	4	8
1	2	9	9
1	4	4	8
1		7	8
			8

The number affected in a group of 200 can be regarded as a proportion, but since the 'chance of success' is clearly small the Poisson distribution should be a good approximation to the binomial so that the square-root transformation should be satisfactory.

24 In a study of the distribution of a certain wild species of marigold, 5 types of habitat were investigated and unit areas taken from several sites within each habitat. The numbers of separate plants in each area were as follows:

		Habitat		
J	K	L	M	N
5	7	55	23	25
3	5	40	22	19
4	14	46	17	25
3	8	41	16	18
3	6	44	31	14
3	6		18	10
	7		20	16
	7			19

25 The number of accidents, r, in the period 4.00 p.m. to 7.00 p.m. on a certain section of road was thought to depend on the mean traffic density, x (vehicles/hour), during this period. Observations on 48 days gave the following results:

r	x	r	x	r	x	r	x
3	2000	4	950	2	1280	1	940
3	1450	1	1160	3	1011	2	2480
2	2110	4	1550	1	1260	1	830
3	640	2	1710	4	1360	0	100
1	530	1	880	1	2010	3	1700
1	190	2	1290	1	1950	4	860
3	1010	1	840	0	670	1	760
3	1980	0	1160	6	2030	4	1420
0	390	3	980	2	1260	2	2400
3	690	6	2130	0	890	1	880
4	1130	3	1440	0	1250	2	1380
2	1620	3	1320	2	1300	2	1220

Use the square-root transformation in investigating the regression on x. Obtain an estimate, with confidence limits, of the average number of accidents to be expected when the mean traffic density is 1700 vehicles/hour.

Exercises 26–9 require tests of goodness of fit, of the type illustrated in 9.5.1. The tests can be carried out by the use of the χ^2-test or the Kolmogorov–Smirnov one-sample test. These tests are very similar in their capabilities, but it can be argued that the Kolmogorov–Smirnov test is more likely to detect a systematic departure of the sample distribution from the null hypothesis form, whilst the χ^2-test may better detect substantial irregular deviations.

26 The distribution of the number (r) of fertile eggs in batches of 10 from a supposedly uniform supply was investigated using 85 batches; the values of r were as follows:

7	7	6	8	6	8	7	6	8	6	10	7	6
19	9	9	7	9	9	8	6	7	7	7	7	8
8	10	8	9	7	6	6	9	7	7	9	6	7
7	6	6	8	5	7	6	6	5	7	9	7	9
8	8	8	7	8	7	9	10	7	9	7	7	9
7	7	9	8	8	5	9	7	9	7	6	8	8
6	6	6	6	8	7	7					

Obtain the frequencies for the various values of r and estimate the proportion fertile from the formula

$$\hat{p} = \frac{\sum r f_r}{10 \sum f_r}.$$

Calculate the frequencies expected for a binomial distribution with $n = 10$ and $p = \hat{p}$, and compare these frequencies with those observed by means of a test of goodness of fit. If the χ^2-test is used the degrees of freedom will be

$$\text{number of classes} - 2$$

since the parameter p is estimated. At first sight it might appear that the number of classes would be the number of possible values of r, that is 11, but in fact this number will need to be reduced because one requirement of the χ^2-test is that no class shall have an expected frequency less than 5; this is achieved by combining the extreme classes, the observed (expected) frequency for a combined class being the sum of the observed (expected) frequencies for the classes that were combined. The number of classes used in the formula above is the number after such combination.

27 The number (r) of mice responding to a standard dose of a hormone was determined on 70 groups of 12 animals, with the following results:

11	7	8	10	9	9	8	10	9	10	10	10	8
10	10	8	11	7	12	11	6	9	10	10	10	8
11	12	11	10	10	11	7	11	11	10	11	9	10
11	12	12	8	10	9	10	11	11	11	9	11	12
11	12	9	12	11	7	7	9	11	8	12	11	11
9	12	11	12	9								

It was important to know whether the distribution of r could be regarded as binomial, because this affected the method of analysis of the assay for which these mice were typical test animals. The analysis should proceed as in exercise 26.

28 The numbers (r) of young born alive in 80 gerbil litters were as follows:

3	5	5	12	3	11	2	5	4	10	6	4	2
4	8	7	5	10	5	12	4	7	5	13	4	2
2	8	2	2	3	8	2	9	2	4	4	5	6
2	5	6	7	3	2	6	6	7	9	3	3	3
2	4	2	5	6	5	8	5	2	1	9	7	5
7	1	6	4	2	2	5	1	3	6	6	2	4
2	5											

Test whether this distribution can reasonably be regarded as Poisson form – the parameter m of the Poisson is estimated by

$$\hat{m} = \bar{r} = \frac{\sum r f_r}{\sum f_r}$$

where f_r is the frequency of the value r.

29 Counts were made of the number of cells on 100 areas of a haemocytometer slide because it was suspected that the cells tended to form clumps and so deviate from a Poisson distribution: the observations were as follows:

11	13	7	11	10	18	21	21	9	13	16	23	14
5	10	27	16	23	32	14	9	13	10	8	22	18
13	8	31	22	13	21	21	14	15	22	11	5	4
10	16	9	20	10	10	16	12	7	19	7	10	22
5	7	30	12	21	16	14	19	20	5	15	18	20
14	8	10	9	19	7	16	6	15	13	15	7	21
12	17	11	14	16	16	37	18	16	12	21	11	9
9	12	4	19	24	17	6	8	23				

Test, as in exercise 28, whether the distribution can reasonably be regarded as Poisson.

FURTHER READING

Whilst there is no shortage of books on more advanced statistical topics, many are unduly mathematical in approach and are not very helpful to the biologist. Three which are, however, in my opinion outstanding are Bliss (1967, 1970), Rayner (1969) and Sokal and Rohlf (1981); all are big books containing careful discussions of many topics important and useful to the biologist.

REFERENCES

Bliss, C. I. (1967, 1970). *Statistics in Biology*, volumes I and II. New York: McGraw-Hill.

Chernoff, H. and Moses, L. E. (1959). *Elementary Decision Theory*. New York: Wiley.

Cochran, W. G. (1977). *Sampling Techniques*, 3rd edition. New York: Wiley.

Federer, W. T. (1955). *Experimental Design*. New York: Macmillan.

Finney, D. J. (1971). *Probit Analysis*, 3rd edition. Cambridge University Press.

Fisher, Sir Ronald A. and Yates, F. (1974). *Statistical Tables for Biological, Agricultural and Medical Research*, 6th edition. Edinburgh and London: Oliver & Boyd.

Galton, Sir Francis (1886). *Proc. Roy. Soc.* A **40**, 42.

Genstat Manual (1988). *Genstat 5: Reference Manual*. Oxford University Press.

Gordon, M. J. (1957). *Proc. Nat. Acad. Sci. Wash.* **43**, 913.

Huff, D. (1986). *How to Lie with Statistics*. London: Penguin Books.

Kendall, Sir Maurice G. (1970). *Rank Correlation Methods*, 4th edition. London: Griffin.

Lane, P. W., Galwey, N. W. and Alvey, N. G. (1988). *Genstat 5: A First Course*. Oxford University Press.

Lehmann, E. L. (1963). *Ann. Math. Statist.* **34**, 1507.

Lindley, D. V. (1962). *Statistician* **12**, 43.

Lindley, D. V. (1980). *Introduction to Probability and Statistics from a Bayesian Viewpoint*: Part I, *Probability*; Part 2, *Inference*. Cambridge University Press.

Lindley, D. V. and Scott, W. F. (1984). *New Cambridge Elementary Statistical Tables*. Cambridge University Press.

Mann, H. B. and Whitney, D. R. (1947). *Ann. Math. Statist.* **18**, 50.

Nair, K. R. (1940). *Sankhyā* **4**, 551.

Norusis, M. J. (1983). SPSS-X *Introductory Statistics Guide*. New York: McGraw-Hill.

Norusis, M. J. (1985). SPSS-X *Advanced Statistics Guide*. New York: McGraw-Hill.

Pearce, S. C. (1965). *Biological Statistics: An Introduction*. New York: McGraw-Hill.

REFERENCES

Pearson, E. S. and Hartley, D. O. (1958). *Biometrika Tables for Statisticians*, volume I. Cambridge University Press.

Rayner, A. A. (1969). *A First Course in Biometry for Agriculture Students*. Pietermaritzburg: University of Natal Press.

Ryan, T. A., Jr, Joiner, B. L. and Ryan, B. F. (1980). Minitab *Reference Manual*. Statistics Department, Pennsylvania State University.

Sampford, M. R. (1962). *An Introduction to Sampling Theory with Applications to Agriculture*. Edinburgh and London: Oliver & Boyd.

Savage, L. J. (1972). *The Foundations of Statistics*. New York: Dover.

Seymour, G. A. and Wiggins, R. A. (1981). PLOTALL *Computer Graphics Language User's Guide*, 2nd edition. University of Akron Computer Centre.

Siegel, S. (1956). *Nonparametric Statistics for the Behavioural Sciences*. New York: McGraw-Hill.

Sokal, R. R. and Rohlf, F. J. (1981). *Biometry*. San Francisco: Freeman.

Thompson, W. R. (1936). *Ann. Math. Statist.* **7**, 122.

Velleman, P. F. and Welsch, R. E. (1981). *Amer. Statist.* **35**, 234.

Yates, F. (1981). *Sampling Methods for Censuses and Surveys*, 2nd edition. London: Griffin.

APPENDIX

Table A1. *Random digits.*†

20 17	42 28	23 17	59 66	38 61	02 10	86 10	51 55	92 52	44 25
74 49	04 19	03 04	10 33	53 70	11 54	48 63	94 60	94 49	57 38
94 70	49 31	38 67	23 42	29 65	40 88	78 71	37 18	48 64	06 57
22 15	78 15	69 84	32 52	32 54	15 12	54 02	01 37	38 37	12 93
93 29	12 18	27 30	30 55	91 87	50 57	58 51	49 36	12 53	96 40
45 04	77 97	36 14	99 45	52 95	69 85	03 83	51 87	85 56	22 37
44 91	99 49	89 39	94 60	48 49	06 77	64 72	59 26	08 51	25 57
16 23	91 02	19 96	47 59	89 65	27 84	30 92	63 37	26 24	23 66
04 50	65 04	65 65	82 42	70 51	55 04	61 47	88 83	99 34	82 37
32 70	17 72	03 61	66 26	24 71	22 77	88 33	17 78	08 92	73 49
03 64	59 07	42 95	81 39	06 41	20 81	92 34	51 90	39 08	21 42
62 49	00 90	67 86	93 48	31 83	19 07	67 68	49 03	27 47	52 03
61 00	95 86	98 36	14 03	48 88	51 07	33 40	06 86	33 76	68 57
89 03	90 49	28 74	21 04	09 96	60 45	22 03	52 80	01 79	33 81
01 72	33 85	52 40	60 07	06 71	89 27	14 29	55 24	85 79	31 96
27 56	49 79	34 34	32 22	60 53	91 17	33 26	44 70	93 14	99 70
49 05	74 48	10 55	35 25	24 28	20 22	35 66	66 34	26 35	91 23
49 74	37 25	97 26	33 94	42 23	01 28	59 58	92 69	03 66	73 82
20 26	22 43	88 08	19 85	08 12	47 65	65 63	56 07	97 85	56 79
48 87	77 96	43 49	76 93	08 79	22 18	54 55	93 75	97 26	90 77
08 72	87 46	75 73	00 11	27 07	05 20	30 85	22 21	04 67	19 13
95 97	98 62	17 27	31 42	64 71	46 22	32 75	19 32	20 99	94 85
37 99	57 31	70 40	46 55	46 12	24 32	36 74	69 20	72 10	95 93
05 79	58 37	85 33	75 18	88 71	23 44	54 28	00 48	96 23	66 45
55 85	63 42	00 79	91 22	29 01	41 39	51 40	36 65	26 11	78 32

† One method of using the digits to select at random from a group of 20 is described in 1.2.2; a more efficient method is as follows. Divide each pair of digits by 20 and take the remainder (remainder zero is for this purpose treated as remainder 20); these remainders are then the required random numbers. Starting at the top of the first column this method gives 20, 14, (14), 2, 13, . . . ; unless sampling is required to be with replacement we would ignore the second 14.

This procedure has to be modified if the size of the group is not an exact submultiple of a power of 10. Thus if the group size is 19, the pairs of digits 1, . . . , 19; 20, . . . , 38; 39, . . . , 57; 58, . . . , 76; 77, . . . , 95 all give remainders 1, . . . , 19, but the pairs 96, . . . , 00 give the incomplete set of remainders 1, . . . , 5 and must therefore be excluded because their use would unbalance the frequencies of the remainders. Starting at the bottom of the first column on page 404 we would thus select from a group of 19 those individuals numbered 18, 8, 15, 14, ...

Table A1 (*continued*)

67 28	96 25	68 36	24 72	03 85	49 24	05 69	64 86	08 19	91 21
85 86	94 78	32 59	51 82	86 43	73 84	45 60	89 57	06 87	08 15
40 10	60 09	05 88	78 44	63 13	58 25	37 11	18 47	75 62	52 21
94 55	89 48	90 80	77 80	26 89	87 44	23 74	66 20	20 19	26 52
11 63	77 77	23 20	33 62	62 19	29 03	94 15	56 37	14 09	47 16
64 00	26 04	54 55	38 57	94 62	68 40	26 04	24 25	03 61	01 20
50 94	13 23	78 41	60 58	10 60	88 46	30 21	45 98	70 96	36 89
66 98	37 96	44 13	45 05	34 59	75 85	48 97	27 19	17 85	48 51
66 91	42 83	60 77	90 91	60 90	79 62	57 66	72 28	08 70	96 03
33 58	12 18	02 07	19 40	21 29	39 45	90 42	58 84	85 43	95 67
52 49	40 16	72 40	73 05	50 90	02 04	98 24	05 30	27 25	20 88
74 98	93 99	78 30	79 47	96 92	45 58	40 37	89 76	84 41	74 68
50 26	54 30	01 88	69 57	54 45	69 88	23 21	05 69	93 44	05 32
49 46	61 89	33 79	96 84	28 34	19 35	28 73	39 59	56 34	97 07
19 65	13 44	78 39	73 88	62 03	36 00	25 96	86 76	67 90	21 68
64 17	47 67	87 59	81 40	72 61	14 00	28 28	55 86	23 38	16 15
18 43	97 37	68 97	56 56	57 95	01 88	11 89	48 07	42 60	11 92
65 58	60 87	51 09	96 61	15 53	66 81	66 88	44 75	37 01	28 88
79 90	31 00	91 14	85 65	31 75	43 15	45 93	64 78	34 53	88 02
07 23	00 15	59 05	16 09	94 42	20 40	63 76	65 67	34 11	94 10
90 08	14 24	01 51	95 46	30 32	33 19	00 14	19 28	40 51	92 69
53 82	62 02	21 82	34 13	41 03	12 85	65 30	00 79	56 30	15 48
98 17	26 15	04 50	76 25	20 33	54 84	39 31	23 33	59 64	96 27
08 91	12 44	82 40	30 62	45 50	64 54	65 17	89 25	59 44	99 95
37 21	46 77	84 87	67 39	85 54	97 37	33 41	11 74	90 50	29 62

Table A2. *This table is used as in 2.5.1.1, in finding confidence limits for the median of one population.*

Sample size n	Approximately 95 per cent confidence		Approximately 99 per cent confidence	
	k	Exact conf. (per cent)	k	Exact conf. (per cent)
6	1	96.88		
7	1	98.44		
8	1	99.22	1	99.22
9	2	96.10	1	99.60
10	2	97.86	1	99.80
11	2	98.82	1	99.90
12	3	96.14	2	99.36
13	3	97.76	2	99.66
14	3	98.70	2	99.82
15	4	96.48	3	99.26
16	4	97.88	3	99.58
17	5	95.10	3	99.76
18	5	96.92	4	99.24
19	5	98.08	4	99.56
20	6	95.86	4	99.74
21	6	97.34	5	99.28
22	6	98.30	5	99.56
23	7	96.54	5	99.74
24	7	97.74	6	99.34
25	8	95.68	6	99.60
26	8	97.10	7	99.06
27	8	98.08	7	99.40
28	9	96.44	7	99.62
29	9	97.58	8	99.18
30	10	95.72	8	99.48
31	10	97.06	8	99.66
32	10	98.00	9	99.30
33	11	96.50	9	99.54
34	11	97.56	10	99.10
35	12	95.90	10	99.40
36	12	97.12	10	99.60
37	13	95.30	11	99.24
38	13	96.64	11	99.50
39	13	97.62	12	99.06
40	14	96.16	12	99.36
41	14	97.24	12	99.52
42	15	95.64	13	99.20
43	15	96.84	13	99.46
44	16	95.12	14	99.04
45	16	96.44	14	99.34

405

Table A2 (*continued*)

Sample size n	Approximately 95 per cent confidence		Approximately 99 per cent confidence	
	k	Exact conf. (per cent)	k	Exact conf. (per cent)
46	16	97.42	14	99.54
47	17	96.00	15	99.20
48	17	97.06	15	99.44
49	18	95.56	16	99.06
50	18	96.72	16	99.34
51	19	95.12	16	99.54
52	19	96.36	17	99.22
53	19	97.30	17	99.46
54	20	95.98	18	99.10
55	20	97.00	18	99.36
56	21	95.60	18	99.54
57	21	96.68	19	99.24
58	22	95.20	19	99.46
59	22	96.36	20	99.14
60	22	97.26	20	99.38
61	23	96.04	21	99.02
62	23	97.00	21	99.28
63	24	95.70	21	99.48
64	24	96.72	22	99.18
65	25	95.36	22	99.40
66	25	96.44	23	99.08
67	26	95.02	23	99.32
68	26	96.16	23	99.50
69	26	97.06	24	99.24
70	27	95.86	24	99.44
71	27	96.80	25	99.14
72	28	95.56	25	99.36
73	28	96.56	26	99.04
74	29	95.26	26	99.30
75	29	96.30	26	99.48
76	29	97.14	27	99.20
77	30	96.04		
78	30	96.92		
79	31	95.78		
80	31	97.70		
81	32	95.52		

Table A3. Values of the Wilcoxon U-statistic for use in obtaining 95 per cent confidence limits for the difference between two medians (3.2) and in testing such a difference at the 5 per cent level of significance (3.2.2).

$n_1 \backslash n_2$	1	2	3	4	5	6	7	8	9	10	11	12	13	14	15	16	17	18	19	20
1																				
2								0	0	0	0	1	1	1	1	1	2	2	2	2
3					0	1	1	2	2	3	3	4	4	5	5	6	6	7	7	8
4				0	1	2	3	4	4	5	6	7	8	9	10	11	11	12	13	13
5			0	1	2	3	5	6	7	8	9	11	12	13	14	15	17	18	19	20
6			1	2	3	5	6	8	10	11	13	14	16	17	19	21	22	24	25	27
7			1	3	5	6	8	10	12	14	16	18	20	22	24	26	28	30	32	34
8		0	2	4	6	8	10	13	15	17	19	22	24	26	29	31	34	36	38	41
9		0	2	4	7	10	12	15	17	20	23	26	28	31	34	37	39	42	45	48
10		0	3	5	8	11	14	17	20	23	26	29	33	36	39	42	45	48	52	55
11		0	3	6	9	13	16	19	23	26	30	33	37	40	44	47	51	55	58	62
12		1	4	7	11	14	18	22	26	29	33	37	41	45	49	53	57	61	65	69
13		1	4	8	12	16	20	24	28	33	37	41	45	50	54	59	63	67	72	76
14		1	5	9	13	17	22	26	31	36	40	45	50	55	59	64	67	74	78	83
15		1	5	10	14	19	24	29	34	39	44	49	54	59	64	70	75	80	85	90
16		1	6	11	15	21	26	31	37	42	47	53	59	64	70	75	81	86	92	98
17		2	6	11	17	22	28	34	39	45	51	57	63	67	75	81	87	93	99	105
18		2	7	12	18	24	30	36	42	48	55	61	67	74	80	86	93	99	106	112
19		2	7	13	19	25	32	38	45	52	58	65	72	78	85	92	99	106	113	119
20		2	8	13	20	27	34	41	48	55	62	69	76	83	90	98	105	112	119	127

APPENDIX

Table A4. *Probabilities associated with the Kruskal–Wallis statistic H used, as in 3.3, for comparing the medians for several populations. If the number of populations exceeds three an approximate test is used, as described in the text.*

n_1	n_2	n_3	H	p	n_1	n_2	n_3	H	p
2	1	1	2.7000	0.500	4	3	1	5.8333	0.021
								5.2083	0.050
2	2	1	3.6000	0.200				5.0000	0.057
								4.0556	0.093
2	2	2	4.5714	0.067				3.8889	0.129
			3.7143	0.200	4	3	2	6.4444	0.008
								6.3000	0.011
3	1	1	3.2000	0.300				5.4444	0.046
								5.4000	0.051
3	2	1	4.2857	0.100				4.5111	0.098
			3.8571	0.133				4.4444	0.102
3	2	2	5.3572	0.029					
			4.7143	0.048	4	3	3	6.7455	0.010
			4.5000	0.067				6.7091	0.013
			4.4643	0.105				5.7909	0.046
								5.7273	0.050
3	3	1	5.1429	0.043				4.7091	0.092
			4.5714	0.100				4.7000	0.101
			4.0000	0.129	4	4	1	6.6667	0.010
3	3	2	6.2500	0.011				6.1667	0.022
			5.3611	0.032				4.9667	0.048
			5.1389	0.061				4.8667	0.054
			4.5556	0.100				4.1667	0.082
			4.2500	0.121				4.0667	0.102
3	3	3	7.2000	0.004	4	4	2	7.0364	0.006
			6.4889	0.011				6.8727	0.011
			5.6889	0.029				5.4545	0.046
			5.6000	0.050				5.2364	0.052
			5.0667	0.086				4.5545	0.098
			4.6222	0.100				4.4455	0.103
4	1	1	3.5714	0.200	4	4	3	7.1439	0.010
								7.1364	0.011
4	2	1	4.8214	0.057				5.5985	0.049
			4.5000	0.076				5.5758	0.051
			4.0179	0.114				4.5455	0.099
								4.4773	0.102
4	2	2	6.0000	0.014	4	4	4	7.6538	0.008
			5.3333	0.033				7.5385	0.011
			5.1250	0.052				5.6923	0.049
			4.4583	0.100				5.6538	0.054
			4.1667	0.105				4.6539	0.097
								4.5001	0.104

Table A4 (*continued*)

n_1	n_2	n_3	H	p	n_1	n_2	n_3	H	p
Sample sizes					**Sample sizes**				
5	1	1	3.8571	0.143	5	4	3	7.4449	0.010
								7.3949	0.011
5	2	1	5.2500	0.036				5.6564	0.049
			5.0000	0.048				5.6308	0.050
			4.4500	0.071				4.5487	0.099
			4.2000	0.095				4.5231	0.103
			4.0500	0.119					
					5	4	4	7.7604	0.009
5	2	2	6.5333	0.008				7.7440	0.011
			6.1333	0.013				5.6571	0.049
			5.1600	0.034				5.6176	0.050
			5.0400	0.056				4.6187	0.100
			4.3733	0.090				4.5527	0.102
			4.2933	0.122					
					5	5	1	7.3901	0.009
5	3	1	6.4000	0.012				6.8364	0.011
			4.9600	0.048				5.1273	0.046
			4.8711	0.052				4.9091	0.053
			4.0178	0.095				4.1091	0.086
			3.8400	0.123				4.0364	0.105
5	3	2	6.9091	0.009	5	5	2	7.3385	0.010
			6.8218	0.010				7.2692	0.010
			5.2509	0.049				5.3385	0.047
			5.1055	0.052				5.2462	0.051
			4.6509	0.091				4.6231	0.097
			4.4945	0.101				4.5077	0.100
5	3	3	7.0788	0.009	5	5	3	7.5780	0.010
			6.9818	0.011				7.5429	0.010
			5.6485	0.049				5.7055	0.046
			5.5152	0.051				5.6264	0.051
			4.5333	0.097				4.5451	0.100
			4.4121	0.109				4.5363	0.102
5	4	1	6.9545	0.008	5	5	4	7.8229	0.010
			6.8400	0.011				7.7914	0.010
			4.9855	0.044				5.6657	0.049
			4.8600	0.056				5.6429	0.050
			3.9873	0.098				4.5229	0.099
			3.9600	0.102				4.5200	0.101
5	4	2	7.2045	0.009	5	5	5	8.0000	0.009
			7.1182	0.010				7.9800	0.010
			5.2727	0.049				5.7800	0.049
			5.2682	0.050				5.6600	0.051
			4.5409	0.098				4.5600	0.100
			4.5182	0.101				4.5000	0.102

Table A5. *Values of the χ^2 distribution for v degrees of freedom and probability level P per cent, for use in one-tailed tests – that is, the majority of uses of this distribution.*†

P	99	95	10	5	1	0.1
v = 1	0.0^3157	0.00393	2.71	3.84	6.63	10.83
2	0.0201	0.103	4.61	5.99	9.21	13.81
3	0.115	0.352	6.25	7.81	11.34	16.27
4	0.297	0.711	7.78	9.49	13.28	18.47
5	0.554	1.15	9.24	11.07	15.09	20.52
6	0.872	1.64	10.64	12.59	16.81	22.46
7	1.24	2.17	12.02	14.07	18.48	24.32
8	1.65	2.73	13.36	15.51	20.09	26.12
9	2.09	3.33	14.68	16.92	21.67	27.88
10	2.56	3.94	15.99	18.31	23.21	29.59
11	3.05	4.57	17.28	19.68	24.73	31.26
12	3.57	5.23	18.55	21.03	26.22	32.91
13	4.11	5.89	19.81	22.36	27.69	34.53
14	4.66	6.57	21.06	23.68	29.14	36.12
15	5.23	7.26	22.31	25.00	30.58	37.70
16	5.81	7.96	23.54	26.30	32.00	39.25
17	6.41	8.67	24.77	27.59	33.41	40.79
18	7.01	9.39	25.99	28.87	34.81	42.31
19	7.63	10.12	27.20	30.14	36.19	43.82
20	8.26	10.85	28.41	31.41	37.57	45.31
21	8.90	11.59	29.62	32.67	38.93	46.80
22	9.54	12.34	30.81	33.92	40.29	48.27
23	10.20	13.09	32.01	35.17	41.64	49.73
24	10.86	13.85	33.20	36.42	42.98	51.18
25	11.52	14.61	34.38	37.65	44.31	52.62
26	12.20	15.38	35.56	38.89	45.64	54.05
27	12.88	16.15	36.74	40.11	46.96	55.48
28	13.56	16.93	37.92	41.34	48.28	56.89
29	14.26	17.71	39.09	42.56	49.59	58.30
30	14.95	18.49	40.26	43.77	50.89	59.70
40	22.16	26.51	51.81	55.76	63.69	73.40
50	29.71	34.76	63.17	67.50	76.15	86.66
60	37.48	43.19	74.40	79.08	88.38	99.61
70	45.44	51.74	85.53	90.53	100.4	112.3
80	53.54	60.39	96.58	101.9	112.3	124.8
90	61.75	69.13	107.6	113.1	124.1	137.2
100	70.06	77.93	118.5	124.3	135.8	149.4

† Values for two-sided procedures, used in obtaining confidence limits for a variance estimate (6.4.1), are given in Table A13.

Table A6. *Values of the Wilcoxon signed-rank statistic T required for rejecting the null hypothesis that the population medians of two paired samples are equal (small values of T are significant). The procedure is described in 3.4.1.*

	Value of T for significance at	
n	5 per cent	1 per cent
6	0	
7	2	
8	4	0
9	6	2
10	8	3
11	11	5
12	15	7
13	17	10
14	21	13
15	25	16
16	30	20
17	35	23
18	40	28
19	46	32
20	52	38
21	59	43
22	66	49
23	73	55
24	81	61
25	89	68

Table A7. *Values of the sign-test statistic R required for rejecting the null hypothesis that two populations are equivalent* (small *values of R are significant*). *The procedure is described in 3.4.1.2.*

	Value of R for significance at	
n	5 per cent	1 per cent
6	0	
7	0	
8	0	0
9	1	0
10	1	0
11	1	0
12	2	1
13	2	1
14	2	1
15	3	2
16	3	2
17	4	2
18	4	3
19	4	3
20	5	3
21	5	4
22	5	4
23	6	4
24	6	5
25	7	5

Table A8. *Values of the Friedmann statistic S for rejecting the null hypothesis that the medians of n populations are equal; the calculation is described in 3.4.2.*[†]

n	3			4			5		
Approx. significance (per cent) *m*	5	1	0.1	5	1	0.1	5	1	0.1
2				20 4.2					
3	18 2.8			37 3.3			64 4.5	76 0.78	86 0.09
4	26 4.2	32 0.46		52 3.6	64 0.69	74 0.09			
5	32 3.9	42 0.85	50 0.08	65 4.4	83 0.87	105 0.06			
6	42 2.9	54 0.81	72 0.01	76 4.3	100 1.00	128 0.09			
7	50 2.7	62 0.84	86 0.03						
8	50 4.7	72 0.99	98 0.09						
9	56 4.8	78 1.00	114 0.07						
10	62 4.6	96 0.75	126 0.08						

† Since S is a discrete variable, the significance levels are only approximate; the exact probability (per cent) is given below each value of S.

APPENDIX

Table A9. *Values of the test statistic required for 5 per cent and 1 per cent significance in the Kolmogorov–Smirnov two-sample test. The procedure is described in 3.5.*

Sample sizes n	Significance level (per cent)	
	5	1
4	4	
5	4	5
6	5	6
7	5	6
8	5	6
9	6	7
10	6	7
11	6	8
12	6	8
13	7	8
14	7	8
15	7	9
16	7	9
17	8	9
18	8	10
19	8	10
20	8	10
21	8	10
22	9	11
23	9	11
24	9	11
25	9	11
26	9	11
27	9	12
28	10	12
29	10	12
30	10	12
35	11	13
40	11	14

414

Table A10. *Probabilities (per cent) for testing the significance of* $|S|$, *obtained in calculating Kendall's coefficient of rank correlations. The procedure is described in 4.3.1.*†

	Values of n					Values of n		
$|S|$	4	5	8	9	$|S|$	6	7	10
2	75	82	90	92	3	72	77	86
4	33	48	72	76	5	47	56	73
6	8	23	55	61	7	27	38	60
8		8	40	48	9	14	24	48
10		1.7	27	36	11	6	14	38
12			18	26	13	1.7	7	29
14			11	18	15	0.3	3.0	22
16			6	12	17		1.0	16
18			3.2	8	19		0.3	11
20			1.4	4.4	21		0.04	7
22			0.6	2.4	23			4.6
24			0.17	1.3	25			2.8
26			0.04	0.6	27			1.7
28			0.0^25	0.24	29			0.9
30				0.09	31			0.5
32				0.02	33			0.2
34				0.0^25	35			0.09
36				0.0^36	37			0.04
					39			0.012
					41			0.0^23
					43			0.0^36
					45			0.0^46

† If there are ties among the observations these probabilities are only approximate (see 4.3.1.1).

APPENDIX

Table A11. *The distribution function $\Phi(x)$ measures the probability that a normal variable will exceed its mean by less than x standard deviations (x positive).*†

x	$\Phi(x)$	P	x	$\Phi(x)$	P
0.00	0.5000	1.0000	1.55	0.9394	0.1212
0.05	0.5199	0.9602	1.60	0.9452	0.1096
0.10	0.5398	0.9204	1.65	0.9505	0.0990
0.15	0.5596	0.8808	1.70	0.9554	0.0892
0.20	0.5793	0.8414	1.75	0.9599	0.0802
0.25	0.5987	0.8026	1.80	0.9641	0.0718
0.30	0.6179	0.7642	1.85	0.9678	0.0644
0.35	0.6368	0.7264	1.90	0.9713	0.0574
0.40	0.6554	0.6892	1.95	0.9744	0.0512
0.45	0.6736	0.6528	2.00	0.9772	0.0456
0.50	0.6915	0.6170	2.05	0.9798	0.0404
0.55	0.7088	0.5824	2.10	0.9821	0.0358
0.60	0.7257	0.5486	2.15	0.9842	0.0316
0.65	0.7422	0.5156	2.20	0.9861	0.0278
0.70	0.7580	0.4840	2.25	0.9878	0.0244
0.75	0.7734	0.4532	2.30	0.9893	0.0214
0.80	0.7881	0.4238	2.35	0.9906	0.0188
0.85	0.8023	0.3954	2.40	0.9918	0.0164
0.90	0.8159	0.3682	2.45	0.9929	0.0142
0.95	0.8289	0.3422	2.50	0.9938	0.0124
1.00	0.8413	0.3174	2.55	0.9946	0.0108
1.05	0.8531	0.2938	2.60	0.9953	0.0094
1.10	0.8643	0.2714	2.65	0.9960	0.0080
1.15	0.8749	0.2502	2.70	0.9965	0.0070
1.20	0.8849	0.2302	2.75	0.9970	0.0060
1.25	0.8944	0.2112	2.80	0.9974	0.0052
1.30	0.9032	0.1936	2.85	0.9978	0.0044
1.35	0.9115	0.1770	2.90	0.9981	0.0038
1.40	0.9192	0.1616	2.95	0.9984	0.0032
1.45	0.9265	0.1470	3.00	0.9986	0.0028
1.50	0.9332	0.1336			

† The probability that such a variable will be more than x standard deviations below its mean is $\{1 - \Phi(x)\}$. The probability P is that of a normal variable deviating from its mean by more than x standard deviations; thus

$$P = 2[1 - \Phi(x)], \quad x > 0.$$

Table A12. Values of the t-statistic for degrees of freedom v and two-sided probability level P per cent. The use of this statistic is described in 6.4 and several later sections.

P	90	80	70	60	50	40	30	20	10	5	2	1	0.1
v = 1	0.158	0.325	0.510	0.727	1.000	1.376	1.936	3.078	6.314	12.706	31.821	63.657	636.619
2	0.142	0.289	0.445	0.617	0.816	1.061	1.386	1.886	2.920	4.303	6.965	9.925	31.598
3	0.137	0.277	0.424	0.584	0.765	0.978	1.250	1.638	2.353	3.182	4.541	5.841	12.924
4	0.134	0.271	0.414	0.569	0.741	0.941	1.190	1.533	2.132	2.776	3.747	4.604	8.610
5	0.132	0.267	0.408	0.559	0.727	0.920	1.156	1.476	2.015	2.571	3.365	4.032	6.869
6	0.131	0.265	0.404	0.553	0.718	0.906	1.134	1.440	1.943	2.447	3.143	3.707	5.959
7	0.130	0.263	0.402	0.549	0.711	0.896	1.119	1.415	1.895	2.365	2.998	3.499	5.408
8	0.130	0.262	0.399	0.546	0.706	0.889	1.108	1.397	1.860	2.306	2.896	3.355	5.041
9	0.129	0.261	0.398	0.543	0.703	0.883	1.100	1.383	1.833	2.262	2.821	3.250	4.781
10	0.129	0.260	0.397	0.542	0.700	0.879	1.093	1.372	1.812	2.228	2.764	3.169	4.587
11	0.129	0.260	0.396	0.540	0.697	0.876	1.088	1.363	1.796	2.201	2.718	3.106	4.437
12	0.128	0.259	0.395	0.539	0.695	0.873	1.083	1.356	1.782	2.179	2.681	3.055	4.318
13	0.128	0.259	0.394	0.538	0.694	0.870	1.079	1.350	1.771	2.160	2.650	3.012	4.221
14	0.128	0.258	0.393	0.537	0.692	0.868	1.076	1.345	1.761	2.145	2.624	2.977	4.140
15	0.128	0.258	0.393	0.536	0.691	0.866	1.074	1.341	1.753	2.131	2.602	2.947	4.073
16	0.128	0.258	0.392	0.535	0.690	0.865	1.071	1.337	1.746	2.120	2.583	2.921	4.015
17	0.128	0.257	0.392	0.534	0.689	0.863	1.069	1.333	1.740	2.110	2.567	2.898	3.965
18	0.127	0.257	0.392	0.534	0.688	0.862	1.067	1.330	1.734	2.101	2.552	2.878	3.922
19	0.127	0.257	0.391	0.533	0.688	0.861	1.066	1.328	1.729	2.093	2.539	2.861	3.883
20	0.127	0.257	0.391	0.533	0.687	0.860	1.064	1.325	1.725	2.086	2.528	2.845	3.850
21	0.127	0.257	0.391	0.532	0.686	0.859	1.063	1.323	1.721	2.080	2.518	2.831	3.819
22	0.127	0.256	0.390	0.532	0.686	0.858	1.061	1.321	1.717	2.074	2.508	2.819	3.792
23	0.127	0.256	0.390	0.532	0.685	0.858	1.060	1.319	1.714	2.069	2.500	2.807	3.767
24	0.127	0.256	0.390	0.531	0.685	0.857	1.059	1.318	1.711	2.064	2.492	2.797	3.745
25	0.127	0.256	0.390	0.531	0.684	0.856	1.058	1.316	1.708	2.060	2.485	2.787	3.725

Table A12 (*continued*)

P	90	80	70	60	50	40	30	20	10	5	2	1	0.1
26	0.127	0.256	0.390	0.531	0.684	0.856	1.058	1.315	1.706	2.056	2.479	2.779	3.707
27	0.127	0.256	0.389	0.531	0.684	0.855	1.057	1.314	1.703	2.052	2.473	2.771	3.690
28	0.127	0.256	0.389	0.530	0.683	0.855	1.056	1.313	1.701	2.048	2.467	2.763	3.674
29	0.127	0.256	0.389	0.530	0.683	0.854	1.055	1.311	1.699	2.045	2.462	2.756	3.659
30	0.127	0.256	0.389	0.530	0.683	0.854	1.055	1.310	1.697	2.042	2.457	2.750	3.646
40	0.126	0.255	0.388	0.529	0.681	0.851	1.050	1.303	1.684	2.021	2.423	2.704	3.551
60	0.126	0.254	0.387	0.527	0.679	0.848	1.046	1.296	1.671	2.000	2.390	2.660	3.460
120	0.126	0.254	0.386	0.526	0.677	0.845	1.041	1.289	1.658	1.980	2.358	2.617	3.373
∞	0.126	0.253	0.385	0.524	0.674	0.842	1.036	1.282	1.645	1.960	2.326	2.576	3.291

Table A13. Values of the χ^2-statistic, with v degrees of freedom and probability level P per cent, for use in two-tailed procedures, such as obtaining confidence limits for a variance estimate (see 6.4.1).

P	5		1	
	$\underline{\chi}^2$	$\bar{\chi}^2$	$\underline{\chi}^2$	$\bar{\chi}^2$
v = 1	0.0^3982	5.02	0.0^4393	7.88
2	0.0506	7.38	0.0100	10.60
3	0.216	9.35	0.0717	12.84
4	0.484	11.14	0.207	14.86
5	0.831	12.83	0.412	16.75
6	1.24	14.45	0.676	18.55
7	1.69	16.01	0.989	20.28
8	2.18	17.53	1.34	21.95
9	2.70	19.02	1.73	23.59
10	3.25	20.48	2.16	25.19
11	3.82	21.92	2.60	26.76
12	4.40	23.34	3.07	28.30
13	5.01	24.74	3.57	29.82
14	5.63	26.12	4.07	31.32
15	6.26	27.49	4.60	32.80
16	6.91	28.85	5.14	34.27
17	7.56	30.19	5.70	35.72
18	8.23	31.53	6.26	37.16
19	8.91	32.85	6.84	38.58
20	9.59	34.17	7.43	40.00
21	10.28	35.48	8.03	41.40
22	10.98	36.78	8.64	42.80
23	11.69	38.08	9.26	44.18
24	12.40	39.36	9.89	45.56
25	13.12	40.65	10.52	46.93
26	13.84	41.92	11.16	48.29
27	14.57	43.19	11.81	49.64
28	15.31	44.46	12.46	50.99
29	16.05	45.72	13.12	52.34
30	16.79	46.98	13.79	53.67
40	24.43	59.34	20.71	66.77
50	32.36	71.42	27.99	79.49
60	40.48	83.30	35.53	91.95
70	48.76	95.02	43.28	104.2
80	57.15	106.6	51.17	116.3
90	65.65	118.1	59.20	128.3
100	74.22	129.6	67.33	140.2

Table A14. *Values of the d-statistic for use, as described in 6.6.2, in comparing the means of two populations which differ in variance.*

		θ						
	v_1	0°	15°	30°	45°	60°	75°	90°
5 per cent points	6	2.447	2.440	2.435	2.435	2.435	2.440	2.447
	8	2.447	2.430	2.398	2.364	2.331	2.310	2.306
$v_2 = 6$	12	2.447	2.423	2.367	2.301	2.239	2.193	2.179
	24	2.447	2.418	2.342	2.247	2.156	2.088	2.064
	∞	2.447	2.413	2.322	2.201	2.082	1.993	1.960
	6	2.306	2.310	2.331	2.364	2.398	2.430	2.447
	8	2.306	2.300	2.294	2.292	2.294	2.300	2.306
$v_2 = 8$	12	2.306	2.292	2.262	2.229	2.201	2.183	2.179
	24	2.306	2.286	2.236	2.175	2.118	2.077	2.064
	∞	2.306	2.281	2.215	2.128	2.044	1.982	1.960
	6	2.179	2.193	2.239	2.301	2.367	2.423	2.447
	8	2.179	2.183	2.201	2.229	2.262	2.292	2.306
$v_2 = 12$	12	2.179	2.175	2.169	2.167	2.169	2.175	2.179
	24	2.179	2.168	2.142	2.112	2.085	2.069	2.064
	∞	2.179	2.163	2.120	2.064	2.011	1.973	1.960
	6	2.064	2.088	2.156	2.247	2.342	2.418	2.447
	8	2.064	2.077	2.118	2.175	2.236	2.286	2.306
$v_2 = 24$	12	2.064	2.069	2.085	2.112	2.142	2.168	2.179
	24	2.064	2.062	2.058	2.056	2.058	2.062	2.064
	∞	2.064	2.056	2.035	2.009	1.983	1.966	1.960
	6	1.960	1.993	2.082	2.201	2.322	2.413	2.447
	8	1.960	1.982	2.044	2.128	2.215	2.281	2.306
$v_2 = ∞$	12	1.960	1.973	2.011	2.064	2.120	2.163	2.179
	24	1.960	1.966	1.983	2.009	2.035	2.056	2.064
	∞	1.960	1.960	1.960	1.960	1.960	1.960	1.960
1 per cent points	6	3.707	3.654	3.557	3.514	3.557	3.654	3.707
	8	3.707	3.643	3.495	3.363	3.307	3.328	3.355
$v_2 = 6$	12	3.707	3.636	3.453	3.246	3.104	3.053	3.055
	24	3.707	3.631	3.424	3.158	2.938	2.822	2.797
	∞	3.707	3.626	3.402	3.093	2.804	2.627	2.576
	6	3.355	3.328	3.307	3.363	3.495	3.643	3.707
	8	3.355	3.316	3.329	3.206	3.239	3.316	3.355
$v_2 = 8$	12	3.355	3.307	3.192	3.083	3.032	3.039	3.055
	24	3.355	3.301	3.158	2.988	2.862	2.805	2.797
	∞	3.355	3.295	3.132	2.916	2.723	2.608	2.576
	6	3.055	3.053	3.104	3.246	3.453	3.636	3.707
	8	3.055	3.039	3.032	3.083	3.192	3.307	3.355
$v_2 = 12$	12	3.055	3.029	2.978	2.954	2.978	3.029	3.055
	24	3.055	3.020	2.938	2.853	2.803	2.793	2.797
	∞	3.055	3.014	2.909	2.775	2.661	2.595	2.576
	6	2.797	2.822	2.938	3.158	3.424	3.631	3.707
	8	2.797	2.805	2.862	2.988	3.158	3.301	3.355
$v_2 = 24$	12	2.797	2.793	2.803	2.853	2.938	3.020	3.055
	24	2.797	2.785	2.759	2.747	2.759	2.785	2.797
	∞	2.797	2.777	2.726	2.664	2.613	2.585	2.576
	6	2.576	2.627	2.804	3.093	3.402	3.626	3.707
	8	2.576	2.608	2.723	2.916	3.132	3.295	3.355
$v_2 = ∞$	12	2.576	2.595	2.661	2.775	2.909	3.014	3.055
	24	2.576	2.585	2.613	2.664	2.726	2.777	2.797
	∞	2.576	2.576	2.576	2.576	2.576	2.576	2.576

Table A15. Two-tailed values of the F-statistic, for degrees of freedom v_1, v_2 and probability level 5 per cent; this form of the statistic is used in a two-sided comparison of two variance estimates (see 6.6.4).

$v_2 \backslash v_1$	1	2	3	4	5	6	7	8	10	12	24	∞
1	0.0^2154	0.0260	0.0575	0.082	0.100	0.114	0.124	0.132	0.144	0.153	0.175	0.199
	648	800	864	900	922	937	948	957	969	977	997	1018
2	0.0^2125	0.0256	0.0625	0.094	0.119	0.138	0.153	0.165	0.183	0.196	0.231	0.271
	38.5	39.0	39.2	39.2	39.3	39.3	39.4	39.4	39.4	39.4	39.5	39.5
3	0.0^2116	0.0255	0.0649	0.100	0.129	0.152	0.170	0.185	0.207	0.224	0.269	0.321
	17.4	16.0	15.4	15.1	14.9	14.7	14.6	14.5	14.4	14.3	14.1	13.9
4	0.0^2111	0.0255	0.0662	0.104	0.135	0.161	0.181	0.198	0.224	0.243	0.296	0.358
	12.22	10.65	9.98	9.60	9.36	9.20	9.07	8.98	8.84	8.75	8.51	8.26
5	0.0^2108	0.0254	0.0671	0.107	0.140	0.167	0.189	0.207	0.236	0.257	0.317	0.389
	10.01	8.43	7.76	7.39	7.15	6.98	6.85	6.76	6.62	6.52	6.28	6.02
6	0.0^2107	0.0254	0.0680	0.109	0.143	0.172	0.195	0.215	0.246	0.268	0.334	0.415
	8.81	7.26	6.60	6.23	5.99	5.82	5.70	5.60	5.46	5.37	5.12	4.85
7	0.0^2106	0.0254	0.0685	0.110	0.146	0.175	0.200	0.221	0.253	0.277	0.348	0.437
	8.07	6.54	5.89	5.52	5.29	5.12	4.99	4.90	4.76	4.67	4.42	4.14
8	0.0^2104	0.0254	0.0690	0.111	0.148	0.179	0.204	0.266	0.260	0.285	0.360	0.457
	7.57	6.06	5.42	5.05	4.82	4.65	4.53	4.43	4.30	4.20	3.95	3.67
10	0.0^2103	0.0254	0.0694	0.113	0.151	0.183	0.210	0.233	0.269	0.297	0.379	0.488
	6.94	5.46	4.83	4.47	4.24	4.07	3.95	3.85	3.72	3.62	3.37	3.08
12	0.0^2102	0.0254	0.0699	0.114	0.153	0.186	0.214	0.238	0.276	0.305	0.394	0.515
	6.55	5.10	4.47	4.12	3.89	3.73	3.61	3.51	3.37	3.28	3.02	2.72
24	0.0^2100	0.0253	0.0709	0.118	0.159	0.195	0.226	0.253	0.297	0.331	0.440	0.610
	5.72	4.32	3.72	3.38	3.15	2.99	2.87	2.78	2.64	2.54	2.27	1.94
∞	0.0^398	0.0253	0.0719	0.121	0.166	0.206	0.242	0.272	0.325	0.368	0.515	1.00
	5.02	3.69	3.12	2.79	2.57	2.41	2.29	2.19	2.05	1.94	1.64	1.00

APPENDIX

Table A16a. *One-tailed values of the F-statistic, for degrees of freedom v_1, v_2 and probability level 5 per cent. The use of the statistic is described in 7.2 and several later sections.*

$v_1 =$	1	2	3	4	5	6	7	8	10	12	24	∞
$v_2 = 1$	161.4	199.5	215.7	224.6	230.2	234.0	236.8	238.9	241.9	243.9	249.0	254.3
2	18.5	19.0	19.2	19.2	19.3	19.3	19.4	19.4	19.4	19.4	19.5	19.5
3	10.13	9.55	9.28	9.12	9.01	8.94	8.89	8.85	8.79	8.74	8.64	8.53
4	7.71	6.94	6.59	6.39	6.26	6.16	6.09	6.04	5.96	5.91	5.77	5.63
5	6.61	5.79	5.41	5.19	5.05	4.95	4.88	4.82	4.74	4.68	4.53	4.36
6	5.99	5.14	4.76	4.53	4.39	4.28	4.21	4.15	4.06	4.00	3.84	3.67
7	5.59	4.74	4.35	4.12	3.97	3.87	3.79	3.73	3.64	3.57	3.41	3.23
8	5.32	4.46	4.07	3.84	3.69	3.58	3.50	3.44	3.35	3.28	3.12	2.93
9	5.12	4.26	3.86	3.63	3.48	3.37	3.29	3.23	3.14	3.07	2.90	2.71
10	4.96	4.10	3.71	3.48	3.33	3.22	3.14	3.07	2.98	2.91	2.74	2.54
11	4.84	3.98	3.59	3.36	3.20	3.09	3.01	2.95	2.85	2.79	2.61	2.40
12	4.75	3.89	3.49	3.26	3.11	3.00	2.91	2.85	2.75	2.69	2.51	2.30
13	4.67	3.81	3.41	3.18	3.03	2.92	2.83	2.77	2.67	2.60	2.42	2.21
14	4.60	3.74	3.34	3.11	2.96	2.85	2.76	2.70	2.60	2.53	2.35	2.13
15	4.54	3.68	3.29	3.06	2.90	2.79	2.71	2.64	2.54	2.48	2.29	2.07
16	4.49	3.63	3.24	3.01	2.85	2.74	2.66	2.59	2.49	2.42	2.24	2.01
17	4.45	3.59	3.20	2.96	2.81	2.70	2.61	2.55	2.45	2.38	2.19	1.96
18	4.41	3.55	3.16	2.93	2.77	2.66	2.58	2.51	2.41	2.34	2.15	1.92
19	4.38	3.52	3.13	2.90	2.74	2.63	2.54	2.48	2.38	2.31	2.11	1.88
20	4.35	3.49	3.10	2.87	2.71	2.60	2.51	2.45	2.35	2.28	2.08	1.84
21	4.32	3.47	3.07	2.84	2.68	2.57	2.49	2.42	2.32	2.25	2.05	1.81
22	4.30	3.44	3.05	2.82	2.66	2.55	2.46	2.40	2.30	2.23	2.03	1.78
23	4.28	3.42	3.03	2.80	2.64	2.53	2.44	2.37	2.27	2.20	2.00	1.76
24	4.26	3.40	3.01	2.78	2.62	2.51	2.42	2.36	2.25	2.18	1.98	1.73
25	4.24	3.39	2.99	2.76	2.60	2.49	2.40	2.34	2.24	2.16	1.96	1.71
26	4.23	3.37	2.98	2.74	2.59	2.47	2.39	2.32	2.22	2.15	1.95	1.69
27	4.21	3.35	2.96	2.73	2.57	2.46	2.37	2.31	2.20	2.13	1.93	1.67
28	4.20	3.34	2.95	2.71	2.56	2.45	2.36	2.29	2.19	2.12	1.91	1.65
29	4.18	3.33	2.93	2.70	2.55	2.43	2.35	2.28	2.18	2.10	1.90	1.64
30	4.17	3.32	2.92	2.69	2.53	2.42	2.33	2.27	2.16	2.09	1.89	1.62
32	4.15	3.29	2.90	2.67	2.51	2.40	2.31	2.24	2.14	2.07	1.86	1.59
34	4.13	3.28	2.88	2.65	2.49	2.38	2.29	2.23	2.12	2.05	1.84	1.57
36	4.11	3.26	2.87	2.63	2.48	2.36	2.28	2.21	2.11	2.03	1.82	1.55
38	4.10	3.24	2.85	2.62	2.46	2.35	2.26	2.19	2.09	2.02	1.81	1.53
40	4.08	3.23	2.84	2.61	2.45	2.34	2.25	2.18	2.08	2.00	1.79	1.51
60	4.00	3.15	2.76	2.53	2.37	2.25	2.17	2.10	1.99	1.92	1.70	1.39
120	3.92	3.07	2.68	2.45	2.29	2.18	2.09	2.02	1.91	1.83	1.61	1.25
∞	3.84	3.00	2.60	2.37	2.21	2.10	2.01	1.94	1.83	1.75	1.52	1.00

Table A16b. *One-tailed values of the F-statistic, for degrees of freedom v_1, v_2 and probability level 1 per cent. The use of the statistic is described in 7.2 and several later sections.*

$v_1 =$	1	2	3	4	5	6	7	8	10	12	24	∞
$v_2 = 1$	4052	5000	5403	5625	5764	5859	5928	5981	6056	6106	6235	6366
2	98.5	99.0	99.2	99.2	99.3	99.3	99.4	99.4	99.4	99.4	99.5	99.5
3	34.1	30.8	29.5	28.7	28.2	27.9	27.7	27.5	27.2	27.1	26.6	26.1
4	21.2	18.0	16.7	16.0	15.5	15.2	15.0	14.8	14.5	14.4	13.9	13.5
5	16.26	13.27	12.06	11.39	10.97	10.67	10.46	10.29	10.05	9.89	9.47	9.02
6	13.74	10.92	9.78	9.15	8.75	8.47	8.26	8.10	7.87	7.72	7.31	6.88
7	12.25	9.55	8.45	7.85	7.46	7.19	6.99	6.84	6.62	6.47	6.07	5.65
8	11.26	8.65	7.59	7.01	6.63	6.37	6.18	6.03	5.81	5.67	5.28	4.86
9	10.56	8.02	6.99	6.42	6.06	5.80	5.61	5.47	5.26	5.11	4.73	4.31
10	10.04	7.56	6.55	5.99	5.64	5.39	5.20	5.06	4.85	4.71	4.33	3.91
11	9.65	7.21	6.22	5.67	5.32	5.07	4.89	4.74	4.54	4.40	4.02	3.60
12	9.33	6.93	5.95	5.41	5.06	4.82	4.64	4.50	4.30	4.16	3.78	3.36
13	9.07	6.70	5.74	5.21	4.86	4.62	4.44	4.30	4.10	3.96	3.59	3.17
14	8.86	6.51	5.56	5.04	4.70	4.46	4.28	4.14	3.94	3.80	3.43	3.00
15	8.68	6.36	5.42	4.89	4.56	4.32	4.14	4.00	3.80	3.67	3.29	2.87
16	8.53	6.23	5.29	4.77	4.44	4.20	4.03	3.89	3.69	3.55	3.18	2.75
17	8.40	6.11	5.18	4.67	4.34	4.10	3.93	3.79	3.59	3.46	3.08	2.65
18	8.29	6.01	5.09	4.58	4.25	4.01	3.84	3.71	3.51	3.37	3.00	2.57
19	8.18	5.93	5.01	4.50	4.17	3.94	3.77	3.63	3.43	3.30	2.92	2.49
20	8.10	5.85	4.94	4.43	4.10	3.87	3.70	3.56	3.37	3.23	2.86	2.42
21	8.02	5.78	4.87	4.37	4.04	3.81	3.64	3.51	3.31	3.17	2.80	2.36
22	7.95	5.72	4.82	4.31	3.99	3.76	3.59	3.45	3.26	3.12	2.75	2.31
23	7.88	5.66	4.76	4.26	3.94	3.71	3.54	3.41	3.21	3.07	2.70	2.26
24	7.82	5.61	4.72	4.22	3.90	3.67	3.50	3.36	3.17	3.03	2.66	2.21
25	7.77	5.57	4.68	4.18	3.86	3.63	3.46	3.32	3.13	2.99	2.62	2.17
26	7.72	5.53	4.64	4.14	3.82	3.59	3.42	3.29	3.09	2.96	2.58	2.13
27	7.68	5.49	4.60	4.11	3.78	3.56	3.39	3.26	3.06	2.93	2.55	2.10
28	7.64	5.45	4.57	4.07	3.75	3.53	3.36	3.23	3.03	2.90	2.52	2.06
29	7.60	5.42	4.54	4.04	3.73	3.50	3.33	3.20	3.00	2.87	2.49	2.03
30	7.56	5.39	4.51	4.02	3.70	3.47	3.30	3.17	2.98	2.84	2.47	2.01
32	7.50	5.34	4.46	3.97	3.65	3.43	3.26	3.13	2.93	2.80	2.42	1.96
34	7.45	5.29	4.42	3.93	3.61	3.39	3.22	3.09	2.90	2.76	2.38	1.91
36	7.40	5.25	4.38	3.89	3.58	3.35	3.18	3.05	2.86	2.72	2.35	1.87
38	7.35	5.21	4.34	3.86	3.54	3.32	3.15	3.02	2.83	2.69	2.32	1.84
40	7.31	5.18	4.31	3.83	3.51	3.29	3.12	2.99	2.80	2.66	2.29	1.80
60	7.08	4.98	4.13	3.65	3.34	3.12	2.95	2.82	2.63	2.50	2.12	1.60
120	6.85	4.79	3.95	3.48	3.17	2.96	2.79	2.66	2.47	2.34	1.95	1.38
∞	6.63	4.61	3.78	3.32	3.02	2.80	2.64	2.51	2.32	2.18	1.79	1.00

Table A16c. *One-tailed values of the F-statistic, for degrees of freedom v_1, v_2 and probability level 0.1 per cent. The use of the statistic is described in 7.2 and several later sections.*

$v_1 =$	1	2	3	4	5	6	7	8	10	12	24	∞
$v_2 = 1$†	4053	5000	5404	5625	5764	5859	5929	5981	6056	6107	6235	6366†
2	998.5	999.0	999.2	999.2	999.3	999.3	999.4	999.4	999.4	999.4	999.5	999.5
3	167.0	148.5	141.1	137.1	134.6	132.8	131.5	130.6	129.2	128.3	125.9	123.5
4	74.14	61.25	56.18	53.44	51.71	50.53	49.66	49.00	48.05	47.41	45.77	44.05
5	47.18	37.12	33.20	31.09	29.75	28.83	28.16	27.65	26.92	26.42	25.14	23.79
6	35.51	27.00	23.70	21.92	20.80	20.03	19.46	19.03	18.41	17.99	16.90	15.75
7	29.25	21.69	18.77	17.20	16.21	15.52	15.02	14.63	14.08	13.71	12.73	11.70
8	25.42	18.49	15.83	14.39	13.48	12.86	12.40	12.05	11.54	11.19	10.30	9.34
9	22.86	16.39	13.90	12.56	11.71	11.13	10.69	10.37	9.87	9.57	8.72	7.81
10	21.04	14.91	12.55	11.28	10.48	9.93	9.52	9.20	8.74	8.44	7.64	6.76
11	19.69	13.81	11.56	10.35	9.58	9.05	8.66	8.35	7.92	7.63	6.85	6.00
12	18.64	12.97	10.80	9.63	8.89	8.38	8.00	7.71	7.29	7.00	6.25	5.42
13	17.82	12.31	10.21	9.07	8.35	7.86	7.49	7.21	6.80	6.52	5.78	4.97
14	17.14	11.78	9.73	8.62	7.92	7.44	7.08	6.80	6.40	6.13	5.41	4.60
15	16.59	11.34	9.34	8.25	7.57	7.09	6.74	6.47	6.08	5.81	5.10	4.31
16	16.12	10.97	9.01	7.94	7.27	6.80	6.46	6.19	5.81	5.55	4.85	4.06
17	15.72	10.66	8.73	7.68	7.02	6.56	6.22	5.96	5.58	5.32	4.63	3.85
18	15.38	10.39	8.49	7.46	6.81	6.35	6.02	5.76	5.39	5.13	4.45	3.67
19	15.08	10.16	8.28	7.27	6.62	6.18	5.85	5.59	5.22	4.97	4.29	3.51
20	14.82	9.95	8.10	7.10	6.46	6.02	5.69	5.44	5.08	4.82	4.15	3.38
21	14.59	9.77	7.94	6.95	6.32	5.88	5.56	5.31	4.95	4.70	4.03	3.26
22	14.38	9.61	7.80	6.81	6.19	5.76	5.44	5.19	4.83	4.58	3.92	3.15
23	14.19	9.47	7.67	6.70	6.08	5.65	5.33	5.09	4.73	4.48	3.82	3.05
24	14.03	9.34	7.55	6.59	5.98	5.55	5.23	4.99	4.64	4.39	3.74	2.97
25	13.88	9.22	7.45	6.49	5.89	5.46	5.15	4.91	4.56	4.31	3.66	2.89
26	13.74	9.12	7.36	6.41	5.80	5.38	5.07	4.83	4.48	4.24	3.59	2.82
27	13.61	9.02	7.27	6.33	5.73	5.31	5.00	4.76	4.41	4.17	3.52	2.75
28	13.50	8.93	7.19	6.25	5.66	5.24	4.93	4.69	4.35	4.11	3.46	2.69
29	13.39	8.85	7.12	6.19	5.59	5.18	4.87	4.64	4.29	4.05	3.41	2.64
30	13.29	8.77	7.05	6.12	5.53	5.12	4.82	4.58	4.24	4.00	3.36	2.59
32	13.12	8.64	6.94	6.01	5.43	5.02	4.72	4.48	4.14	3.91	3.27	2.50
34	12.97	8.52	6.83	5.92	5.34	4.93	4.63	4.40	4.06	3.83	3.19	2.42
36	12.83	8.42	6.74	5.84	5.26	4.86	4.56	4.33	3.99	3.76	3.12	2.35
38	12.71	8.33	6.66	5.76	5.19	4.79	4.49	4.26	3.93	3.70	3.06	2.29
40	12.61	8.25	6.59	5.70	5.13	4.73	4.44	4.21	3.87	3.64	3.01	2.23
60	11.97	7.77	6.17	5.31	4.76	4.37	4.09	3.86	3.54	3.32	2.69	1.89
120	11.38	7.32	5.78	4.95	4.42	4.04	3.77	3.55	3.24	3.02	2.40	1.54
∞	10.83	6.91	5.42	4.62	4.10	3.74	3.47	3.27	2.96	2.74	2.13	1.00

† Values for $v_2 = 1$ should be multiplied by 100.

Table A17. *Values of the test statistic required for 5 per cent and 1 per cent significance in the Kolmogorov–Smirnov one-sample test. The procedure is described in 7.2.3.1.*

Sample size n	Significance level (per cent)	
	5	1
1	0.975	0.995
2	0.842	0.929
3	0.708	0.828
4	0.624	0.733
5	0.565	0.669
6	0.521	0.618
7	0.486	0.577
8	0.457	0.543
9	0.432	0.514
10	0.410	0.490
11	0.391	0.468
12	0.375	0.450
13	0.361	0.433
14	0.349	0.418
15	0.338	0.404
16	0.328	0.392
17	0.318	0.381
18	0.309	0.371
19	0.301	0.363
20	0.294	0.356
25	0.27	0.32
30	0.24	0.29
35	0.23	0.27
Over 35	$\dfrac{1.36}{\sqrt{n}}$	$\dfrac{1.63}{\sqrt{n}}$

APPENDIX

Table A18. *Values of* r, *the estimate on* v *degrees of freedom of the product-moment correlation coefficient, required for significance at the conventional levels. The calculation of* r *is described in 8.2.1.*

	Significance level (per cent)		
v	5	1	0.1
1	0.99692	0.999877	0.9999988
2	0.95000	0.990000	0.99900
3	0.8783	0.95873	0.99116
4	0.8114	0.91720	0.97406
5	0.7545	0.8745	0.95074
6	0.7067	0.8343	0.92493
7	0.6664	0.7977	0.8982
8	0.6319	0.7646	0.8721
9	0.6021	0.7348	0.8471
10	0.5760	0.7079	0.8233
11	0.5529	0.6835	0.8010
12	0.5324	0.6614	0.7800
13	0.5139	0.6411	0.7603
14	0.4973	0.6226	0.7420
15	0.4821	0.6055	0.7246
16	0.4683	0.5879	0.7084
17	0.4555	0.5751	0.6932
18	0.4438	0.5614	0.6787
19	0.4329	0.5487	0.6652
20	0.4227	0.5368	0.6524
25	0.3809	0.4869	0.5974
30	0.3494	0.4487	0.5541
35	0.3246	0.4182	0.5189
40	0.3044	0.3932	0.4896
45	0.2875	0.3721	0.4648
50	0.2732	0.3541	0.4433
60	0.2500	0.3248	0.4078
70	0.2319	0.3017	0.3799
80	0.2172	0.2830	0.3568
90	0.2050	0.2673	0.3375
100	0.1946	0.2540	0.3211

Table A19. *Confidence limits for the mean of a Poisson distribution given a single observation r from the distribution (see 9.3.2).*

Confidence level	95 per cent		99 per cent	
r	Lower	Upper	Lower	Upper
0	0.0000	3.69	0.00000	5.30
1	0.0253	5.57	0.00501	7.43
2	0.242	7.22	0.103	9.27
3	0.619	8.77	0.338	10.98
4	1.09	10.24	0.672	12.59
5	1.62	11.67	1.08	14.15
6	2.20	13.06	1.54	15.66
7	2.81	14.42	2.04	17.13
8	3.45	15.76	2.57	18.58
9	4.12	17.08	3.13	20.00
10	4.80	18.39	3.72	21.40
11	5.49	19.68	4.32	22.78
12	6.20	20.96	4.94	24.14
13	6.92	22.23	5.58	25.50
14	7.65	23.49	6.23	26.84
15	8.40	24.74	6.89	28.16
16	9.15	25.98	7.57	29.48
17	9.90	27.22	8.25	30.79
18	10.67	28.45	8.94	32.09
19	11.44	29.67	9.64	33.38
20	12.22	30.89	10.35	34.67
21	13.00	32.10	11.07	35.95
22	13.79	33.31	11.79	37.22
23	14.58	34.51	12.52	38.48
24	15.38	35.71	13.25	39.74
25	16.18	36.90	14.00	41.00
26	16.98	38.10	14.74	42.25
27	17.79	39.28	15.49	43.50
28	18.61	40.47	16.24	44.74
29	19.42	41.65	17.00	45.98
30	20.24	42.83	17.77	47.21
35	24.38	48.68	21.64	53.32
40	28.58	54.47	25.59	59.36
45	32.82	60.21	29.60	65.34
50	37.11	65.92	33.66	71.27

APPENDIX

Table A20. *Values v such that if the smaller of the two Poisson variables, r_1 and r_2, is less than or equal to v then the null hypothesis that the means of the distributions are equal should be rejected at the stated nominal significance level (see 9.3.4).*

	Nominal significance level			Nominal significance level	
$r_1 + r_2$	0.05	0.01	$r_1 + r_2$	0.05	0.01
1			41	14	12
2			42	15	13
3			43	15	13
4			44	16	13
5	0		45	16	14
6	0		46	16	14
7	0	0	47	17	15
8	1	0	48	17	15
9	1	0	49	18	15
10	1	0	50	18	16
11	2	1	51	19	16
12	2	1	52	19	17
13	3	1	53	20	17
14	3	2	54	20	18
15	3	2	55	20	18
16	4	2	56	21	18
17	4	3	57	21	19
18	5	3	58	22	19
19	5	4	59	22	20
20	5	4	60	23	20
21	6	4	61	23	20
22	6	5	62	24	21
23	7	5	63	24	21
24	7	5	64	24	22
25	7	6	65	25	22
26	8	6	66	25	23
27	8	7	67	26	23
28	9	7	68	26	23
29	9	7	69	27	24
30	10	8	70	27	24
31	10	8	71	28	25
32	10	8	72	28	25
33	11	9	73	28	26
34	11	9	74	29	26
35	12	10	75	29	26
36	12	10	76	30	27
37	13	10	77	30	27
38	13	11	78	31	28
39	13	11	79	31	28
40	14	12	80	32	29

Table A21. *The transformation of a proportion* (*here expressed as a percentage*) *to degrees by the relationship* $\theta = \sin^{-1}\sqrt{(r/n)}$ (*see* 9.4.1).

$100r/n$	0	1	2	3	4	5	6	7	8	9
0	0	5.7	8.1	10.0	11.5	12.9	14.2	15.3	16.4	17.5
10	18.4	19.4	20.3	21.1	22.0	22.8	23.6	24.4	25.1	25.8
20	26.6	27.3	28.0	28.7	29.3	30.0	30.7	31.3	31.9	32.6
30	33.2	33.8	34.4	35.1	35.7	36.3	36.9	37.5	38.1	38.6
40	39.2	39.8	40.4	41.0	41.6	42.1	42.7	43.3	43.9	44.4
50	45.0	45.6	46.1	46.7	47.3	47.9	48.4	49.0	49.6	50.2
60	50.8	51.4	51.9	52.5	53.1	53.7	54.3	54.9	55.6	56.2
70	56.8	57.4	58.1	58.7	59.3	60.0	60.7	61.3	62.0	62.7
80	63.4	64.2	64.9	65.6	66.4	67.2	68.0	68.9	69.7	70.6
90	71.6	72.5	73.6	74.7	75.8	77.1	78.5	80.0	81.9	84.3

Table A22. *Random observations from a normal distribution for use in sampling experiments.*

17.9	18.7	11.3	17.3	17.5	14.9	12.3	13.6	12.1	16.1
11.0	16.4	20.7	16.4	12.2	18.4	12.2	13.6	17.1	17.2
18.2	13.8	13.5	16.9	17.0	8.7	20.2	17.0	18.0	18.1
16.7	24.1	19.1	15.8	16.4	17.7	13.6	16.0	14.0	10.3
15.4	18.2	10.7	17.4	12.1	13.3	16.3	19.1	9.9	18.3
10.4	16.7	15.0	18.7	16.0	16.0	11.2	12.1	11.0	18.4
12.2	15.6	13.9	19.6	16.3	16.2	14.5	14.4	9.2	10.8
18.9	14.7	16.0	19.1	15.9	17.0	18.5	15.9	17.2	6.9
15.1	20.9	16.6	14.3	10.2	11.2	17.7	19.3	8.6	15.3
14.1	16.0	16.0	22.0	14.0	17.3	19.0	12.0	10.0	12.6
19.4	13.9	13.5	20.4	14.7	13.8	17.8	18.1	18.1	15.9
15.9	14.5	16.1	13.5	13.1	15.8	15.7	17.8	17.6	14.7
16.8	16.0	18.4	17.1	13.8	17.8	13.7	14.7	11.9	15.1
18.6	20.2	14.2	18.1	14.4	17.2	16.8	13.7	9.1	14.8
17.5	15.3	21.2	14.9	15.7	16.0	13.6	16.0	13.4	16.0
18.5	21.4	13.4	18.3	11.4	15.6	14.6	17.8	21.6	15.9
15.0	11.5	17.4	12.6	16.4	15.5	16.0	17.5	14.3	15.2
13.5	13.6	13.4	14.8	13.6	11.9	12.4	12.7	17.4	15.9
12.1	14.5	19.1	18.2	15.8	14.1	13.9	10.9	12.4	18.6
10.2	18.1	16.1	15.6	17.8	14.4	18.5	16.5	13.4	16.1
16.9	18.6	17.5	14.1	16.5	10.4	19.6	16.2	11.4	16.4
16.2	14.6	11.3	12.3	16.9	13.9	22.0	15.7	16.0	12.6
10.7	18.4	13.6	17.1	11.3	13.8	14.8	16.5	14.7	19.0
16.5	11.6	15.3	16.9	11.3	11.8	12.9	10.6	16.0	14.3
17.1	14.7	13.5	16.0	15.3	10.6	17.4	15.1	22.1	17.5
15.6	11.3	16.3	15.9	18.5	16.4	10.3	9.0	13.3	16.4
12.8	11.8	16.9	14.2	15.7	16.2	11.3	18.3	16.9	18.6
9.8	11.6	12.3	19.1	8.2	16.3	14.2	9.9	15.8	14.4
13.7	18.8	16.7	7.8	12.6	14.3	10.1	13.9	14.4	18.3
14.2	16.4	17.8	16.6	16.1	15.2	15.4	15.9	12.4	20.3
20.9	12.9	11.8	9.8	10.4	15.0	11.0	13.0	19.3	10.2
15.2	16.0	15.1	12.5	13.6	21.7	16.1	10.3	10.9	17.9
18.3	7.4	11.0	11.1	11.8	16.3	13.4	13.9	15.5	10.1
13.8	12.4	13.9	20.7	16.0	13.6	19.2	14.0	16.0	14.4
22.1	16.9	8.6	15.3	16.2	15.4	14.7	15.3	11.4	20.0
12.5	17.8	10.7	12.7	16.9	16.2	17.7	11.2	17.5	16.5
13.4	9.6	15.8	11.7	14.8	11.2	15.5	15.1	15.5	15.6
10.0	13.9	16.4	11.5	18.4	14.7	19.3	17.6	13.0	10.1
18.1	15.1	10.1	12.8	16.9	16.2	14.2	18.4	14.2	18.6
17.3	18.2	13.8	16.9	15.6	13.0	18.4	19.4	13.1	10.8
13.0	20.4	9.2	18.3	15.0	22.1	13.7	15.1	12.8	15.2
19.1	14.9	13.9	9.4	8.6	17.9	19.4	14.0	12.1	10.2
14.1	14.0	16.5	13.7	14.1	7.5	17.3	19.5	13.3	17.7
15.3	14.2	16.1	16.8	8.6	15.3	14.0	11.8	16.0	15.8
18.3	13.6	20.7	14.2	12.0	15.7	16.7	11.4	19.0	17.8

Table A22 (*continued*)

11.7	12.0	16.6	18.0	14.1	17.6	14.0	13.7	18.4	15.7
19.8	13.2	14.3	14.7	18.1	13.6	16.6	17.7	15.4	9.9
14.0	19.5	14.3	15.3	11.7	11.1	8.6	14.2	8.7	12.1
14.2	16.4	13.8	13.2	17.7	20.8	15.0	10.9	14.7	12.6
11.4	18.4	20.8	8.7	16.6	11.1	13.9	14.2	17.3	12.7
12.0	14.3	19.3	12.9	17.9	14.0	13.3	12.7	11.7	12.3
12.0	15.0	15.3	16.3	16.0	15.4	16.3	17.7	19.1	18.4
17.4	17.3	11.2	12.4	14.9	11.0	17.6	13.2	16.8	14.5
14.3	11.1	6.4	12.0	14.5	12.6	15.8	17.0	14.6	13.4
17.8	14.3	10.2	12.4	15.9	12.8	15.7	12.5	20.2	18.7
16.7	13.2	17.2	14.8	21.2	13.3	15.4	16.8	17.5	10.7
13.6	18.0	12.5	7.9	16.5	10.4	11.6	19.0	14.6	9.2
16.4	12.2	10.5	11.8	17.6	14.2	19.0	15.4	18.2	16.0
15.2	13.5	16.2	20.9	19.8	10.9	17.7	17.2	14.5	19.3
14.2	14.7	16.4	17.9	19.9	6.7	12.7	14.1	17.4	19.4
14.9	16.1	10.4	18.2	14.4	11.6	18.1	13.8	16.7	13.8
14.7	15.7	13.2	12.6	16.4	20.4	12.9	16.8	21.2	18.7
13.9	16.8	13.9	12.8	16.0	16.3	13.7	16.2	12.7	10.3
8.9	16.0	16.5	17.4	18.4	17.0	14.1	14.4	15.8	14.7
16.7	12.6	16.9	8.7	17.1	18.7	14.7	15.0	16.0	8.9
14.1	17.0	14.1	17.2	16.7	9.6	14.5	15.0	12.3	19.2
15.2	14.1	16.8	20.5	15.5	12.5	13.1	15.2	14.1	21.3
10.8	17.1	18.7	18.6	11.8	10.8	14.1	17.9	16.9	18.8
17.4	19.3	13.2	18.1	12.9	18.8	14.1	15.4	13.7	16.9
11.3	13.9	15.0	11.0	22.7	11.7	12.8	17.6	13.7	14.2
16.4	18.5	21.3	15.5	15.7	17.9	12.8	11.8	14.9	15.1
18.5	11.5	15.6	22.4	16.5	16.6	14.3	19.4	17.8	18.2
11.3	16.3	14.4	16.3	13.6	10.9	18.3	11.7	16.5	12.2
13.9	14.7	10.9	11.9	11.5	17.1	12.9	16.7	13.1	13.6
13.4	15.1	14.4	12.1	12.1	15.9	16.5	13.6	14.3	16.9
14.6	20.2	16.8	9.3	10.5	16.6	17.2	15.0	18.2	16.8
10.1	12.0	18.0	16.9	8.3	13.2	18.4	14.1	17.1	18.7
16.2	21.1	15.7	12.3	12.2	13.4	14.1	14.6	12.3	15.0
12.8	13.9	17.6	20.6	16.3	18.4	18.2	15.8	17.1	16.2
14.3	16.0	11.8	9.1	14.2	15.7	19.1	13.5	14.9	15.4
16.9	17.2	11.0	16.5	11.9	11.0	17.3	9.6	14.2	19.5
15.0	12.0	13.2	14.5	18.9	16.2	14.5	11.5	15.3	15.0
12.4	13.5	16.2	13.9	15.5	17.0	17.4	14.2	18.6	9.5
20.8	17.3	15.2	13.6	15.8	19.8	11.7	14.9	18.2	15.9
15.0	11.6	7.4	15.2	15.4	15.9	15.2	16.6	12.3	17.6
9.4	11.7	17.8	8.4	14.4	17.3	17.0	18.0	17.9	11.7
11.6	18.2	20.6	11.3	14.1	15.9	17.2	13.3	15.2	16.3
12.2	19.4	14.4	13.2	10.3	16.2	17.5	18.7	12.4	16.4
13.5	19.1	13.5	13.6	9.3	12.6	11.7	16.7	17.5	17.4
17.2	14.5	15.7	15.7	14.2	12.9	15.8	19.6	13.1	12.3

Table A22 (*continued*)

15.0	13.1	16.5	15.7	16.8	16.6	17.6	11.9	12.7	13.9
13.6	21.8	16.6	14.7	15.7	15.5	8.1	15.8	17.4	16.3
8.7	13.7	12.2	14.5	13.9	16.7	12.0	9.8	10.9	13.8
16.7	17.0	18.1	16.3	18.5	17.9	18.8	14.3	19.0	18.2
17.9	15.5	13.4	15.3	11.8	13.4	16.1	14.1	20.3	6.8
15.0	16.6	17.9	17.0	14.5	16.7	12.4	20.4	10.4	21.2
14.5	12.9	16.8	12.8	16.0	9.2	16.8	19.4	17.5	19.1
6.7	20.0	12.1	9.6	12.1	15.7	16.3	11.8	15.0	18.7
14.9	19.5	11.2	19.2	12.1	7.6	9.1	16.6	12.5	13.1
11.2	16.9	13.5	17.9	11.1	13.2	17.4	13.7	17.3	16.1
11.0	14.6	16.7	13.3	9.4	18.3	17.8	9.6	15.1	21.1
15.4	16.1	16.3	11.0	14.4	14.8	18.4	6.3	5.2	15.1
10.0	15.1	11.5	15.2	14.5	16.1	13.3	20.0	19.5	16.3
16.4	10.9	17.3	9.1	13.7	13.9	15.8	17.6	14.1	12.2
15.2	14.3	8.0	18.3	21.9	15.4	14.3	12.7	13.7	8.0
17.3	11.6	17.1	13.7	13.9	17.1	18.0	15.4	15.7	17.7
18.6	13.9	14.2	11.7	16.6	17.7	15.9	12.5	10.4	19.8
13.9	12.8	18.8	12.0	11.0	18.4	16.6	17.7	16.3	9.4
16.9	18.2	20.9	17.1	16.6	13.2	12.6	13.8	21.5	13.9
12.1	10.4	17.5	19.4	17.6	17.1	14.9	17.0	12.6	13.3
17.8	13.0	9.6	11.7	15.5	22.3	14.8	12.4	20.1	15.0
14.1	10.9	13.9	11.8	16.3	19.5	14.1	15.6	12.8	15.4
18.6	16.1	17.7	18.3	7.1	19.4	18.5	13.2	16.9	16.4
14.6	18.3	16.1	14.4	18.1	13.9	11.7	17.9	14.7	16.4
21.4	12.7	15.5	19.1	14.5	23.9	19.7	13.5	18.0	17.6
15.5	14.9	16.2	11.0	11.4	16.0	14.1	14.8	20.5	19.1
17.3	17.4	13.8	15.1	18.8	10.0	14.1	16.1	14.8	12.0
11.1	15.6	17.9	19.3	19.5	10.4	17.1	18.9	11.9	16.6
16.8	17.4	11.8	14.8	12.0	11.9	15.5	14.4	15.8	17.6
16.7	15.3	16.3	14.4	16.4	11.9	18.2	10.4	11.5	16.2
15.4	12.8	8.6	13.4	16.5	15.5	21.8	15.0	19.0	17.9
17.7	10.4	11.8	16.3	14.9	12.3	11.2	16.0	18.8	14.5
13.8	21.7	18.1	19.0	15.9	13.3	17.4	13.6	20.9	12.8
15.6	14.6	17.8	20.5	20.2	18.7	17.4	18.5	13.4	16.9
18.8	20.0	14.8	16.1	10.2	13.8	18.5	14.0	16.3	16.7
18.0	16.9	15.2	16.5	17.2	17.8	15.0	13.6	11.2	11.1
16.9	18.2	10.1	16.9	18.2	12.2	16.0	12.9	15.2	19.0
16.2	12.4	20.4	18.3	17.2	12.7	14.1	17.3	17.5	10.2
12.2	18.3	16.0	18.2	16.8	17.6	9.9	18.9	17.7	18.0
19.8	15.4	15.8	10.7	13.0	14.9	17.1	13.4	10.8	12.6
16.3	12.9	10.4	10.7	17.0	14.8	14.8	11.9	14.2	17.0
20.3	14.0	15.7	18.6	8.8	15.4	15.9	17.0	14.2	18.3
12.3	13.8	15.2	9.2	11.3	14.8	17.3	12.5	17.1	13.5
20.0	17.2	14.5	14.3	20.8	14.0	14.9	15.4	16.7	17.1
11.5	16.5	18.7	12.2	14.8	15.1	17.0	20.8	16.1	14.5

Table A22 (*continued*)

16.9	9.9	11.0	16.2	12.1	11.0	17.1	18.3	16.7	13.3
9.8	15.1	12.6	12.0	15.8	14.0	14.6	14.7	16.5	17.7
12.5	13.3	15.9	8.7	15.3	12.7	15.6	14.4	16.0	15.5
11.1	16.1	10.4	13.4	15.6	15.4	12.1	14.4	20.0	19.1
14.2	15.9	17.6	12.8	14.6	14.1	19.9	10.3	18.5	16.4
10.6	12.8	14.8	14.6	6.2	14.2	12.3	12.2	13.8	15.1
9.2	14.9	13.1	14.2	17.7	16.6	11.1	16.9	14.2	20.3
15.7	16.0	17.8	16.2	17.9	11.8	11.2	14.2	12.9	17.4
13.3	15.0	14.9	17.3	17.3	13.6	17.3	18.1	16.1	15.4
15.2	11.2	15.9	11.9	14.3	18.7	15.0	16.8	13.1	15.3
15.2	19.1	16.1	17.7	10.1	13.0	17.5	13.5	14.6	14.7
13.9	14.5	18.0	18.3	14.5	16.1	15.7	11.4	12.4	12.0
18.4	21.3	13.6	13.0	13.4	13.8	16.6	15.0	18.9	16.8
18.5	16.7	17.5	10.2	16.8	13.7	19.6	17.0	17.3	11.3
16.4	11.5	16.2	10.6	16.7	12.4	14.7	12.9	16.3	16.3

APPENDIX

Table A23. *Observations in the survey in exercise 7.5(21).*

	Area 1	Breed A				
	955	938	1084	994	1004	903
	943	992	949	1053	1038	995
	930	965	1106	1064	921	963
	1007	947	1018	1027	1025	988
	983	882	1016	963	868	1090
	991	910	1119	968	939	931
	1036	978	1102	994	923	978
	994	1007	1184	1138	985	971
	994	1080	978	1002	943	1032
	971	932	1090	1040	903	1011
	985	1021	1280	1096	1030	1039
	1017	990	1142	957	891	1054
	1008	938	1039	996	1024	963
	955	991	1048	1027	1015	999
	1036	1130	1076	1013	972	935
	1004	982	1110	977	959	998
	1010	954	1118	984	1062	934
	1078	1023	1083	1012	926	988
	990	966	1134	908	984	964
	983	1092	998	965	1022	1022

Area 1	Breed B							
745	847	739	733	743	694	789	841	777
725	777	692	873	826	788	857	762	793
760	804	765	857	895	727	750	784	815
739	829	815	851	825	886	742	774	753
859	770	777	729	811	787	841	787	863
777	760	790	872	819	864	896	848	742
836	778	776	949	895	792	738	761	716
782	795	781	809	832	842	822	814	758
714	901	784	896	874	826	736	745	843
811	871	758	803	827	774	836	820	686
815	803	751	743	840	790	781	825	887
780	792	724	799	796	772	762	736	843
795	755	776	865	907	771	819	817	829
767	822	816	750	857	790	870	757	833
814	789	721	875	891	814	786	860	865
762	842	848	798	750	813	827	739	776
788	914	800	776	845	793	730	686	719
823	797	727	751	828	762	796	813	888
777	832	802	808	848	724	769	743	791
807	797	730	841	776	788	739	770	849
718	835	778	929	753	838	863	858	804
739	743	847	800	853	835	773	781	740
804	756	818	819	844	822	870	814	701
717	761	848	795	841	810	849	794	790
816	834	691	877	880	784	837	738	786
742	804	802	775	731	716	773	862	808
758	790	798	830	813	846	878	822	832
887	897	757	891	820	818	779	814	836
804	846	767	880	739	768	899	745	796
804	855	741	861	880	859	859	873	852

Table A23 (*continued*)

Area 1	Breed C					
687	704	686	731	684	797	710
627	727	652	723	824	743	740
667	733	742	716	680	634	623
671	590	620	753	712	763	610
615	685	741	735	753	698	715
644	668	681	603	730	724	755
645	706	705	707	760	716	668
734	739	605	659	626	621	628
652	587	684	650	683	637	711
739	712	762	720	756	720	670
705	687	681	771	726	613	598
626	675	755	670	677	725	689
731	703	766	709	693	672	741
634	693	753	677	680	809	669
620	640	719	788	767	652	637
607	702	765	733	794	678	731
625	736	641	681	676	635	705
602	642	729	701	654	693	775
702	770	797	623	656	713	643
602	583	710	683	747	703	731
770	674	647	631	732	625	735
661	705	702	761	646	674	635
558	672	625	677	730	706	658
617	620	693	765	601	670	743
710	724	706	744	785	764	649

Area 2	Breed A		
952	816	863	920
966	888	842	974
957	919	830	980
848	950	837	1032
992	909	938	916
915	943	935	873
901	839	845	915
786	898	842	984
861	855	875	1061
907	861	956	954
923	887	937	979
861	875	935	886
866	885	861	899
994	875	812	936
846	964	842	972
906	938	962	964
917	941	932	931
939	930	867	941
915	921	869	976
897	986	943	958

Table A23 (*continued*)

Area 2	Breed B		
749	762	734	654
715	625	686	722
696	662	797	679
745	693	777	755
695	696	801	698
675	610	741	651
634	692	766	673
650	715	561	741
712	673	699	605
698	693	743	796
653	680	564	677
662	671	683	642
758	680	778	657
739	664	610	751
746	685	780	714
682	737	666	689
746	674	715	678
739	694	593	689
740	759	650	693
641	651	699	692

Area 2	Breed C			
546	593	587	522	650
655	566	592	575	594
609	524	636	605	665
585	535	727	562	676
644	475	645	531	637
564	450	581	622	639
577	611	607	589	571
637	584	526	561	600
672	642	692	688	658
630	591	642	631	563
619	577	596	681	639
617	561	562	514	592
621	570	557	518	652
615	646	629	679	587
614	443	654	724	624
519	557	590	672	579
524	654	672	614	622
521	577	614	532	627
502	462	646	678	600
625	583	627	639	668

In each section of the Table, columns correspond to herds.

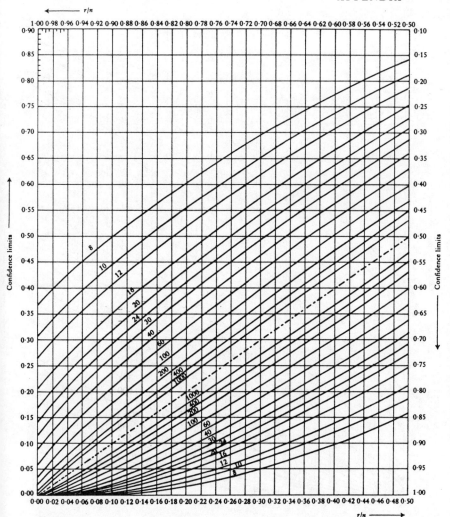

Fig. A1 (a). Chart for finding 95% confidence limits for binomial p given a single observation from the distribution: the method of using the chart is described in 9.3.1.

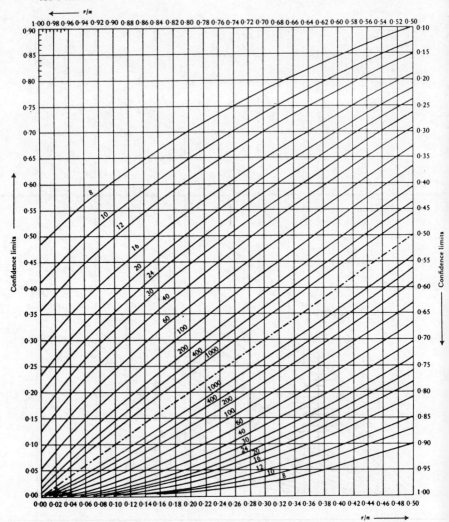

Fig. A1 (*b*). Chart for finding 99% confidence limits for binomial *p* given a single observation from the distribution: the method of using the chart is described in 9.3.1.

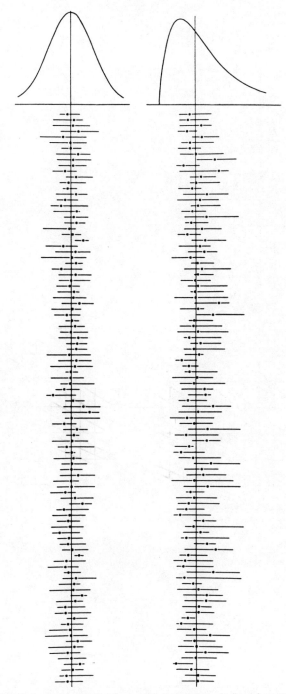

Fig. A2. Point and interval estimates for the median obtained from 100 samples of 17 observations from each of 2 populations. The forms of the distributions are shown, with the true values of the medians marked.

INDEX